Advances in

Pharmacology

Volume 26

Cyclic GMP: Synthesis, Metabolism, and Function

Advances in ____________

Pharmacology

Volume 26

Cyclic GMP: Synthesis, Metabolism, and Function

Edited by

Ferid Murad
Molecular Geriatrics Corporation
Lake Bluff, Illinois

Academic Press
A Division of Harcourt Brace & Company
San Diego New York Boston London Sydney Tokyo Toronto

Cover photograph: Photomicrographs of the crystalline structure of cyclic GMP. Kindly provided by Raymond Zinkowski of Molecular Geriatrics Corporation.

This book is printed on acid-free paper. ∞

Academic Press, Inc.
525 B Street, Suite 1900, San Diego, California 92101-4495

United Kingdom Edition published by
Academic Press Limited
24–28 Oval Road, London NW1 7DX

International Standard Serial Number: 0065-3144

International Standard Book Number: 0-12-032926-3

PRINTED IN THE UNITED STATES OF AMERICA

94 95 96 97 98 99 QW 9 8 7 6 5 4 3 2 1

Contents

Contributors xiii
Preface xv

Introduction and Some Historical Comments 1
Ferid Murad

References 5

Cloning of Guanylyl Cyclase Isoforms
Masaki Nakane and Ferid Murad

I. Introduction 7
II. Cloning of Particulate Guanylyl Cyclases 7
III. Cloning of Soluble Guanylyl Cyclases 11
IV. Summary 13
References 15

Regulation of Cytosolic Guanylyl Cyclase by Nitric Oxide: The NO–Cyclic GMP Signal Transduction System
Ferid Murad

I. Introduction 19
II. Effects of Nitric Oxide on Cyclic GMP Synthesis and Smooth Muscle Relaxation 21
III. Effects of Endothelial-Derived Relaxing Factor on Cyclic GMP Formation 26
IV. Nitric Oxide Formation 27
V. Summary 30
References 30

Regulation of Cytosolic Guanylyl Cyclase by Porphyrins and Metalloporphyrins
Louis J. Ignarro

I. Introduction and Overview 35

II. Studies Leading to the Discovery That Protoporphyrin IX Activates Guanylate Cyclase 37
A. Early Studies 37
B. Influence of Thiols on Guanylate Cyclase Activity 38
C. Requirement of Heme for Guanylate Cyclase Activation by NO 40
III. Kinetic Mechanisms by Which Protoporphyrin IX Activates Guanylate Cyclase and Similarity to Nitric Oxide–Heme Complex 43
IV. Role of Copper Bound to Guanylate Cyclase 45
V. Mechanism by Which Phenylhydrazine Activates Guanylate Cyclase 46
VI. Structure–Activity Relationships Involving Porphyrins and Metalloporphyrins 50
VII. Regulation of Guanylate Cyclase Activity by Porphyrins and Metalloporphyrins 54
VIII. Nitric Oxide–Heme Exchange between Hemoproteins and Guanylate Cyclase 58
IX. Summary and Conclusions 61
References 62

Regulation of Particulate Guanylate Cyclase by Natriuretic Peptides and *Escherichia coli* Heat-Stable Enterotoxin

Dale C. Leitman, Scott A. Waldman, and Ferid Murad

I. Introduction 67
II. Atrial Natriuretic Peptide (ANP) Receptors 69
A. Pharmacological Heterogeneity of ANP Receptors 69
B. Heterogeneity of ANP Receptor Subunit Structure 70
C. ANP-R1 and Particulate Guanylate Cyclase Reside on the Same Transmembrane Protein 70
D. Cloning of ANP Receptors 71
E. Coupling of ANP-R1 Receptor to the Activation of Particulate Guanylate Cyclase 72
III. Heat-Stable Enterotoxin (ST) Receptor 74
A. Pharmacological Heterogeneity of ST Receptors 74
B. Heterogeneity of ST Receptor Subunit Structure 76
C. Heterogeneity of ST Receptor Subcellular Distribution 76
D. Cloning of ST Receptors 77
E. Purification of ST Receptors from Intestinal Mucosa 78
F. Relationship of ST Receptors and Particulate Guanylate Cyclase in Intestinal Cells 79
G. Coupling of ST Receptors and Activation of Particulate Guanylate Cyclase 80

IV. Particulate Guanylate Cyclase–Cyclic GMP Second Messenger System 81
References 82

Cyclic GMP and Regulation of Cyclic Nucleotide Hydrolysis
William K. Sonnenburg and Joseph A. Beavo

I. Introduction 87
A. Scope of Chapter 87
B. General Control of Cyclic Nucleotide Steady-State Levels 88
C. Multiple PDEs Control Cyclic GMP Hydrolysis 88
D. Basis for PDE Family Designation, General Domain Organization, and Conserved Motifs 89
II. Ca^{2+}/CaM-Dependent PDE Family 91
A. Multiple Isoforms 91
B. Kinetic Properties 92
C. Structure and Domain Organization 93
D. Tissue and Cellular Distribution 94
E. Calcium/CaM-Dependent PDE Regulation of Cyclic GMP 95
III. Cyclic GMP-Specific PDE Family 96
A. General Properties 97
B. Cellular Distribution and Functions 97
C. Regulation of Activity 98
D. Structural Features 99
IV. Cyclic GMP-Stimulated Phosphodiesterase Family 99
A. Multiple Isoforms 99
B. Kinetic Properties 99
C. Domain Organization 101
D. Tissue Distribution 101
E. Regulation 102
V. Cyclic GMP-Inhibited PDE Family 104
A. General Properties and Multiple Isoforms within Family 104
B. Tissue and Cellular Distribution 105
C. Regulation 105
D. Structural Features 106
VI. Summary 107
References 107

Progress in Understanding the Mechanism and Function of Cyclic GMP-Dependent Protein Kinase
Sharron H. Francis and Jackie D. Corbin

I. Introduction 115
II. Tissue Distribution of Cyclic GMP Kinase 117

III. Isozymes 118
A. Type I Isoforms 120
B. Type II Isoform 120
C. mRNA Size and Distribution 121
D. Chromosomal Location/Exon–Intron Organization 123
IV. Nonmammalian Cyclic GMP Kinases 124
V. General Structure 124
VI. Microheterogeneity 126
VII. Domain Structures and Functions 128
A. Dimerization Domain 129
B. Autoinhibitory Domain 130
C. Cyclic GMP-Binding Domains 136
D. Catalytic Domain 144
E. Carboxyl-Terminal Domain 149
VIII. Physiological Function 149
A. Regulation of Smooth Muscle Tone 150
B. Inhibition of Platelet Aggregation 152
C. Regulation of Intracellular Calcium Levels 152
D. Other Possible Functions 155
IX. Cross-Activation 156
X. Concluding Remarks 159
References 159

Effects of Cyclic GMP on Smooth Muscle Relaxation
Timothy D. Warner, Jane A. Mitchell, Hong Sheng, and Ferid Murad

I. Introduction 171
II. Isoforms of Guanylyl Cyclase Present in Smooth Muscle 172
A. Particulate Guanylyl Cyclase 172
B. Cytosolic Guanylyl Cyclase 174
III. Mechanism of Cyclic GMP-Mediated Smooth Muscle Relaxation 180
A. Correlation between Cyclic GMP Levels and Relaxation of Smooth Muscle 180
B. Cyclic GMP-Dependent Protein Kinase 182
C. Cyclic GMP, Calcium, and Intracellular Signaling 182
D. Other Possible Mechanisms 184
IV. Summary 184
References 185

Interrelationships of Cyclic GMP, Inositol Phosphates, and Calcium
Masato Hirata and Ferid Murad

I. Introduction 195
II. Cyclic GMP Effects on Calcium 196

A. Plasmalemmal Ca^{2+} Pump 196
B. Endoplasmic Reticulum Ca^{2+} Pump 200
C. Na^+–Ca^{2+} Exchange Mechanisms 201
D. Voltage-Dependent Ca^{2+} Channels 203
E. Receptor-Operated Ca^{2+} Entry 205
III. Cyclic GMP Effect on Ins(1,4,5)P_3 Production 207
A. Mechanisms of Ins(1,4,5)P_3 Production 207
B. Mode of Cyclic GMP Inhibition of Ins(1,4,5)P_3 Production 208
References 209

Cyclic GMP Regulation of Calcium Slow Channels in Cardiac Muscle and Vascular Smooth Muscle Cells

Nicholas Sperelakis, Noritsugu Tohse, Yusuke Ohya, and Hiroshi Masauda

I. Introduction and Overview 217
II. Cardiac Muscle 220
A. Cyclic AMP Stimulation of Slow Ca^{2+} Channels 220
B. Phosphorylation Hypothesis 222
C. Cyclic GMP Inhibition of Slow Ca^{2+} Current 224
D. Calmodulin–Protein Kinase and Protein Kinase C 229
E. Summary and Conclusions 230
III. Vascular Smooth Muscle Cells 230
A. Inhibition of Ca^{2+} Slow Channels by Cyclic AMP and Cyclic GMP 230
B. Regulation of Ca^{2+} Channels by ATP 234
C. Modulation of Ca^{2+} Channels by Agonists 239
D. Summary and Conclusions 245
IV. Skeletal Muscle Fibers 245
V. Summary 246
References 246

Effect of Cyclic GMP on Intestinal Transport

Arie B. Vaandrager and Hugo R. DeJonge

I. Introduction 253
II. Synthesis of Cyclic GMP in the Intestine 254
A. Intestinal Form of Guanylyl Cyclase (Type C) 254
B. Localization of Guanylyl Cyclases in the Intestine 255
III. Effects of Cyclic GMP on Intestinal Transport in Mammals 257
A. Effects of Cyclic GMP in Epithelial Cells 257
B. Effects of Cyclic GMP in Nonepithelial Cell Types 262
IV. Effects of Cyclic GMP on Intestinal Transport in Winter Flounder 263

V. Mechanisms of Cyclic GMP Action in the Intestine 263
A. General Mechanisms of Cyclic GMP Action 263
B. Mechanisms of Inhibition of Na and Cl Absorption by Cyclic GMP 265
C. Mechanisms of Stimulation of Cl Secretion by Cyclic GMP 267
VI. Intestinal Form of Cyclic GMP-Dependent Protein Kinase (Type II) 267
VII. The CFTR Cl Channel and Its Regulation by Cyclic GMP 270
VIII. Function of Cyclic GMP-Induced Electrolyte and Fluid Secretion in the Intestine 273
References 275

Cyclic GMP in Lower Forms

Joachim E. Schultz and Susanne Klumpp

I. Introduction 285
II. Bacteria 286
III. Slime Molds and Fungi 287
A. *Dictyostelium discoideum* 287
B. Others 291
IV. Cyclic GMP in the Protozoans *Tetrahymena* and *Paramecium* 292
A. Regulation of Intracellular Cyclic GMP Formation 292
B. Guanylyl Cyclases 295
C. Phosphodiesterase and Cyclic GMP-Dependent Protein Kinase 297
V. Cyclic GMP in Other Lower Forms 297
References 298

Clinical Relationships of Cyclic GMP

Jean R. Cusson, Johanne Tremblay, Pierre Larochelle, Ernesto L. Schiffrin, Jolanta Gutkowska, and Pavel Hamet

I. Introduction 305
II. Cyclic GMP Measurements in Healthy Humans 305
A. Measurement Conditions 306
B. Exogenous Blood Volume Expansion 309
C. Endogenous Blood Volume Expansion 310
D. Effect of Pressor Doses of Phenylephrine 311
E. Effect of Intravenous ANP Administration: The Nonlinear ANP–Cyclic GMP Relationship 311
III. Cyclic GMP Measurements in Essential Hypertension 313
A. Baseline Values 313
B. Effect of Blood Volume Expansion 314

C. Effect of Pressor Doses of Phenylephrine 314
D. Effect of Intravenous ANP Administration 314
IV. Use of Cyclic GMP Measurements as a Tool in Other Diseases 315
V. Conclusions 316
References 317

Future Directions 321
Ferid Murad

References 324

Index 325
Contents of Previous Volumes 331

Contributors

Numbers in parentheses indicate the pages on which the authors' contributions begin.

Joseph A. Beavo (87), Department of Pharmacology, University of Washington, Seattle, Washington 98195

Jackie D. Corbin (115), Department of Molecular Physiology and Biophysics, Vanderbilt University School of Medicine, Nashville, Tennessee 37232

Jean R. Cusson (305), Centre de Recherche, Hôtel-Dieu de Montréal, Montréal, Québec, Canada H2W 1T8

Hugo R. DeJonge (253), Department of Biochemistry, School of Medicine, Erasmus University, Rotterdam, 3000 DR Rotterdam, The Netherlands

Sharron H. Francis (115), Department of Molecular Physiology and Biophysics, Vanderbilt University School of Medicine, Nashville, Tennessee 37232

Jolanta Gutkowska (305), Centre de Recherche, Hôtel-Dieu de Montréal, Montréal, Québec, Canada H2W 1T8, and Institut de Recherches Cliniques de Montréal, Montréal, Québec, Canada H2W 1R7

Pavel Hamet (305), Centre de Recherche, Hôtel-Dieu de Montréal, Montréal, Québec, Canada H2W 1T8

Masato Hirata (195), Department of Biochemistry, Faculty of Dentistry, Kyushu University, Fukuoka 812, Japan

Louis J. Ignarro (35), Department of Pharmacology, Center for the Health Sciences, University of California, Los Angeles, School of Medicine, Los Angeles, California 90024

Susanne Klumpp (285), Abteilung Biochemie, Pharmazeutisches Institut der Universität, 72076 Tübingen, Germany

Pierre Larochelle (305), Centre de Recherche, Hôtel-Dieu de Montréal, Montréal, Québec, Canada H2W 1T8

Dale C. Leitman (67), Metabolic Research Unit, University of California, San Francisco, School of Medicine, San Francisco, California 94143

Hiroshi Masuda (217), Department of Pediatrics, School of Medicine, University of Hiroshima, Hiroshima 734, Japan

Jane A. Mitchell (171), William Harvey Research Institute, St. Bartholomew's Hospital Medical College, London EC1M 6BQ, United Kingdom

Ferid Murad (5, 7, 19, 67, 171, 195, 324), Molecular Geriatrics Corporation, Lake Bluff, Illinois 60044

Masaki Nakane (7), Pharmaceutical Products Division, Abbott Laboratories, Abbott Park, Illinois 60064

Yusuke Ohya (217), Second Department of Internal Medicine, School of Medicine, Kyushu University, Fukuoka 812, Japan

Ernesto L. Schiffrin (305), Institut de Recherches Cliniques de Montréal, Montréal, Québec, Canada H2W 1R7

Joachim E. Schultz (285), Abteilung Biochemie, Pharmazeutisches Institut der Universität, 72076 Tübingen, Germany

Hong Sheng (171), Department of Pharmacology, University of California, Los Angeles, School of Medicine, Los Angeles, California 90024

William K. Sonnenburg (87), Department of Pharmacology, University of Washington, Seattle, Washington 98195

Nicholas Sperelakis (217), Department of Physiology and Biophysics, College of Medicine, University of Cincinnati, Cincinnati, Ohio 45267

Noritsugu Tohse (217), Department of Pharmacology, School of Medicine, Hokkaido University, Sapporo 060, Japan

Johanne Tremblay (305), Centre de Recherche, Hôtel-Dieu de Montréal, Montréal, Québec, Canada H2W 1T8

Arie B. Vaandrager (253), Department of Biochemistry, School of Medicine, Erasmus University, Rotterdam, 3000 DR Rotterdam, The Netherlands

Scott A. Waldman (67), Division of Clinical Pharmacology, Departments of Medicine and Pharmacology, Thomas Jefferson University, Philadelphia, Pennsylvania 19107

Timothy D. Warner (171), William Harvey Research Institute, St. Bartholomew's Hospital Medical College, London EC1M 6BQ, United Kingdom

Preface

The field of cyclic GMP research has expanded considerably in recent years to warrant a volume dedicated to the numerous advances in the field. The interrelationships of cyclic GMP metabolism to the effects of numerous agents, such as nitric oxide and other nitrovasodilators, endothelium-dependent vasodilators, atrial natriuretic peptides, and *Escherichia coli* heat-stable enterotoxin, have led to many new pharmacological and biochemical models and reagents that aid in investigating the role of cyclic GMP in signal transduction and information transfer. Cyclic GMP is one of but a handful of small second messenger molecules that regulate many physiological and pathophysiological processes. Cyclic GMP is a very important member of this class of compounds. In recent years numerous meetings and symposia have been held dealing with different aspects of this field. For example, the discovery of some of the important effects of nitric oxide were a result of cyclic GMP research.

However, to date, there has not been a volume dedicated to the topic of cyclic GMP. This volume is intended to summarize much, but not all, of the work in the field for students, trainees, and others not directly involved in the area of cyclic GMP research. Presumably, the cyclic GMP addict will be familiar with most of the work summarized. Chapters include those discussing the guanylyl cyclase and phosphodiesterase isoenzyme families for cyclic GMP synthesis and hydrolysis, a chapter on cyclic GMP-dependent protein kinase, and several chapters on various hormones and ligands that regulate cyclic GMP formation and/or metabolism. Several chapters deal with some of the effects of cyclic GMP on other second messengers such as calcium, ion transport, and smooth muscle relaxation. Clinical studies with cyclic GMP and atrial natriuretic peptide are also discussed. Some areas have been intentionally omitted, such as the role of cyclic GMP in the central nervous system, because the information is incomplete, the model systems are complex and inadequate, or it is too early to determine which directions the work will go. Space constraints also limited what could be covered. Recent review articles are available for some of these topics.

In editing this volume, numerous interesting and important experiments yet to be performed have become obvious to me and have convinced me that, despite the growing information base in recent years, many important experiments remain to be done. Some of these questions and thoughts are discussed in the last chapter entitled "Future Directions." I hope the readers not only gain a general overview of the field from this volume, but also recognize additional important experiments that may apply to their work in related fields.

I thank the authors, collaborators, and trainees who provided the chapters and encouragement to prepare this book, and Darlene Rentschler and Janice Rundgren for their secretarial assistance.

Ferid Murad

Introduction and Some Historical Comments

Ferid Murad
Molecular Geriatrics Corporation
Lake Bluff, Illinois 60044

The identification of cyclic GMP as an endogenous substance occurred about three decades ago and about 6 years after the discovery of cyclic AMP by Sutherland and Rall. The perceived significance of cyclic AMP as a ubiquitous intracellular second messenger in signal transduction led to the search for analogous cyclic nucleotide monophosphates of other purines and pyrimidines. Several naturally occurring compounds were subsequently characterized and/or identified. However, only cyclic AMP and cyclic GMP continue to attract the attention of numerous laboratories.

Cyclic GMP has always been suspected by some investigators to play an important role as a second messenger. The interests and enthusiasm for cyclic GMP have grown considerably in recent years, enough to warrant the first monograph in this area. Unfortunately, cyclic GMP was viewed by some as a lesser cousin to cyclic AMP in signal transduction for many years. However, cyclic GMP has clearly joined the ranks of the limited number of small molecules that act as intracellular messengers to mediate the diverse effects of various hormones, neurotransmitters, and agonists in tissues. The relatively short list of small molecular messengers include cyclic AMP, cyclic GMP, Ca^{2+}, inositol phosphates, diacyl glycerol, eicosanoids nitric oxide, and a few other candidates.

The work with cyclic GMP can be viewed in several major phases in the past three decades. The 1960s represent the period in which the enzymes for its synthesis (guanylyl cyclase) and hydrolysis (cyclic nucleotide phosphodiesterase) and its target in most cells and tissues (cyclic GMP-

Advances in Pharmacology, Volume 26

dependent protein kinase) were described within a few years of the discovery of GMP. The apparent rapid progress in these early years fostered even greater interests and expectations for this new second messenger. During the early 1970s there was considerable enthusiasm for the possible role of this new cyclic nucleotide and second messenger. Numerous descriptive studies were performed by many laboratories that led to many hypotheses about its possible physiological role. Some hypotheses and proposed functions were overly simplistic and, unfortunately, incorrect.

In the mid and late 1970s it became apparent that the problem was much more complex than originally anticipated, which is often the case in most fields. It was learned that there were multiple isoenzymes for cyclic GMP syntheses and hydrolysis. This work supported the view that there would be different intracellular compartments of cyclic GMP, different mechanisms to regulate cyclic GMP accumulation in these compartments, and perhaps different functions to be associated with these intracellular pools of cyclic GMP. Clearly, the regulation of the various isoforms for either synthesis or hydrolysis were different, making it quite difficult to correlate cycle GMP synthesis, metabolism, and accumulation to various hormonal responses in different physiological models. Unlike the cyclic AMP field where hormones and agonists could regulate cyclic AMP synthesis and metabolism in cell free systems with some degree of fidelity to the behavior of intact tissues, most agents that altered cyclic GMP in intact cells failed to exert effects in cell extracts. Thus, the characterization of the components participating in hormone–receptor coupling to cyclic GMP accumulation and function and opportunities to reconstitute the components did not seem feasible in the foreseeable future. The lack of progress with the molecular mechanisms of hormone–receptor coupling to altered cyclic GMP synthesis and metabolism and some premature conclusions regarding cyclic GMP functions resulted in many investigators turning their interests and attention to other readily resolvable problems. The 1970s can be viewed as a decade of early excitement followed by subsequent disenchantment and decreased interest in cyclic GMP. Some subtle artifacts in published data became apparent and some problems were not so subtle. As a result, few laboratories continued their work with cycle GMP. It was obviously time for some detailed biochemical characterization and purification of the components of the cyclic GMP system.

Many of us have tried to answer two major questions in the area of cyclic GMP: (*a*) how precisely is cyclic GMP synthesis regulated after hormone–receptor interactions (i.e., what are the molecular events in this signal transduction cascade) and (*b*) what are some of the physiological and biochemical functions regulated by cyclic GMP. Some important observations were made in the 1970s that permitted cyclic GMP synthesis and metabolism to be perturbed in intact and cell-free systems and, thus, it became possible to begin to address these questions. First, nitric oxide, nitrate, hydroxylamine, and other "nitrovasodilators" (a term that we coined for this broad class of agents and prodrugs) that lead to the generation of nitric oxide enzymatically or nonenzymatically were found to activate guanylyl cyclase and increase cyclic GMP in intact tissues as well as tissue extracts. These observations, reviewed in Chapters 3 and 4, not only lead to an understanding of the mechanism of action of this class of drugs on smooth muscle relaxation, but also provided the first group of molecules that increased cyclic GMP syntheses in both cell-free preparations and intact tissues. Thus, these reagents became a prototype for understanding hormonal regulation of cyclic GMP synthesis in future studies.

After we found the effects of nitric oxide on cyclic GMP syntheses, we proposed that this free radical could function as an endogenous second messenger to explain hormonal regulation of cyclic GMP synthesis (Murad *et al.*, 1978a,b). However, work from several independent directions was required before this hypothesis was proven (see Chapter 3 and references therein). This work also led to our subsequent work and expectations that endothelium-dependent vasodilators would work through increased cyclic GMP synthesis (reviewed in Chapters 3, 4, and 8).

Second, *Escherichia coli* heat stable enterotoxin (ST) was found to activate guanylyl cyclase in both intestinal mucosa homogenates and intact cell preparations and cyclic GMP mediated the effects of this peptide toxin on intestinal ion transport and the induced diarrhea. Thus, the first naturally occurring peptide ligand was identified that served as a "hormone prototype" to perturb this second messenger system in some select tissues. This area is reviewed in Chapters 5 and 11. Our earlier studies with the effects of nitrovasodilators on cyclic GMP and smooth muscle relaxation and the effects of *E. coli* enterotoxin on intestinal ion transport set the stage for our subsequent work with atrial natriuretic peptides (ANP). This newly described family of peptide hormones in the early 1980s was found to relax vascular smooth muscle and cause renal natriuresis. Since ANP had some physiological effects similar to those of nitrovasodilators and ST on smooth muscle relaxation and ion transport, we suspected that these newly described hormones may also mediate their effects through cyclic GMP. This was indeed the case and the work with ANP is discussed in Chapters 5 and 13.

The 1980s was a period of many rapid developments in the area of cyclic GMP. Definitive studies from our laboratory and other laboratories demonstrated that cyclic GMP mediated the effects of nitrovasodilators, endothelium-dependent vasodilators, natriuretic peptides, and some enterotoxins. The functions for cyclic GMP expanded from smooth muscle relaxation and intestinal secretion to phototransduction, platelet aggregation, platelet adhesion, neurotransmission, hormonal secretion, and many other processes.

In the mid and late 1980s many laboratories joined or rejoined the ranks of cyclic GMP and/or nitric oxide converts and the research interests and commitments have grown logrithmically since. The potential for a role of cyclic GMP and/or nitric oxide in biological regulation is quite vast and currently unknown. These systems are quite ubiquitous and much work remains to be done. If one assumes that there are a limited number of small second messengers such as cyclic GMP and nitric oxide, then one can expect that 5 to 10% of all of the diverse physiological and biochemical processes in cell biology will be regulated by these messengers. Obviously, a large task is before us and much is yet to be done. This monograph will briefly review the field of cyclic GMP and the early observations with nitric oxide, natriuretic factors, and related agents. Numerous enzyme isoforms for cyclic GMP synthesis and hydrolysis and nitric oxide synthesis are now apparent and, undoubtedly, more gene products and post-translational modifications will be described in the future. Many of these isoenzymes have been characterized, purified, and cloned and these discussions can be found in Chapters 2, 3, 6, and 7. Understanding their transcriptional and post-translational regulations is in its infancy. Although it has been suspected that most, or all, of the physiological effects of cyclic GMP will be mediated through cyclic GMP-dependent protein kinase and phosphorylation of various endogenous macromolecular targets, this is not the case, as reviewed in Chapters 6 through 11.

Some areas of cyclic GMP have been reviewed very briefly in this monograph or have been omitted because of limited space or an inability to obtain commitments from some authors due to their schedules. In most cases, review articles on these topics are included in the chapter references to lead the reader to these other areas of interest. When this project began about 2½ years ago, I had hoped that this monograph could be broader and more extensive. However, the rapid developments in this field in the past several years have been extraordinary and a more complete monograph on the topic is no longer possible in a reasonable time frame. For example, in the area of nitric oxide and its biochemical and biological effects, there were more than 1900 publications during 1992. In recent years, the information in the field of cyclic GMP and/or nitric oxide has grown more rapidly than any other second messenger system to date.

References

Murad, F., Mittal, C. K., Arnold, W. P., and Braughler, J. M. (1978a). Effect of nitro-compound smooth muscle relaxants and other materials on cyclic GMP metabolism. *Adv. Pharmacol. Ther.* **3,** 123–132.

Murad, F., Mittal, C. K., Arnold, W. P., Katsuki, S., and Kimura, H. (1978b). Guanylate cyclase: Activation by azide, nitro compounds, nitric oxide and hydroxyl radical and inhibition by hemoglobin and myoglobin. *Adv. Cyclic Nucleotide Res.* **9,** 145–158.

Cloning of Guanylyl Cyclase Isoforms

Masaki Nakane* and Ferid Murad†

**Pharmaceutical Products Division*
Abbott Laboratories
Abbott Park, Illinois 60064

†*Molecular Geriatrics Corporation*
Lake Bluff, Illinois 60044

I. Introduction

The synthesis of cyclic GMP is catalyzed by guanylyl cyclase [GTP pyrophosphate-lyase (cyclizing), EC 4.6.1.2], and the enzyme is present in several particulate and soluble isoforms, which are structurally different proteins with different properties. The particulate isoforms can be activated by natriuretic peptides and *Escherichia coli* heat-stable enterotoxin, whereas the soluble enzyme can be activated by nitric oxide and some porphyrins. Recently, the structures of some particulate and soluble guanylyl cyclases have been described from various sources, and an activation mechanism has been proposed. We describe here the cDNA cloning and expression studies for particulate and soluble guanylyl cyclases that have taken place in several laboratories.

II. Cloning of Particulate Guanylyl Cyclases

It was known for some time that particulate guanylyl cyclase is distinct from the soluble isoform because of its kinetic, physicochemical, and antigenic properties (Kimura and Murad, 1975; Chrisman *et al.*, 1975; Nakane and Deguchi, 1980, 1982; Brandwein *et al.*, 1981). Various pep-

Advances in Pharmacology, Volume 26

tides, including sea urchin chemotactic peptides, atrial and brain natriuretic peptide (ANP and BNP), and heat-stable enterotoxin form *E. coli* (STa), are known to activate particulate guanylyl cyclase (Waldman *et al.,* 1984; Kuno *et al.,* 1986a; Garbers, 1989). In rat lung, this enzyme was suggested to represent a cell surface receptor for ANP receptor (ANP-R1) (Kuno *et al.,* 1986b; Takayanagi *et al.,* 1987; Paul *et al.,* 1987).

The first cDNA isolated for particulate guanylyl cyclase was from sea urchin (*Arbacia punctulata*). Sea urchin sperm has an exceptionally high particulate guanylyl cyclase activity and it had been purified to homogeneity using GTP affinity column chromatography (Garbers, 1978). Singh *et al.* (1988) succeeded in the isolation of a cDNA for the particulate guanylyl cyclase from sea urchin testis and found a single transmembrane domain and a region in the cytoplasmic domain homologous to the protein kinase family. When the cDNA was transfected in mammalian cells, neither enzyme activity nor peptide binding was observed. Later, another cDNA for particulate guanylyl cyclase was isolated from the sea urchin *Strongylocentrotus purpuratus,* which evolved much later than *A. punctulata* (Thorpe and Garbers, 1989). The amino acid sequence of the enzyme is highly conserved in most regions when compared to the enzyme of *A. punctulata*. However, the carboxy tail of the protein completely diverges from that of *A. punctulata,* where it is highly homologous with a region of soluble guanylyl cyclase (Nakane *et al.,* 1988; Koesling *et al.,* 1988).

By low-stringency hybridization with this sea urchin particulate guanylyl cyclase cDNA probe, cDNA clones encoding particulate guanylyl cyclase that possesses ANP receptor activity were isolated from a human kidney cDNA library (Lowe *et al.,* 1989). The ANP receptor/guanylyl cyclase has a signal sequence followed by an extracellular domain homologous to the ANP-R2 receptor, which has already been cloned (Fuller *et al.,* 1988). The ANP-R2 receptor has a lower molecular mass (60–70 kDa) and is not coupled to guanylyl cyclase. Readers are also referred to the discussion in Chapter 5. A single transmembrane domain precedes a cytoplasmic domain with homology to the protein kinase family and a catalytic domain with homology to soluble guanylyl cyclase (Nakane *et al.,* 1988; Koesling *et al.,* 1988). COS-7 cells transfected with the cDNA showed specific ANP binding and ANP stimulated cGMP production.

This human particulate guanylyl cyclase cDNA was in turn used as a probe to isolate a cDNA coding for particulate guanylyl cyclase from a rat brain cDNA library (Chinkers *et al.,* 1989). The deduced sequence of the ANP receptor/guanylyl cyclase is consistent with the notion that the ANP-binding portion is on the exterior surface and the homologous region to the protein kinase family and the catalytic site on the cytosolic surface. A rat brain cDNA clone encoding the particulate guanylyl cyclase also expressed both guanylyl cyclase and ANP-binding activity when transfected into COS-7 cells. The ANP-binding characteristics by competition studies with ANP analogs were found to be those expected for the high-molecular-weight receptor for ANP (ANP-R1 receptor). Cross-linking studies with the transfected cells demonstrated a major ANP receptor band of 130 kDa on SDS–polyacrylamide gel electrophoresis. Although these data were definitive evidence that particulate guanylyl cyclase functions as an ANP receptor, earlier studies with purification of particulate guanylyl cyclase and the ANP receptor indicated that both activities resided in the same macromolecule with presumably an extracellular receptor domain and an intracellular catalytic domain (Kuno *et al.*, 1986b; Takayanagi *et al.*, 1987; Paul *et al.*, 1987).

Murine ANP receptor/guanylyl cyclase cDNA was isolated from a cDNA library constructed from poly(A)$^+$ RNA of a Leydig tumor cell line (Pandey and Singh, 1990). The predicted protein structure was basically the same as that of rat brain ANP receptor/guanylyl cyclase. In addition, the expression of the cDNA transfected in rat Leydig tumor cells stimulated the production of testosterone and intracellular cGMP after treatment with ANP, indicating that ANP can regulate the testicular steroidogenic responsiveness.

By low-stringency screening using a cDNA of ANP receptor/guanylyl cyclase, then designated GC-A, a second ANP receptor/guanylyl cyclase cDNA was isolated from rat brain (Schultz *et al.*, 1989) and human placenta (Chang *et al.*, 1989), and designated GC-B. The deduced amino acid sequence of GC-B is 74 and 78% identical with GC-A within the intracellular domain, but 43 and 44% identical within the extracellular domain, respectively. GC-B is preferentially activated by brain natriuretic peptide (BNP) in the micromolar range rather than ANP, whereas GC-A responds to ANP better than to BNP in the nanomolar range. The relatively high BNP and ANP concentration required for GC-B activation suggested the presence of a more potent natural ligand for GC-B. Recently, it was shown that the affinity of C-type natriuretic peptide (CNP) (Sudoh *et al.*, 1990) for GC-B is 50- or 500-fold higher than that of ANP or BNP, respectively, suggesting that CNP may be the physiological ligand for GC-B (Koller *et al.*, 1991; Ohyama *et al.*, 1992).

Maximal activation of particulate guanylyl cyclase by ANP requires the presence of ATP, which can be mimicked by nonhydrolyzable ATP analogs (Kurose *et al.*, 1987; Chang *et al.*, 1990). By serial cDNA clonings of particulate guanylyl cyclase, the intracellular region was found to contain the sequence similar to that of protein kinase catalytic domains, and GC-A expressed in insect cells using a baculovirus vector was absolutely dependent on the presence of ATP for activation by ANP (Chinkers *et al.*, 1991). Chinkers and Garbers (1989) also reported that when the kinase-like domain was removed by deletion mutagenesis, the resulting ANP receptor/guanylyl cyclase expressed several-fold higher enzyme activity than wild-type, but the activity was independent of ANP and the stimulation by ATP was completely lost. Recently, a GC-B cDNA clone with a 75-bp deletion at the 3′-flanking region of the putative transmembrane domain, the shorter form lacking the nucleotide binding site, was isolated from rat brain (Ohyama *et al.*, 1992). Expression of the cDNA in mammalian cells revealed that this deleted GC-B could not induce cGMP production by the binding of CNP. These data suggested that the binding of ligands to the extracellular domain of the receptor initiates a conformational change that affects interactions between the protein kinase-like (ATP binding) and the catalytic domain, resulting in de-repression of guanylyl cyclase activity.

The approximately 300 amino acid portion of the particulate guanylyl cyclase carboxy terminal has been predicted as the catalytic domain, based on the homology with soluble guanylyl cyclase (Nakane *et al.*, 1988, 1990; Koesling *et al.*, 1988, 1990) and adenylyl cyclase (Krupinski *et al.*, 1989; Bakalyar and Reed, 1990). When this carboxy terminal of particulate guanylyl cyclase was subcloned and expressed in *E. coli,* the extract had substantial guanylyl cyclase activity, providing direct evidence that the carboxy portion of the particulate guanylyl cyclase contains a catalytic domain. No associated adenylyl cyclase activity was found in the extract (Thorpe and Morkin, 1990).

Recently, a third class of particulate guanylyl cyclase was cloned from a small intestine cDNA library of rat and human (Schultz *et al.*, 1990; de Sauvage *et al.*, 1991; Singh *et al.*, 1991). In intestine, STa has been shown to bind to a cell surface receptor, which subsequently leads to activation of guanylyl cyclase (Hughes *et al.*, 1978; Field *et al.*, 1978; Guerrant *et al.*, 1980). The enzyme in intestine is mainly a particulate form and is insoluble in various detergents, suggesting a different isoform of the enzyme from GC-A and GC-B (Waldman and Murad, 1987). A cDNA clone that encodes an STa receptor/guanylyl cyclase (GC-C) contains an extracellular region divergent from that of GC-A and GC-B. However, the intracellular region retains the protein kinase-like and catalytic domains. Expression in COS-7 cells results in not only high guanylyl cyclase activity, but also the specific binding of STa and the cyclic GMP elevation by STa, but not by ANP. These results show that the STa receptor is also a particulate guanylyl cyclase. It has been argued that the STa receptor and particulate guanylyl cyclase were distinct proteins (Kuno *et al.*, 1986a; Waldman *et al.*, 1986), but the cloning work shows otherwise. The idea that multiple STa receptors might exist, like ANP receptors (Fuller *et al.*,

1988), at least one being a low-molecular-weight receptor that does not possess a guanylyl cyclase domain, may explain the discrepancy. In fact, Kuno *et al.* (1986a) reported that the main STa-binding proteins in rat intestine are smaller (80, 68, and 60 kDa) than those of GC-C (121 kDa). Readers are also referred to Chapter 5 for additional discussion of STa effects on guanylyl cyclase.

III. Cloning of Soluble Guanylyl Cyclases

Soluble guanylyl cyclase has been purified to apparent homogeneity and shown to exist as a heterodimer (Kamisaki *et al.*, 1986; Humbert *et al.*, 1990). The higher-molecular-weight subunit is designated α (82 kDa from rat lung and 73 kDa from bovine lung in SDS–polyacrylamide gel) and the smaller subunit, β (70 kDa from rat and bovine lung in SDS–polyacrylamide gel).

The first soluble guanylyl cyclase cDNAs isolated were for the β subunit from rat lung (Nakane *et al.*, 1988) and bovine lung (Koesling *et al.*, 1988). The mRNA for the β subunit is most abundant in lung and brain, and the carboxy terminal region of the β subunit protein is highly homologous with a carboxy-terminal region of all the particulate guanylyl cyclases. However, when the β-subunit cDNA was transfected in mammalian cells, no enzyme activity was observed (Nakane *et al.*, 1988).

Later, the cDNAs for the α subunit were isolated from rat lung (Nakane *et al.*, 1990) and from bovine lung (Koesling *et al.*, 1990). A comparison of the amino acid sequences of the α and β subunits revealed about 32% homology between the α and the β subunits over the whole sequence. The amino-terminal part shows relatively low homology (about 20%), whereas the carboxy-terminal region shows a high degree of homology (about 40%) and also has homology with the catalytic domain of particulate guanylyl cyclase (Thorpe and Garbers, 1989; Lowe *et al.*, 1989; Chinkers *et al.*, 1989; Pandey and Singh, 1990; Schultz *et al.*, 1989, 1990; Chang *et al.*, 1989) and adenylyl cyclase (Krupinski *et al.*, 1989; Bakalyar and Reed, 1990), suggesting that both subunits have catalytic domains.

Again, when the α-subunit cDNA was permanently transfected to L cells, expression did not yield catalytically active enzyme. However, coexpression of both subunits (α and β) yielded significant guanylyl cyclase activity that was activated markedly by sodium nitroprusside, a potent activator of soluble guanylyl cyclase (Nakane *et al.*, 1990). This observation was confirmed in transient expression experiments using COS-7 cells (Harteneck *et al.*, 1990; Buechler *et al.*, 1991). In addition, cotransfection with either of the antisense oligonucleotide complementary to the α- or the β-subunit mRNA inhibited the expression of soluble guanylyl cyclase activity (Buechler *et al.*, 1991), suggesting that both subunits are required to be expressed and interactive to permit synthesis of cGMP and activation by nitrovasodilator or nitric oxide.

Although the relative quantity of mRNAs for the α and β subunits from most tissues of rat correlates with the distribution of soluble guanylyl cyclase activity, some inconsistencies in the quantities of mRNA and catalytic activity in liver and kidney have been reported (Nakane *et al.*, 1990). In addition, monoclonal antibody against soluble guanylyl cyclase precipitated most soluble enzyme activity in brain and lung, but only 40% of the activity in kidney and liver (Brandwein *et al.*, 1981), suggesting another soluble isoenzyme in the latter tissues. This new subunit was isolated using the amino acid sequence of the putative catalytic domain of soluble guanylyl cyclase in the polymerase chain reaction (Yuen *et al.*, 1990). The sequence revealed more similarity toward the known β subunit than the α subunit, and was subsequently designated the β_2 subunit. This subunit is preferentially expressed in kidney and liver, whereas the mRNA for both α and β subunits (now designated the α_1 and β_1 subunits) are most abundant in lung and brain (Nakane *et al.*, 1990; Yuen *et al.*, 1990). The β_2 subunit lacks 62 amino-terminal amino acids of the lung β_1 subunit, but at the carboxy terminal extends with 86 amino acids beyond that of the lung β_1 subunit. In addition, it contains a consensus sequence for isoprenylation/carboxymethylation. The β_2 subunit shows a higher similarity at the amino-terminal region to the β_1 subunit than the α_1 subunit. However, the putative catalytic domain resembles GC-A more closely than the β_1 subunit. The expression of the β_2 subunit by itself did not yield an enzymatically active guanylyl cyclase (Yuen *et al.*, 1990).

Harteneck *et al.* (1991) isolated a cDNA coding for a new subunit of soluble guanylyl cyclase from human fetal brain library with a calculated molecular mass of 82 kDa. Because this subunit shows a higher degree of homology toward the α_1 subunit, and was able to replace the α_1 but not the β_1 subunit in co-expression experiments, this new subunit appears to be an isoform of the α subunit and was, thus, designated as α_2.

Recently, Giuili *et al.* (1992) isolated cDNAs corresponding to the 70- and 82-kDa subunits of soluble guanylyl cyclase from human adult brain, and they designated these subunits α_3 and β_3. The comparison of the sequence of various subunits revealed that the α_3 and β_3 subunits are more homologous to the α_1 and β_1 subunits than to the α_2 and β_2 subunits, indicating that these are the human α_1 and β_1 subunits of soluble guanylyl cyclase.

IV. Summary

The cloning of particulate and soluble guanylyl cyclases is summarized in Table I. With respect to transmembrane signal transduction systems, guanylyl and adenylyl cyclases can be grouped together with some protein tyrosine kinases and protein tyrosine phosphatases to form a diverse protein family with various structural and functional similarities (Garbers, 1989, 1991, 1992; Koesling *et al.*, 1991; Chinkers and Garbers, 1991; Fig. 1). Particulate guanylyl cyclase contains a single transmembrane domain, and the peptide-binding portion (ligand receptor) is on the exterior surface and the catalytic region on the interior, similar to the protein tyrosine kinase/receptor and the protein tyrosine phosphatase/receptor families (Yarden *et al.*, 1986; Charbonneau *et al.*, 1988; Tonks *et al.*, 1988). Protein tyrosine kinases and phosphatases are also activated by ligand binding to the extracellular domain, which in turn results in phosphorylation or dephosphorylation. On the other hand, soluble guanylyl cyclase exists as a heterodimer with two putative catalytic domains, and both subunits are

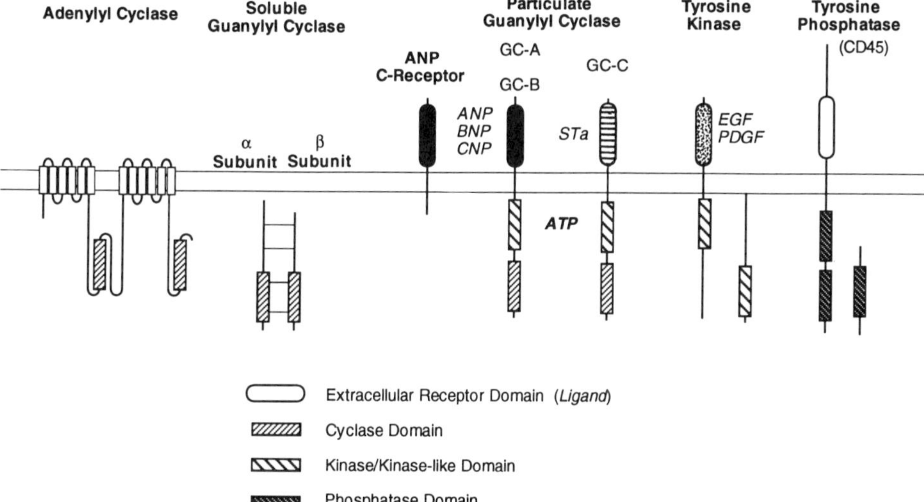

Fig. 1 Members of the cyclase–kinase–phosphatase family. Particulate and soluble guanylyl cyclases, adenylyl cyclase, ANP-R2 receptor, receptor-linked particulate and soluble protein tyrosine kinases, and protein tyrosine phosphatases form a family due to their amino acid homology and structural similarity. Labeled boxes indicate sequence homologies. ANP, atrial natriuretic peptide; BNP, brain natriuretic peptide; CNP, C-type natriuretic peptide; STa, heat-stable enterotoxin from *E. coli;* EGF, epidermal growth factor; PDGF, platelet-derived growth factor; CD45, leukocyte common antigen.

Table I

Structures of Cloned Guanylyl Cyclases

Tissue (abbreviation)	Activator on expression	Molecular mass (kDa)		Reference
		SDS–PAGE	cDNA	
Particulate guanylyl cyclase (homodimer?)				
Sea urchin, *Arbacia punctulata* testis	—	160	106	Singh *et al.* (1988)
Strongylocentrotus purpuratus testis	—	135	124	Thorpe and Garbers (1989)
Human kidney (GC-A)	ANP>BNP	—	115	Lowe *et al.* (1989)
Rat brain (GC-A)	ANP>BNP	120	116	Chinkers *et al.* (1989)
	ANP>BNP>>CNP	—	—	Ohyama *et al.* (1992)
Murine Leydig cells (GC-A)	ANP	—	119	Pandey and Singh (1990)
Rat brain (GC-B)	BNP>ANP	120?	115	Schultz *et al.* (1989)
	CNP>ANP=BNP	—	—	Ohyama *et al.* (1992)
Human placenta (GC-B)	BNP>ANP	—	115	Chang *et al.* (1989)
Rat small intestine (GC-C)	STa	130	121	Schultz *et al.* (1990)
Human ileum (GC-C)	STa	120	121	de Sauvage *et al.* (1991)
	STa	—	121	Singh *et al.* (1991)
Soluble guanylyl cyclase (heterodimer)				
Rat lung (α_1, β_1)	Nitroprusside	70 + 82	70 + 78	Nakane *et al.* (1988, 1990)
Bovine lung (α_1, β_1)	—	70 + 73	70 + 78	Koesling *et al.* (1988, 1990)
	Nitroprusside	—	—	Harteneck *et al.* (1991)
Rat kidney (β_2)	—	—	76	Yuen *et al.* (1990)
Human fetal brain (α_2)	Nitroprusside	—	82	Harteneck *et al.* (1991)
Human brain (α_3, β_3)	—	—	70 + 81	Giuili *et al.* (1992)

essential for enzyme activity and activation by nitric oxide. It is thus particularly interesting that adenylyl cyclase also contains two catalytic domains, which are both necessary for catalytic activity (Tang *et al.*, 1991). It is possible that particulate guanylyl cyclase may also dimerize on hormonal stimulation and two catalytic domains from two monomers form a functional catalytic center capable of forming cyclic GMP. The catalytic core of GC-A expressed in bacteria was shown to form a homodimer with positively cooperative kinetics (Thorpe *et al.*, 1991). The physiological significance of the existence of multiple forms of soluble guanylyl cyclase subunits remains unclear. Future studies should reveal the differences in tissue distribution and activation by nitrovasodilators in various heterodimers of soluble guanylyl cyclase.

References

Bakalyar, H. A., and Reed, R. R. (1990). Identification of a specialized adenylyl cyclase that may mediate odorant detection. *Science* **250,** 1403–1406.

Brandwein, H., Lewicki, J., and Murad, F. (1981). Production and characterization of monoclonal antibodies to soluble rat lung guanylate cyclase. *Proc. Natl. Acad. Sci. U.S.A.* **78,** 4241–4245.

Buechler, W. A., Nakane, M., and Murad, F. (1991). Expression of soluble guanylyl cyclase activity requires both enzyme subunits. *Biochem. Biophys. Res. Commun.* **174,** 351–357.

Chang, C.-H., Kohse, K. P., Chang, B., Hirata, M., Jiang, B., Douglas, J. E., and Murad, F. (1990). Characterization of ATP-stimulated guanylate cyclase activation in rat lung membranes. *Biochim. Biophys. Acta* **1052,** 159–165.

Chang, M., Lowe, D. G., Lewis, M., Hellmiss, R., Chen, E., and Goeddel, D. V. (1989). Differential activation by atrial and brain natriuretic peptides of two different receptor guanylate cyclases. *Nature (London)* **341,** 68–72.

Charbonneau, H., Tonks, N. K., Walsh, K. A., and Fischer, E. H. (1988). The leukocyte common antigen (CD45): A putative receptor-linked protein tyrosine phosphatase. *Proc. Natl. Acad. Sci. U.S.A.* **85,** 7182–7186.

Chinkers, M., and Garbers, D. L. (1989). The protein kinase domain of the ANP receptor is required for signaling. *Science* **245,** 1392–1394.

Chinkers, M., and Garbers, D. L. (1991). Signal transduction by guanylyl cyclases. *Annu. Rev. Biochem.* **60,** 553–575.

Chinkers, M., Garbers, D. L., Chang, M.-S., Lowe, D. G., Chin, H., Goeddel, D. V., and Schulz, S. (1989). A membrane form of guanylate cyclase is an atrial natriuretic peptide receptor. *Nature (London)* **338,** 78–83.

Chinkers, M., Singh, S., and Garbers, D. L. (1991). Adenine nucleotides are required for activation of rat atrial natriuretic peptide receptor/guanylyl cyclase expressed in a baculovirus system. *J. Biol. Chem.* **266,** 4088–4093.

Chrisman, T. D., Garbers, D. L., Parks, M. M., and Hardman, J. G. (1975). Characterization of particulate and soluble guanylate cyclase from rat lung. *J. Biol. Chem.* **250,** 374–381.

de Sauvage, F. J., Camerato, T. R., and Goeddel, D. V. (1991). Primary structure and functional expression of the human receptor for *Escherichia coli* heat-stable enterotoxin. *J. Biol. Chem.* **266,** 17912–17918.

Field, M., Graf, L. H., Laird, W. J., and Smith, P. L. (1978). Heat stable enterotoxin of *Escherichia coli:* In vitro effects on guanylate cyclase activity, cyclic GMP concentration, and ion transport in small intestine. *Proc. Natl. Acad. Sci. U.S.A.* **75,** 2800–2804.

Fuller, F., Porter, J. G., Arfsten, A. E., Miller, J., Schilling, J. W., Scarborough, R. M., Lewicki, J. A., and Schenk, D. B. (1988). Atrial natriuretic peptide clearance receptor. *J. Biol. Chem.* **263,** 9395–9401.

Garbers, D. L. (1978). Sea urchin sperm guanylate cyclase antibody. Cross reactivity with various rat tissue guanylate cyclases. *J. Biol. Chem.* **253,** 1898–1901.

Garbers, D. L. (1989). Guanylate cyclase, a cell surface receptor. *J. Biol. Chem.* **264,** 9103–9106.

Garbers, D. L. (1991). Guanylyl cyclase-linked receptors. *Pharmacol. Ther.* **50,** 337–345.

Garbers, D. L. (1992). Guanylyl cyclase receptors and their endocrine, paracrine, and autocrine ligands. *Cell (Cambridge, Mass.)* **71,** 1–4.

Giuili, G., Scholl, U., Bulle, F., and Guellaën, G. (1992). Molecular cloning of the cDNAs coding for the two subunits of soluble guanylyl cyclase from human brain. *FEBS Lett.* **304,** 83–88.

Guerrant, R. L., Hughes, J. M., Chang, B., Robertson, D. C., and Murad, F. (1980). Activation of intestinal guanylate cyclase by heat-stable enterotoxin of *Escherichia coli:* Studies of tissue specificity, potential receptors, and intermediates. *J. Infect. Dis.* **142,** 220–228.

Harteneck, C., Koesling, D., Söling, A., Schultz, G., and Böhme, E. (1990). Expression of soluble guanylyl cyclase. *FEBS Lett.* **272,** 221–223.

Harteneck, C., Wedel, B., Koesling, D., Malkewitz, J., Böhme, E., and Schultz, G. (1991). Molecular cloning and expression of a new α-subunit of soluble guanylyl cyclase. *FEBS Lett.* **292,** 217–222.

Hughes, J., Murad, F., Chang, B., and Guerrant, R. (1978). The role of cyclic GMP in the mechanism of action of the heat-stable enterotoxin of *E. coli. Nature (London)* **271,** 755–756.

Humbert, P., Niroomand, F., Fischer, G., Mayer, B., Koesling, D., Hinsch, K.-H., Gausephol, H., Frank, R., Schultz, G., and Böhme, E. (1990). Purification of soluble guanylate cyclase form bovine lung by a new immunoaffinity chromatographic method. *Eur. J. Biochem.* **190,** 273–278.

Kamisaki, Y., Saheki, S., Nakane, M., Palmieri, J., Kuno, T., Chang, B., Waldman, S. A., and Murad, F. (1986). Soluble guanylate cyclase from rat lung exists as a heterodimer. *J. Biol. Chem.* **261,** 7236–7241.

Kimura, H., and Murad, F. (1975). Localization of particulate guanylate cyclase in plasma membranes and microsomes of rat liver. *J. Biol. Chem.* **250,** 4810–4817.

Koesling, D., Herz, J., Gausepohl, H., Nirromand, F., Hinsch, K.-D., Mülsch, A., Böhme, E., Schultz, G., and Frank, R. (1988). The primary structure of the 70 kDa subunit of bovine soluble guanylate cyclase. *FEBS Lett.* **239,** 29–34.

Koesling, D., Harteneck, C., Humbert, P., Bosserhoff, A., Frank, R., Schultz, G., and Böhme, E. (1990). The primary structure of the larger subunit of soluble guanylyl cyclase from bovine lung. *FEBS Lett.* **266,** 128–132.

Koesling, D., Böhme, E., and Schultz, G. (1991). Guanylyl cyclases, a growing family of signal-transducing enzymes. *FASEB J.* **5,** 2785–2791.

Koller, K. J., Lowe, D. G., Bennett, G. L., Minamino, N., Kangawa, K., Matsuo, H., and Goeddel, D. V. (1991). Selective activation of the B natriuretic peptide receptor by C-type natriuretic peptide (CNP). *Science* **252,** 120–123.

Krupinski, J., Coussen, F., Bakalyar, H. A., Tang, W.-J., Feinstein, P. G., Orth, K.,

Slaughter, C., Reed, R. R., and Gilman, A. G. (1989). Adenylyl cyclase amino acid sequence: Possible channel- or transporter-like structure. *Science* **244,** 1558–1564.

Kuno, T., Kamisaki, Y., Waldman, S. A., Gariepy, J., Schoolnik, G., and Murad, F. (1986a). Characterization of the receptor for heat-stable enterotoxin from *Escherichia coli* in rat intestine. *J. Biol. Chem.* **261,** 1470–1476.

Kuno, T., Andresen, J. W., Kamisaki, Y., Waldman, S. A., Chang, L. Y., Saheki, S., Leitman, D. C., Nakane, M., and Murad, F. (1986b). Co-purification of an atrial natriuretic factor receptor and particulate guanylate cyclase from rat lung. *J. Biol. Chem.* **261,** 5817–5823.

Kurose, H., Inagami, T., and Ui, M. (1987). Participation of adenosine 5′-triphosphate in the activation of membrane-bound guanylate cyclase by the atrial natriuretic factor. *FEBS Lett.* **219,** 375–379.

Lowe, D. G., Chang, M.-S., Hellmiss, R., Chen, E., Singh, S., Garbers, D. L., and Goeddel, D. V. (1989). Human atrial natriuretic peptide receptor defines a new paradigm for second messenger signal transduction. *EMBO J.* **8,** 1377–1384.

Nakane, M., and Deguchi, T. (1980). Production and properties of antibody to soluble guanylate cyclase purified from bovine brain. *Biochim. Biophys. Acta* **631,** 20–27.

Nakane, M., and Deguchi, T. (1982). Monoclonal antibody to soluble guanylate cyclase of rat brain. *FEBS Lett.* **140,** 89–92.

Nakane, M., Saheki, S., Kuno, T., Ishii, K., and Murad, F. (1988). Molecular cloning of a cDNA coding for 70 kilodalton subunit of soluble guanylate cyclase from rat lung. *Biochem. Biophys. Res. Commun.* **157,** 1139–1147.

Nakane, M., Arai, K., Saheki, S., Kuno, T., Buechler, W., and Murad, F. (1990). Molecular cloning and expression of cDNAs coding for soluble guanylate cyclase form rat lung. *J. Biol. Chem.* **265,** 16841–16845.

Ohyama, Y., Miyamoto, K., Saito, Y., Minamino, N., Kangawa, K., and Matsuo, H. (1992). Cloning and characterization of two forms of C-type natriuretic peptide receptor in rat brain. *Biochem. Biophys. Res. Commun.* **183,** 743–749.

Pandey, K. N., and Singh, S. (1990). Molecular cloning and expression of murine guanylate cyclase/atrial natriuretic factor receptor cDNA. *J. Biol. Chem.* **265,** 12342–12348.

Paul, A. K., Marala, R. B., Jaiswal, R. K., and Sharma, R. K. (1987). Coexistence of guanylate cyclase and atrial natriuretic factor receptor in a 180-kDa protein. *Science* **235,** 1224–1226.

Schultz, S., Singh, S., Bellet, R. A., Singh, G., Tubb, D. J., Chin, H., and Garbers, D. L. (1989). The primary structure of a plasma membrane guanylate cyclase demonstrates diversity within this receptor family. *Cell (Cambridge, Mass.)* **58,** 1155–1162.

Schultz, S., Green, C. K., Yuen, P. S. T., and Garbers, D. L. (1990). Guanylyl cyclase is a heat stable enterotoxin receptor. *Cell (Cambridge, Mass.)* **63,** 941–948.

Singh, S., Lowe, D. G., Thorpe, D. S., Rodriguez, H., Kuang, W.-J., Dangott, L. J., Chinkers, M., Goeddel, D. V., and Garbers, D. L. (1988). Membrane guanylate cyclase is a cell-surface receptor with homology to protein kinases. *Nature (London)* **334,** 708–712.

Singh, S., Singh, G., Heim, J.-M., and Gerzer, R. (1991). Isolation and expression of a guanylate cyclase-coupled heat stable enterotoxin receptor cDNA from a human colonic cell line. *Biochem. Biophys. Res. Commun.* **179,** 1455–1463.

Sudoh, T., Minamino, N., Kangawa, K., and Matsuo, H. (1990). C-type natriuretic peptide (CNP): A new member of natriuretic peptide family identified in porcine brain. *Biochem. Biophys. Res. Commun.* **168,** 863–870.

Takayanagi, R., Inagami, T., Snajdar, R. M., Imada, T., Tamura, M., and Misono, K. S. (1987). Two distinct forms of receptors for atrial natriuretic factor in bovine adrenocortical cells. *J. Biol. Chem.* **262,** 12104.

Tang, W.-J., Krupinski, J., and Gilman, A. G. (1991). Expression and characterization of calmodulin-activated (type I) adenylyl cyclase. *J. Biol. Chem.* **266,** 8595–8603.

Thorpe, D. S., and Garbers, D. L. (1989). The membrane form of guanylate cyclase: Homology with a subunit of the cytoplasmic form of the enzyme. *J. Biol. Chem.* **264,** 6545–6549.

Thorpe, D. S., and Morkin, E. (1990). The carboxy region contains the catalytic domain of the membrane form of guanylate cyclase. *J. Biol. Chem.* **265,** 14717–14720.

Thorpe, D. S., Niu, S., and Morkin, E. (1991). Overexpression of dimeric guanylyl cyclase cores of an atrial natriuretic peptide receptor. *Biochem. Biophys. Res. Commun.* **180,** 538–544.

Tonks, N. K., Charbonneau, H., Diltz, C. D., Fischer, E. H., and Walsh, K. A. (1988). Demonstration that the leukocyte common antigen CD45 is a protein tyrosine phosphatase. *Biochemistry* **27,** 8695–8701.

Waldman, S. A., and Murad, F. (1987). Cyclic GMP synthesis and function. *Pharmacol. Rev.* **39,** 163–196.

Waldman, S. A., Rapoport, R. S., and Murad, F. (1984). Atrial natriuretic factor selectively activates particulate guanylate cyclase and elevates cyclic GMP in rat tissues. *J. Biol. Chem.* **259,** 14332–14334.

Waldman, S. A., Kuno, T., Kamisaki, Y., Chang, L. Y., Gariepy, J., O'Hanley, P., Schoolnik, G., and Murad, F. (1986). Intestinal receptor for heat-stable enterotoxin of *E. coli* is tightly coupled to a novel form of particulate guanylate cyclase. *Infect. Immun.* **51,** 320–326.

Yarden, Y., Escobedo, J. A., Kuang, W.-J., Yang-Feng, T. L., Daniel, T. O., Tremble, P. M., Chen, E. Y., Ando, M. E., Harkins, R. N., Francke, U., Fried, V. A., Ullrich, A., and Williams, L. T. (1986). Structure of the receptor for platelet-derived growth factor helps define a family of closely related growth factor receptors. *Nature (London)* **323,** 226–232.

Yuen, P. S. T., Potter, L. R., and Garbers, D. L. (1990). A new form of soluble guanylyl cyclase is preferentially expressed in rat kidney. *Biochemistry* **29,** 10872–10878.

Regulation of Cytosolic Guanylyl Cyclase by Nitric Oxide: The NO–Cyclic GMP Signal Transduction System

Ferid Murad

Molecular Geriatrics Corporation
Lake Bluff, Illinois 60044

I. Introduction

The nitric oxide–cyclic GMP signal transduction system has emerged in recent years as a very ubiquitous pathway for intracellular and intercellular communication. This review is intended to describe and summarize some of our observations and those of other laboratories that have helped lead us and others to our present understanding of the nitric oxide–cyclic GMP signal transduction system. Readers are also referred to some of our earlier reviews for references and information (Murad, 1986, 1989a,b; Murad *et al.*, 1988, 1990a; Waldman and Murad, 1987) as well as some discussion and references in Chapters 4 and 8.

Although cyclic GMP was considered a potentially important second messenger in hormonally induced effects for many years, there was considerable disappointment with much of the early work in defining a definitive role for this cyclic nucleotide in signal transduction. The significance of cyclic GMP in cellular regulation comes from our work and that of others with nitrovasodilators, endothelial–dependent vasodilators, atrial natriuretic peptides (ANP) and *Escherichia coli* heat-stable enterotoxin (ST). Although some of these studies will be reviewed in this chapter, readers are also referred to Chapters 4, 5, 8, and 11. As is often the case, some of the important early observations were serendipitous and accidental as

Advances in Pharmacology, Volume 26

was the case with our early work with azide and nitric oxide. This work will be briefly reviewed later.

Nitric oxide is a simple, but unique, gaseous molecule and free radical that can serve many diverse functions including an intracellular second messenger as well as an intercellular messenger (paracrine substance, autacoid, or hormone) to regulate neighboring and perhaps distant cells. The important interrelationships of nitric oxide and cyclic GMP that began in the mid and late 1970s and that have expanded remarkably in recent years have led to our present understanding of a fundamentally ubiquitous and important signal transduction system. In addition to its function as an intracellular second messenger and local extracellular agent for intercellular communication, I would like to suggest that nitric oxide might also be viewed as a more classical humoral substance. Effects of nitric oxide at some distant target site could occur if there were complexes or carrier states for nitric oxide. For example, if a carrier(s) or complex(es) of nitric oxide was formed that was inactive and that could release nitric oxide at a distant target site (perhaps selectively with specific uptake or transport), then nitric oxide could be viewed also as a classical humoral substance. Many hormones are inactive when bound or complexed with carrier proteins and become active in information transfer after their release or dissociation at their distant target. With the rapid recirculation times in the cardiovasculature, nitric oxide or a complex of nitric oxide would only need to survive for seconds to serve as a humoral substance downstream from its formation or at a distant target site. However, this hypothesis will be rather difficult to prove definitely. Because of the ubiquity of nitric oxide in most cell types, it will be virtually impossible to prove that nitric oxide generated at one site is the same molecule that functions at a distant site, particularly since radionuclides of NO are not available. Proof will probably require isotopic labeling and nuclear magnetic resonance detection at concentrations that are well above the physiologically relevant concentrations. The ubiquity and reactivity of nitric oxide with thiols, proteins, sugars, metals, heme proteins, etc., permit us to predict with some degree of certainty that nitric oxide complexes and adducts will, undoubtedly, be present in various extracellular fluids. The question is, do any of these complexes or free nitric oxide serve a humoral function. We suspect that they will, considering the very low concentrations (nanamolar) of nitric oxide required to activate guanylyl cyclase and elevate cyclic GMP levels in tissues (see below).

Nitric oxide is formed by most but not all cells. Its formation and release by central and peripheral neurons permit the molecule to function as a neurotransmitter of "nitrinergic" neurons. Thus, nitric oxide may function as an intracellular second messenger and an intercellular messenger (autacoid, neurotransmitter, or hormone). Such a diverse role for a single molecule has not been described previously. Perhaps the agents that come closest to fulfilling all of these roles are some of the eicosanoids. However, multiple members of this molecular class together share these diverse roles in signal transduction.

The discoveries of the effects of nitric oxide and particularly how they relate to cyclic GMP synthesis will be reviewed below.

II. Effects of Nitric Oxide on Cyclic GMP Synthesis and Smooth Muscle Relaxation

Like many laboratories working with cyclic GMP in the late 1960s and early 1970s, we were adding various hormones and drugs to different intact cells and tissues and attempting to correlate cyclic GMP accumulation in these tissues with some possible physiological and biochemical functions. Frankly, this descriptive cataloging, although leading to a number of publications by us and others, provided no significant insight into the mechanisms of humoral regulation of cyclic GMP synthesis or possible biochemical or physiological functions of cyclic GMP. We then turned to the characterization of the enzyme that synthesized cyclic GMP from GTP, guanylyl cyclase (Kimura and Murad, 1974, 1975a,b,c). We had hoped that biochemical characterization of the enzyme would provide us with insights regarding mechanisms of hormonal regulation of guanylyl cyclase and cyclic GMP functions. This approach, needless to say, has certainly paid off with regard to our current understanding of mechanisms of hormone action, role of nitric oxide, and some cyclic GMP functions. We quickly learned that there were soluble and particulate isoforms of guanylyl cyclase in most tissues (Kimura and Murad, 1974, 1975a,b,c). The kinetic and physiochemical properties of the cytosolic (soluble) guanylyl cyclase were quite different from the membrane associated (particulate) guanylyl cyclase. With regard to the substrate GTP, the crude soluble isoform gave typical and linear Michaelis–Menton kinetics, whereas the crude particulate isoform showed curvilinear plots or cooperativity. We thought initially that the apparent cooperativity in these crude preparations could be attributable to contaminating ATPases and phosphatases that modified the GTP concentrations in incubations. Therefore, we added various inhibitors to our crude guanylyl cyclase incubations such as fluoride, pyrophosphate, and azide. We accidentally found that azide activated some, but not all, preparations of guanylyl cyclase (see Kimura *et al.*, 1975a,b). In addition to azide, nitrite and hydroxylamine also activated our preparations (Kimura *et al.*, 1975a,b). This was an exciting turn of events for us because it was the first group of agents that would activate the enzyme in both intact cell and cell-free preparations. We reasoned that if we understood this activation mechanism, perhaps we would someday be clever enough to reconstitute a hormone effect on GMP synthesis in cell-free systems (Murad, 1986, 1989a,b; Murad *et al.,* 1988, 1990b; Waldman and Murad 1987).

The effects of azide in some, but not all, preparations permitted us to develop assay systems and various classical biochemical mixing experiments to help us understand the mechanisms of activation. The lag time for azide activation and the absence of azide effects in all preparations convinced us that another intermediate was being generated in our experiments that was responsible for guanylyl cyclase activation, and we were committed to identifying this active intermediate. We also found that the azide effect was dependent on the presence of some heme-containing proteins such as catalase, peroxidase, or cytochromes in our preparation that probably convert azide to the active intermediate (see Mittal *et al.,* 1975, 1977). In addition, other heme proteins such as hemoglobin and myoglobin blocked the azide activation of guanylyl cyclase (Mittal *et al.,* 1978). The requirements for some heme proteins to see the azide activation and the inhibitory effects of hemoglobin and myoglobin could explain the apparent tissue selectivity for the effects of azide (i.e., some tissue extracts lacked the required proteins for azide conversion to the active molecule while other tissue extracts contained large amounts of inhibitors of the azide effect).

Fortunately, at the same time our laboratory was working with tracheal smooth muscle and gastrointestinal smooth muscle preparations. We suspected that cyclic GMP might contract smooth muscle and we set out to test the hypothesis. After finding the stimulatory effects of azide, nitrite, and hydroxylamine on guanylyl cyclase, we added these agents to various intact tissues including brain and our smooth muscle preparations. These agents elevated cyclic GMP and caused smooth muscle relaxation rather than contraction (see Katsuki *et al.,* 1977b,c). We logically tried other smooth muscle relaxants such as nitroprusside and nitroglycerin and found that they too activated guanylyl cyclase and elevated cyclic GMP (see Katsuki *et al.,* 1977a,b,c). We coined the term "nitrovasodilators" for this broad class of guanylyl cyclase activators and found that their effects were mediated by the formation of nitric oxide (see Katsuki *et al.,* 1977a; Arnold *et al.,* 1977). Thus, it appeared that azide, hydroxylamine, nitrate, and other nitrovasodilators such as nitroglycerin and nitroprusside could be converted enzymatically or nonenzymatically, based on the prodrug used, to the reactive intermediate. We also learned that nitric oxide formation could explain the effects of this broad class of nitrovasodilators (see Fig. 1). Furthermore, the redox state of tissues and extracts could have

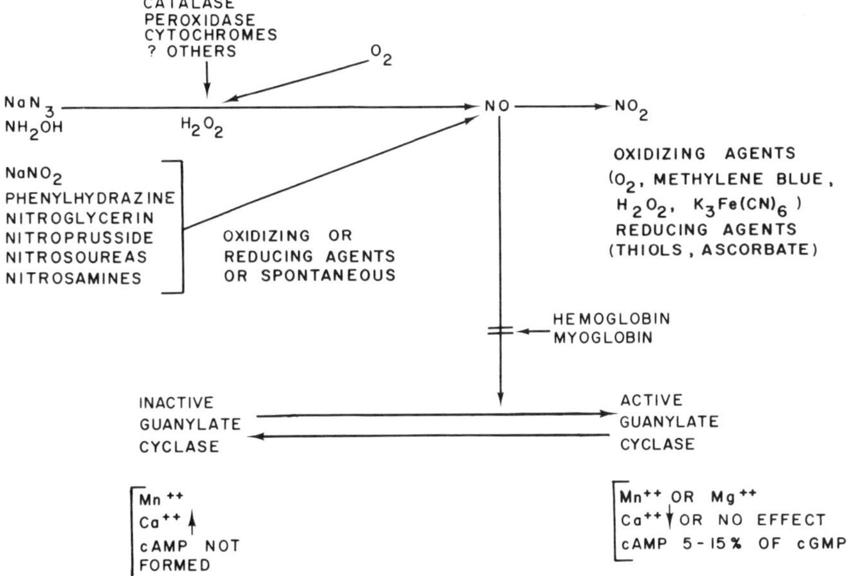

Fig. 1 Conversion of azide and other nitrovasodilators to nitric oxide and their effects on cyclic GMP formation. From Murad *et al.* (1978).

a profound effect on the formation of nitric oxide from some prodrugs and/or the inactivation of nitric oxide by its oxidation to nitrogen dioxide, which was inactive. The effects of oxidizing and reducing agents and nitric oxide scavengers, such as methylene blue, potassium ferricyanide, and hemoglobin, subsequently became important research tools in characterizing the formation and effects of endothelial-derived relaxing factor (see below).

To our knowledge, guanylyl cyclase has been the first and only enzyme whose activity is increased with a free radical activation mechanism. Nitric oxide can activate homogeneous preparations of guanylyl cyclase in the absence of other macromolecules (Braughler *et al.,* 1979a,b). Furthermore, the activation was reversible, as would be expected if this regulatory mechanism were physiologically relevant. Although the precise mechanisms of nitric oxide activation are not totally understood, it is apparent from the work in Bohme's and Ignarros' laboratories (Gerzer *et al.,* 1981; Ignarro *et al.,* 1986) that heme functions as a required prosthetic group. Although several hypotheses have been offered to explain the mechanism of activation, large quantities of purified enzyme (milligrams) are required for detailed ESR/EPR studies with simultaneous monitoring of catalytic activity. Until recently, such studies have not been possible and this was one of the reasons we initiated our cloning and expression studies with the enzyme (see Chapter 2). To date, such studies have not been conducted by us or other laboratories. The possible mechanisms of activation are discussed in greater detail in Chapter 4. It may be that other mechanisms also participate in the activation process since thiol groups in the enzyme also appear to be critical for activation (Brandwein *et al.*, 1981).

Although the soluble isoform of guanylyl cyclase is clearly activated in a reversible manner with nitric oxide under physiological conditions at low concentrations (K_a is about 1–10 nM), other isoforms of guanylyl cyclase are also activated by azide, nitroprusside, or nitric oxide. Our earliest studies in this regard were with particulate guanylyl cyclase from rat intestinal mucosa, a tissue in which the enzyme is predominantly or exclusively particulate. With washed, high-speed particulate fractions from rat intestinal mucosa, nitrovasodilators activated guanylyl cyclase. Most tissues, however, contain greater quantities of soluble isoenzyme, However, many crude particulate guanylyl cyclase preparations from these tissues are often contaminated with entrapped soluble enzyme. Unfortunately, detergents, which are required for the solublization and purification of the particulate enzyme, inhibit the activation. Therefore, we prepared tryptic fragments of particulate guanylyl cyclase from rat liver membranes and also found that these preparations were activated with nitric oxide (Waldman *et al.*, 1982). Recently, we also purified the cytoskeletal isoform of guanyl cyclase from bovine rod outer segments and also found that this isoform could be activated with nitruprusside (Horio and Murad, 1991). Thus, several isoforms of guanylyl cyclase can be activated with nitric oxide and perhaps these isoforms also contain heme prosthetic groups. Obviously, a number of additional studies are required.

When guanylyl cyclase is activated by nitrovasodilators, the properties of the enzyme change dramatically. The V_{max} may be increased as much as 100- to 200-fold under some conditions. Although the native, basal enzyme prefers Mn^{2+}as its cation cofactor, the activated enzyme can utilize Mn^{2+} or Mg^{2+} equally well, which makes more sense considering the K_a values and cellular concentrations of these cations (Kimura *et al.*, 1976). The K_m for GTP is markedly decreased and the enzyme generally becomes more labile to storage. The enzyme can also synthesize cyclic AMP from ATP (Mittal *et al.*, 1979). Although the activated enzyme prefers to make cyclic GMP from GTP, the formation of cyclic AMP from ATP can be appreciable (as much as 5 to 15% of the rate of cyclic GMP formation under some conditions). This alternate pathway for cyclic AMP synthesis has undoubtedly led to misinterpretations of some experiments and data where adenylyl cyclase regulation was expected.

In many cell types, including vascular smooth muscle, the increases in cyclic GMP lead to cyclic GMP-dependent protein kinase activation and altered phosphorylation of many endogenous proteins, including the dephosphorylation of myosin light chain and relaxation (Rapaport *et al.*, 1983; Rapaport and Murad, 1983b; Murad, 1986; Waldman and Murad, 1987) (see Fig. 2). The decreased phosphorylation of myosin light chain probably occurs with cyclic GMP inhibition of phospholipase C activity, which appears to be cyclic GMP-dependent protein kinase mediated (Hirata *et al.*, 1990). Decreased phospholipase C activity and decreased inositol tris-phosphate formation lead to decreased cytosolic calcium, which is required for myosin light-chain kinase activity. These areas are reviewed in greater detail in Chapters 7, 8, 9, and 10.

Although most of the cyclic GMP mediated effects to date are probably mediated through increased cyclic GMP-dependent protein kinase activity (see Chapter 7), other effects of cyclic GMP on phototransduction and phosphodiesterase (Chapter 6) regulation appear to be independent of the

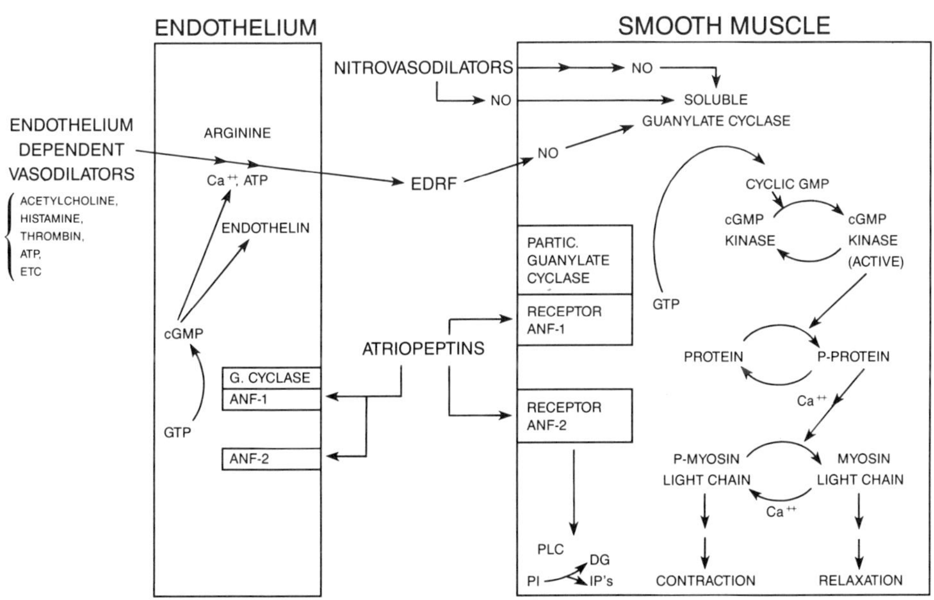

Fig. 2 The effects of some vasodilators on cyclic GMP formation and vascular relaxation. Modified from Murad (1986).

kinase. It is expected that other processes such as transcriptional regulation could be kinase dependent or independent. Again, much is yet to be done.

III. Effects of Endothelial-Derived Relaxing Factor on Cyclic GMP Formation

The studies by Furchgott's laboratory described relaxation of vascular preparations with a variety of agents when the endothelium was intact but not after the endothelium was damaged or removed (Furchgott and Zawodski, 1980). These endothelium-dependent vasodilators produced an endothelial-derived relaxant factor (EDRF) required for the relaxant effects. The lability of this factor and the similarities of these effects to those of nitrovasodilators and *E. coli* heat-stable enterotoxin suggested to us that these effects may also be mediated through cyclic GMP formation. This was indeed the case. We found that a variety of endothelium-dependent vasodilators increased cyclic GMP accumulation in the smooth muscle compartment of vascular preparations (Rapaport and Murad, 1983a). The effects of endothelium-dependent vasodilators on cyclic GMP accumulations were also dependent on the integrity and/or presence of the endothelium in preparations. Furthermore, the effects of EDRF were virtually identical to those of nitrovasodilators with regard to cyclic GMP-dependent protein kinase activation and altered protein phosphorylation including the dephosphorylation of myosin light chain and relaxation (Rapaport and Murad, 1983b; Draznin *et al.*, 1986) (see Fig. 2). Due to the similar biochemical and pharmacological effects of EDRF and nitrovasodilators, we viewed EDRF as the "endogenous nitrovasodilator" (Murad, 1986). Subsequently, Ignarro and Furchgott suggested that EDRF was nitric oxide (Ignarro *et al.*, 1987; Furchgott, 1988). From the reactivity of nitric oxide and the low concentrations of nitric oxide required for guanylyl cyclase activation, we have always suspected that EDRF is a nitric oxide complex(es) or adduct(s) that can liberate nitric oxide. Although this view has subsequently been shared by some other laboratories, the chemical identity of EDRF cannot be proven with current technologies because of the lability of EDRF and the apparent low concentrations in tissues. Nevertheless, most investigators today would agree that EDRF is either nitric oxide or a nitric oxide-like complex and that nitric oxide certainly mediates the effects of EDRF via guanylyl cyclase activation and cyclic GMP formation. It has also been shown that EDRF/NO can also cause other vascular and nonvascular effects such as platelet adhesion, platelet aggregation, and decreased vascular smooth muscle proliferation. The vascular effects of EDRF/NO are discussed further in Chapter 5.

IV. Nitric Oxide Formation

Working with brain preparations and neuroblastoma cell cultures, DeGuchi's laboratory found that an endogenous substance could activate guanylyl cyclase preparations (DeGuchi and Yoshiaka, 1982). This endogenous material was identified as L-arginine. The activation of guanylyl cyclase by L-arginine was similar in many respects to the activation observed with various nitrovasodilators.

Subsequent studies from the laboratories of Hibbs found that murine macrophages could form nitrite and nitrate and the apparent precursor of the synthesis was arginine (Hibbs *et al.,* 1987). Furthermore, arginine analogs such as L-methyl arginine blocked the pathway. Subsequently, Moncada's laboratory extended these observations to endothelial preparations (Palmer *et al.,* 1988). They found, similar to the studies of Hibbs, that one of the guanidino nitrogens of L-arginine could be oxidized and converted to nitric oxide. The other product of the reaction was citrulline. We, and others, found that numerous cell types, in addition to macrophages and endothelial cells, could also carry out the same reactions (Murad *et al.,* 1990a,b; Ishii *et al.,* 1989). To date, there are few cell types that do not possess this enzyme pathway (see Fig. 3).

The enzymes that catalyze this reaction are nitric oxide synthases (NOS); they have also been called EDRF synthase or guanylyl cyclase-activating factor synthase (GAF synthase). These studies rapidly led to the characterization, purification, and cloning of the family of enzymes from many tissues and cell types by many laboratories. Today three gene products from cloning studies have been described (Table I). From purification and characterization studies, additional isoforms and gene products seem likely. Furthermore, there are various post-translational modifications of the enzyme(s), including myristylation and phosphorylation (Pollock *et al.,* 1992; Nakane *et al.,* 1991). Post-translational modifications with myristylation and/or phosphorylation can alter enzyme activity and/or subsellular localization in the membranous or cytosolic compartments. These are currently very active areas of investigation. It seems likely that while six isoforms have been described to date, with at least three different gene products, more isoforms will undoubtedly be described in the near future. Generally, within a given species, the isoforms are 50 to 70%

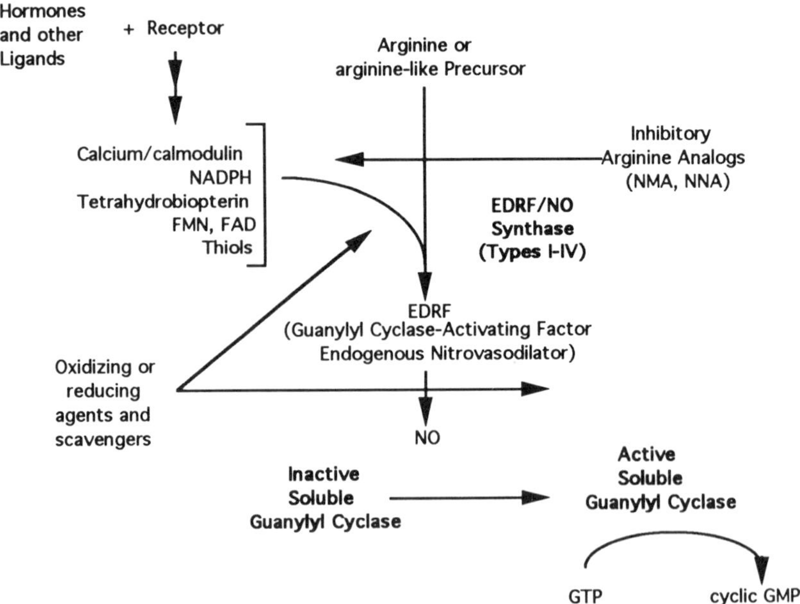

Fig. 3 The nitric oxide–cyclic GMP signal transduction system. Modified from Murad *et al.* (1992).

homologous from cloning studies and deduced amino acid sequencing. However, a specific isoform can be highly homologous between species. For example, the human and rat brain enzymes (Type 1A) are more than 90% identical in sequence . Some isoforms are soluble or particulate and some can be induced with endotoxin and/or cytokines. Some isoforms are also calcium- and calmodulin-dependent (Bredt and Snyder, 1990; Förstermann *et al.,* 1990a,b, 1991a,c). All isoforms to date require NADPH, tetrahydrobiopterin, FMN, and FAD as cofactors. Some isoforms appear to be monomeric, whereas others appear to be active as homodimers.

Polyclonal and/or monoclonal antibodies developed to the various isoforms in our laboratory and other laboratories have been useful in immunohistochemical localization of the various isoforms in different cell types and tissues (Snyder and Bredt, 1991; Schmidt *et al.,* 1992; Pollock *et al.,* 1992). For example, neuronal Type 1A isoform can be found in discrete central and peripheral neurons where nitric oxide may function as a neurotransmitter of these "nitrinergic neurons." NOS-containing neurons include the nonadrenergic–noncholinergic (NANC) neurons in airway smooth muscle, gastrointestinal tract, and corpus cavernosum. Type 1A

Table I
Isoforms of NO Synthase

Type	Cosubstrates cofactors	Regulated by	M_r (kDa)	Present in
I a (soluble)	NADPH, BH_4, FAD/FMN,	Ca^{2+}/calmodulin	155	Brain, cerebellum N1E-115 neuroblastoma cells
I b (soluble)	NADPH	Ca^{2+}/calmodulin	135	Endothelial cells
I c (soluble)	NADPH, BH_4, FAD	Ca^{2+} (*not* calmodulin)	150	Neutrophils
II (soluble)	NADPH, BH_4, FAD/FMN,	unknown (induced by endotoxin/ cytokines)	125	Macrophages smooth muscle, liver, lung, etc.
III (particulate)	NADPH, BH_4, FAD/FMN	Ca^{2+}/calmodulin	135	Endothelial cells
IV (particulate)	NADPH,	unknown (induced by endotoxin/ cytokines)	?	Macrophages and many other induced tissues

Note. All isoenzymes use L-arginine as a substrate and all are inhibited by N^G-methyl-L-arginine and N^G-nitro-L-arginine. The Type Ia, II, and III have been cloned by several laboratories, show about 50 to 60% homology, and obviously represent separate gene products. Some isoforms may also represent post-translational modifications. From Forstermann *et al.* (1991c) and Murad *et al.* (1992).

isoform is also found in pancreatic islets and endometrium (Schmidt *et al.,* 1992). The Type III endothelial isoform has only been reported in endothelial cells, to date, and is the enzyme responsible for EDRF/NO production in vascular preparations (Forstermann *et al.,* 1990b, 1991a,b,c; Pollock *et al.,* 1991). Both Type IA and III isoforms are calcium- and calmodulin-dependent. Type II isoform is inducible with endotoxin and/ or various cytokines and is present in a variety of cell types, including macrophages, smooth muscle, fibroblasts, and liver after appropriate induction. The inducible Type II isoform is calcium–calmodulin independent. The independency of the inducible Type II NOS to calcium and calmodulin is probably incorrect in that recent studies have demonstrated that calmodulin is tightly bound to purified preparations of enzyme (Marletta, personal communications). Perhaps the inducible enzyme is already in the active state with tightly associated calmodulin. Although Type I and III isoforms are thought to be constitutive or housekeeping enzymes, we have found that their expression can also be regulated.

The different physiochemical properties, cofactor requirements, and regulation of the various isoforms of NOS have provided some hope that selective pharmacological regulation of specific isoforms may lead to useful, new therapeutic agents. For example, selective inhibition of the inducible isoforms is expected to be therapeutically useful in managing hypotension in endotoxin-induced shock with sepsis. Selective inhibition of the endothelial Type III isoform should be useful in managing migraine. Thus, the work with nitric oxide synthases is expected to expand dramatically. A literature search has revealed more than 1900 publications on nitric oxide in 1992, an exponential number of publications annually since our first publications on the effects of nitric oxide in 1977.

V. Summary

The understanding of guanylyl cyclase regulation by nitrovasodilators has provided a great deal of information explaining the mechanisms of action of these cardiovascular drugs that have been in clinical use for the past century. The biochemical characterization of guanylyl cyclases and their regulation by NO have also permitted us, and others, to understand the mechanism of action of endothelium-dependent vasodilators and, subsequently, the roles for the nitric oxide–cyclic GMP signal transduction system in numerous cells and tissues. The potential importance of this signal transduction cascade is probably not fully appreciated since numerous additional studies obviously need to be performed. Also, as in many areas of science, serendipitous experiments and observations have added critical data to our present understanding in this field.

Acknowledgments

I thank the numerous trainees and collaborators who, over the years, have been critical to this laboratory's productivity, and many different funding agencies who provided the support for this work during the past two decades.

References

Arnold, W. P., Mittal, C. K., Katsuki, S., and Murad, F. (1977). Nitric oxide activates guanylate cyclase and increases guanosine 3′, 5′-monophosphate levels in various tissue preparations. *Proc. Natl. Acad. Sci. U.S.A.* **74,** 3203–3207.

Brandwein, H. J., Lewicki, J. A., and Murad, F. (1981). Reversible inactivation of guanylate cyclase by mixed disulfide formation. *J. Biol. Chem.* **256,** 2958–2962.

Braughler, J. M., Mittal, C. K., and Murad, F. (1979a). Purification of soluble guanylate cyclase from rat liver. *Proc. Natl. Acad. Sci. U.S.A.* **76,** 219–222.

Braughler, J. M., Mittal, C. K., and Murad, F. (1979b). Effects of thiols, sugars and proteins on nitric oxide activation of guanylate cyclase. *J. Biol. Chem.* **254,** 12450–12454.

Bredt, D. S., and Snyder, S. H. (1990). Isolation of nitric oxide synthetase, a calmodulin-requiring enzyme. *Proc. Natl. Acad. Sci. U.S.A.* **85,** 682–685.

DeGuchi, T., and Yoshiaka M. (1982). L-Arginine identified as an endogenous activator for soluble guanylate cyclase from neuroblastoma cells. *J. Biol. Chem.* **257,** 10147–10157.

Draznin, M. B., Rapoport, R. M., and Murad, F. (1986). Myosin light chain phosphorylation in contraction and relaxation of intact rat thoracic aorta. *Int. J. Biochem.* **18,** 917–928.

Förstermann, U., Gorsky, L., Pollock, J., Ishii, K., Schmidt, H. H. H. W., Heller, M., and Murad, F. (1990a). Hormone induced biosynthesis of endothelium-derived ralaxing factor-nitric oxide-like material in N1E-115 neuroblastoma cells requires calcium and calmodulin. *Mol. Pharmacol.* **38,** 7–13.

Förstermann, U., Gorsky, L., Pollock, J. S., Schmidt, H. H. H. W., Ishii, K., Heller, M., and Murad, F. (1990b). Subcellular localization and regulation of the enzymes responsible for EDRF synthesis in endothelial cells and N1E 115 neuroblastoma cells. *Eur. J. Pharmacol.* **183,** 1625–1626.

Förstermann, U., Pollock, J., Schmidt, H. H. H. W., Heller, M., and Murad, F. (1991a). Calmodulin-dependent endothelium-derived relaxing factor/nitric oxide synthase activity is present in the particulate and cytosolic fractions of bovine aortic endothelial cells. *Proc. Natl. Acad. Sci. U.S.A.* **88,** 1788–1792.

Förstermann, U., Schmidt, H. H. H. W., Pollock, J. S., Heller, M., and Murad, F. (1991b). Enzymes synthesizing guanylyl cyclase activating factor (GAF) in endothelial cells, neuroblastoma cells and rat brain. *J. Cardiovasc. Pharmacol.* **17,** Suppl. 3, 557–564.

Förstermann, U., Schmidt, H. H. H. W., Pollock, J. S., Sheng, H., Mitchell, J. A., Warner, T. D., Nakane, M., and Murad, F. (1991c). Isoforms of EDRF/NO synthase: Characterization and purification from different cell types. *Biochem. Pharmacol.* **42,** 1849–1857.

Furchgott, R., and Zawodski, J. (1980). The obligatory roll of endothelial cells in the relaxation of arterial smooth muscle to acetylcholine. *Nature (London)* **288,** 373–376.

Furchgott, R. F. (1988). Studies on relaxation of rabbit aorta by sodium nitrate: The basis for the proposal that acid-activatable inhibitory factor from bovine retractor penis is organic nitrate and EDRF is nitric oxide. *In* "Vasodilation: Vascular Smooth Muscle Peptides, Autonomic Nerves and Endothelium" (P. M. Vanhoutte, ed.), pp. 401–414. Raven Press, New York.

Gerzer, R., Bohme, E., Hoffman, F., and Schultz, G. (1981). Soluble guanylate cyclase purified from bovine lunc contains heme and copper. *FEBS Lett.* **132,** 71–74.

Hibbs, J. R., Traintor, R. R., and Varrin, Z. (1987). Macrophage cytotoxicity: Role for L-arginine deiminase and imino nitrogen oxidation to nitrive. *Science* **235,** 473–476.

Hirata, M., Kohse, K., Chang, C. H., Ikebe, T., and Murad, F. (1990). Mechanism of cyclic GMP inhibition of inositol phosphate formation in rat aorta segments and cultured bovine aortic smooth muscle cells. *J. Biol. Chem.* **265,** 1268–1273.

Horio, Y., and Murad, F. (1991). Solubilization of guanylate cyclase from bovine rod outer segments and effects of Ca+ + and nitro compounds. *J. Biol. Chem.* **266,** 3411–3415.

Ignarro, L. J., Adams, J., Horwitz, P., and Wood, K. S. (1986). Activation of soluble cyclase by NO-hemeprogeins involves NO-heme exchange: Comparison of heme containing and heim deficient enzymes. *J. Biol. Chem.* **261,** 4997–5002.

Ignarro, L. J., Buga, G. M., Wood, K. S., Byrnes, R.E., and Chaudhuri, G. (1987). Endothelium-derived relaxing factor produced and released from artery and vein is nitri oxide. *Proc. Natl. Acad. Sci. U.S.A.* **84,** 9265–9269.

Ishii, K., Gorsky, L., Förstermann, U., and Murad, F. (1989). Endothelium-derived relaxing factor (EDRF): The endogenous activator of soluble guanylate cyclase in various types of cells. *J. Appl. Cardiol.* **4,** 505–512.

Katsuki, S., Arnold, W., Mittal, C. K., and Murad, F. (1977a). Stimulation of guanylate cyclase by sodium nitroprusside, nitroglycerin and nitric oxide in various tissue preparations and comparison to the effects of sodium azide and hydroxylamine. *J. Cyclic Nucleotide Res.* **3,** 23–35.

Katsuki, S., Arnold, W. P., Mittal, C. K., and Murad, F. (1977b). Stimulation of formation and accumulation of cyclic GMP by smooth muscle relaxing agents. *Proc. Jpn. Cyclic Nucleotide Conf., 2nd, 1977,* pp. 44–50.

Katsuki, S., Arnold, W. P., and Murad, F. (1977c). Effect of sodium nitroprusside, nitroglycerin and sodium azide on levels of cyclic nucleotides and mechanical activity of various tissues. *J. Cyclic Nucleotide Res.* **3,** 239–247.

Kimura, H., and Murad, F. (1974). Evidence for two different forms of guanylate cyclase in rat heart. *J. Biol. Chem.* **249,** 6910–6919.

Kimura, H., and Murad, F. (1975a). Two forms of guanylate cyclase in mammalian tissues and possible mechanisms for their regulation. *Metab., Clin. Exp.* **24,** 439–445.

Kimura, H., and Murad, F. (1975b). Localization of particulate guanylate cyclase in plasma membranes and microsomes of rat liver. *J. Biol. Chem.* **250,** 4810–4817.

Kimura, H., and Murad, F. (1975c). Increased particulate and decreased soluble guanylate cyclase activity in regenerating liver, fetal liver, and hepatoma. *Proc. Natl. Acad. Sci. U.S.A.* **72,** 1965–1969.

Kimura, H., Mittal, C. K., and Murad, F. (1975a). Activation of guanylate cyclase from rat liver and other tissues with sodium azide. *J. Biol. Chem.* **250,** 8016–8022.

Kimura, H., Mittal, C. K., and Murad, F. (1975b). Increases in cyclic GMP levels in brain and liver with sodium azide, an activator of guanylate cyclase. *Nature (London)* **257,** 700–702.

Kimura, H., Mittal, C. K., and Murad, F. (1976). Appearance of magnesium guanylate cyclase activity in rat liver with sodium-azide activation. *J. Biol. Chem.* **251,** 7769–7773.

Mittal, C. K., Kimura, H., and Murad, F. (1975). Requirement for a macromolecular factor for sodium azide activation of guanylate cyclase. *J. Cyclic Nucleotide Res.* **1,** 261–269.

Mittal, C. K., Kimura, H., and Murad, F. (1977). Purification and properties of a protein required for sodium azide activation of guanylate cyclase. *J. Biol. Chem.* **252,** 4348–4390.

Mittal, C. K., Arnold, W. P., and Murad, F. (1978). Characterization of protein inhibitors of guanylate cyclase activation from rat heart and bovine lung. *J. Biol. Chem.* **253,** 1266–1271.

Mittal, C. K., Braughler, J. M., Ichihara, K., and Murad, F. (1979). Synthesis of adenosine 3′, 5′-monophosphate by guanylate cyclase a new pathway for its formation. *Biochim. Biophys. Acta* **585,** 333–342.

Murad, F. (1986). Cyclic guanosine monophosphate as a mediator of vasodilation. *J. Clin. Invest.* **78,** 1–5.

Murad, F. (1989a). Modulation of the guanylate cyclase-cGMP system by vasodilators and the role of free radicals as second messengers. *In* "Vascular Endothelium" (J. D. Catravas, C. N. Gillis, and U. S. Ryan, eds.), pp. 157–164. Plenum, New York.

Murad, F. (1989b). Mechanisms for hormonal regulation of the different isoforms of guanylate cyclase. *In* "Molecular Mechanisms of Hormone Action" (Y. Gehring, E. Helmreich, and G. Schultz, eds.), pp. 186–194. Springer-Verlag, Heidelberg.

Murad, F., Mittal, C. K., Arnold, W. P., Katsuki, S., and Kimura, H. (1978). Guanylate cyclase: Activation by azide, nitro compounds, nitric oxide, and hydroxyl radical and inhibition by hemoglobin and myoglobin. *Adv. Cyclic Nucleotide Res.* **9,** 145–158.

Murad, F., Leitman, D., Waldman, S. A., Chang, C. H., Hirata, M., and Kohse, K. (1988). Effects of nitrovasodilators, endothelium-dependent vasodilators and atrial peptides on cGMP. *Cold Spring Harbor Symp. Quant. Biol., Signal Transduction* **53,** 1005–1009.

Murad, F., Ishii, K., Gorsky, L., Förstermann, U., Kerwin, J. F., and Heller, M. (1990a). Endothelium-derived relaxing factor is a ubiquitous intracellular second messenger and extracellular paracrine substance for cyclic GMP synthesis. *In* "Nitric Oxide from L-Arginine: A Bioregulatory System" (S. Moncada and E. A. Higgs, eds.), Chapter 32, pp. 301–315.

Murad, F., Ishii, K., Förstermann, U., Gorsky, L., Kerwin, J., Pollock, J., and Heller, M. (1990b). EDRF is an intracellular second messenger and autacoid to regulate cyclic GMP synthesis in many cells. *Adv. Cyclic Nucleotide Res.* **24,** 441–448.

Murad, F., Förstermann, U., Nakane, M., Pollock, J., Schmidt, H. H. H. W., Matsumoto, T., Tracey, W. R., and Buechler, W. (1992). Isoforms of nitric oxide synthase and the nitric oxide-cyclic GMP signal transduction system. *Proc. NATO-ASI Conf. Vascular Endothelium,* Rhodes, Greece, 1992.

Nakane, M., Mitchell, J. A., Förstermann, U., and Murad, F. (1991). Phosphorylation by calcium calmodulin-dependent protein kinase II and protein kinase C modulates the activity of nitric oxide synthase. *Biochem. Biophys. Res. Commun.* **180,** 1396–1402.

Palmer, R., Ashton, D., and Moncado, S. (1988). Vascular endothelial cells synthesize nitric oxide from L-arginine. *Nature (London)* **333,** 664–665.

Pollock, J. S., Förstermann, U., Mitchell, J. A., Warner, T. D., Schmidt, H. H. H. W., Nakane, M., and Murad, F. (1991). Purification and characterization of particulate EDRF synthase from cultured and native bovine aortic endothelial cells. *Proc. Natl. Acad. Sci. U.S.A.* **88,** 10480–10484.

Pollock, J., Klinghofer, V., Förstermann, U., and Murad, F. (1992). Endothelial nitric oxide synthase is myristylated. *FEBS Lett.* **309,** 402–404.

Rapoport, R. M., and Murad, F. (1983a). Agonist-induced endothelial-dependent relaxation in rat thoracic aorta may be mediated through cyclic GMP. *Circ. Res.* **52,** 352–357.

Rapoport, R. M., and Murad, F. (1983b). Endothelium-dependent and nitrovasodilator-induced relaxation of vascular smooth muscle: Role for cyclic GMP. *J. Cyclic Nucleotide Protein Phosphorilation Res.* **9,** 281–296.

Rapoport, R. M., Draznin, M. B., and Murad, F. (1983). Endothelium-dependent vasodilator- and nitrovasodilator-induced relaxation may be mediated through cyclic GMP formation and cyclic GMP-dependent protein phosphorylation. *Trans. Assoc. Am. Physicians* **96,** 19–30.

Schmidt, H. H. H. W., Gagne, J., Nakane, M., Pollock, J., Miller, M., and Murad F. (1992). Mapping of neural NO synthase in the rat suggests frequent colocalization with NADPH diaphorase but not soluble guanylyl cyclase and novel paraneural functions for nitrinergic signal transduction. *J. Histochem. Cytochem.* **40,** 1439–1456.

Snyder, S. H., and Bredt, D. S. (1991). Nitric oxide as a neuronal messenger. *Trends Pharmacol. Sci.* **12,** 125–130.

Waldman, S. A., and Murad, F. (1987). Cyclic GMP synthesis and function. *Pharmacol. Rev.* **39,** 163–196.

Waldman, S. A., Lewicki, J. A., Brandwein, H. J., and Murad, F. (1982). Partial purification and characterization of particulate guanylate cyclase from rat liver after solubization with trypsin. *J. Cyclic Nucleotide Res.* **8,** 359–370.

Regulation of Cytosolic Guanylyl Cyclase by Porphyrins and Metalloporphyrins

Louis J. Ignarro
Department of Pharmacology
Center for the Health Sciences
University of California, Los Angeles
School of Medicine
Los Angeles, California 90024

I. Introduction and Overview

The discovery that protoporphyrin IX activates the soluble or cytoplasmic isoform of guanylate cyclase was made in an experiment designed to determine whether protoporphyrin IX, like heme, could bind to a purified preparation of guanylate cyclase that was rendered deficient in heme. In experiments focused on the purification of guanylate cyclase from bovine lung and rat liver, we noted that some partially purified and completely purified preparations of guanylate cyclase required the addition of heme to enzyme reaction mixtures for the full expression of guanylate cyclase activation by nitric oxide (NO). The use of an isoelectric precipitation step early during enzyme purification resulted in guanylate cyclase preparations that were devoid of or deficient in heme, as assessed spectrophotometrically. Certain anion exchange resins were more prone than others to aid in the detachment of heme from guanylate cyclase during enzyme purification. This was particularly true for rat liver guanylate cyclase.

Using a heme-free preparation of bovine lung guanylate cyclase, a 10-fold molar excess of heme was added to a small aliquot of enzyme and the mixture was chromatographed on a gel filtration column to remove all excess unbound heme. The enzyme protein that eluted from the column

Advances in Pharmacology, Volume 26

was activated 60- to 80-fold by added NO, whereas the starting heme-free enzyme preparation was not activated by NO (less than 2-fold activation). Moreover, the protein eluting from the column displayed spectral properties of a hemoprotein. In order to ascertain whether the iron of heme was required for binding of heme to guanylate cyclase, an experiment similar to that described above for heme was conducted with protoporphyrin IX, which is heme without iron. We noted that the enzyme protein eluting from the gel filtration column after reaction with protoporphyrin IX displayed spectral properties of a protein-bound porphyrin, thus indicating that protoporphyrin IX, like heme, could easily bind to guanylate cyclase.

In the same study, an experiment was conducted to determine whether bound protoporphyrin IX, like heme, could restore enzyme activation by NO. Much to our astonishment, however, we found that the protoporphyrin IX-containing guanylate cyclase was already in the maximally activated state (60- to 80-fold) and could not be further activated by NO. Subsequent experiments using heme-free, heme-deficient, and heme-containing preparations of guanylate cyclase revealed that protoporphyrin IX activated all enzyme preparations, whereas enzyme activation by NO was proportional to the amount of heme bound to guanylate cyclase in the various enzyme preparations. Whereas the addition of heme to guanylate cyclase reaction mixtures restored enzyme activation by NO, the addition of heme inhibited enzyme activation by added protoporphyrin IX, and the inhibition was competitive with protoporphyrin IX.

A kinetic analysis of the interaction between guanylate cyclase and NO or protoporphyrin IX revealed that both interactions were indistinguishable. That is, the activation of heme-containing enzyme by NO was characterized by a marked increase in V_{max} and a decrease in the K_m for enzyme substrate. The identical observation was made with either heme-deficient or heme-containing guanylate cyclase activated by protoporphyrin IX. Since NO activates guanylate cyclase by forming a nitrosyl complex with heme (NO–heme), preformed NO–heme complex was prepared, tested, and found to activate either heme-deficient or heme-containing guanylate cyclase by mechanisms that were indistinguishable from those found for protoporphyrin IX. These observations indicated that the mechanism by which NO–heme activates guanylate cyclase is the same as that for protoporphyrin IX, and led to the hypothesis that NO activates guanylate cyclase by a protoporphyrin IX-like binding interaction with the enzyme.

In a search for other chemical agents that activate guanylate cyclase by a protoporphyrin IX-like binding interaction, we found that phenylhydrazine causes enzyme activation by heme-dependent but NO-independent mechanisms. Phenylhydrazine was found to react chemically with heme to yield an iron–phenyl–heme complex that activates heme-deficient guanylate cyclase by mechanisms that are indistinguishable from those for the NO–heme complex or protoporphyrin IX. Thus, both NO and phenylhydrazine require heme for activation of guanylate cyclase, and the corresponding heme complex behaves similarly to protoporphyrin IX.

This chapter describes the relevant studies that led to the finding that protoporphyrin IX activates and heme inhibits cytosolic guanylate cyclase. In addition, the mechanism of guanylate cyclase activation by NO and phenylhydrazine and the modulation of enzyme activity by porphyrins and metalloporphyrins are discussed.

II. Studies Leading to the Discovery That Protoporphyrin IX Activates Guanylate Cyclase

A. Early Studies

The first studies that addressed the activation of cytosolic guanylate cyclase by nitrogen-containing compounds that could be converted into NO were those of Murad and colleagues. Azide, hydroxylamine, and nitrite activated unpurified preparations of guanylate cyclase and stimulated cyclic GMP accumulation in tissues (Kimura *et al.*, 1975a,b; Katsuki *et al.*, 1977). Similar observations were made by DeRubertis and Craven (1976). Nitroglycerin and nitroprusside were shown to stimulate cyclic GMP accumulation in various smooth muscle preparations (Diamond and Holmes, 1975; Diamond and Blisard, 1976; Schultz *et al.*, 1977; Katsuki and Murad, 1977).

The release of NO from solutions of certain nitrosoguanidines had been known since the 1960s (Schoental and Rive, 1965; McCalla *et al.*, 1968; Schulz and McCalla, 1969; Lawley and Thatcher, 1970). The decomposition of inorganic nitrite in aqueous solution to nitrous acid (HONO) and NO is a well-known chemical reaction. Subsequent studies on guanylate cyclase activation by nitroso compounds, nitrite, azide, and hydroxylamine led to the development of the hypothesis that NO is responsible for enzyme activation by these nitrogen-containing compounds (Kimura *et al.*, 1975a,b; DeRubertis and Craven, 1976; Arnold *et al.*, 1977; Katsuki *et al.*, 1977; Craven and DeRubertis, 1978a; Craven *et al.*, 1979; Murad *et al.*, 1978). Subsequent studies from this laboratory revealed the direct release of NO from a variety of nitrogen oxide-containing compounds (Ignarro *et al.*, 1980a,b).

The knowledge that NO is responsible for guanylate cyclase activation elicited by many of the compounds described above, including nitroglycerin and nitroprusside, prompted us to ascertain whether NO and cyclic GMP accounted for the vascular smooth muscle relaxant action of such chemical agents. The test of this hypothesis was to determine whether NO causes vascular smooth muscle relaxation, and in 1979 we reported that NO relaxes precontracted strips of bovine coronary artery and activates cytosolic guanylate cyclase prepared from the same tissue (Gruetter *et al.*, 1979). After making this observation, we focused our research objectives on (*a*) the mechanism of guanylate cyclase activation by NO and (*b*) the mechanism of conversion of organic nitrate and nitrite esters and nitroso compounds to NO.

B. Influence of Thiols on Guanylate Cyclase Activity

The major study that motivated this laboratory to ascertain the mechanism of activation of guanylate cyclase by NO was that of Craven and DeRubertis (1978a), which showed that relatively crude soluble preparations of hepatic guanylate cyclase required the addition of heme or reduced hemoproteins in order to observe enzyme activation by NO and certain nitroso compounds. We were unable to confirm those observations at first with crude enzyme preparations from liver, lung, brain, heart, and platelets. We did find, however, that heme and hemoglobin inhibited guanylate cyclase activation elicited by nitrosoguanidines and nitroprusside, and that the inhibition could be largely overcome by addition of dithiothreitol to enzyme reaction mixtures (Ohlstein *et al.*, 1979). The inhibition by hemoproteins of guanylate cyclase activation by NO and related agents had been reported earlier (Murad *et al.*, 1978). We noted that the more contaminated the soluble fraction with hemoglobin, the greater the requirement of dithiothreitol to unmask guanylate cyclase activation by labile nitroso compounds. Dithiothreitol and other thiols or free sulfhydryls were found to liberate NO gas from nitrosoguanidines and to a lesser extent from nitroprusside, and thereby account for the capacity of the thiols to unmask guanylate cyclase activation by nitrosoguanidines and nitroprusside (Ignarro *et al.*, 1980a). Although thiols enhanced enzyme activation by inorganic nitrite, liberation of NO gas could not be detected. Some thiols enhanced activation of crude guanylate cyclase by azide and other agents in the earliest reports (Kimura *et al.*, 1975b).

Upon investigating the chemical mechanisms by which thiols liberate NO from, or unmask guanylate cyclase activation by, certain nitroso compounds, we found that thiols react chemically with the nitroso compounds to yield the corresponding *S*-nitrosothiols, represented by the general formula R-SNO (Ignarro *et al.*, 1980a,b; Ignarro and Gruetter, 1980). *S*-Nitrosothiols were found to be more resistant than nitrosoguanidines and nitroprusside to the inhibitory action of hemoproteins on guanylate cyclase activation, thereby providing an explanation of our previous observation that thiols unmasked guanylate cyclase activation by the nitroso compounds.

In studies designed to examine the effects of various thiols on the capacity of nitrogen oxide-containing compounds to activate guanylate cyclase, we found that nitroglycerin could not activate guanylate cyclase unless cysteine was added to enzyme reaction mixtures (Ignarro and Gruetter, 1980). Although sodium nitrite was capable of activating crude soluble fractions of guanylate cyclase at very high concentrations (exceeding 10 m*M*), lower concentrations were inactive unless a thiol was added to enzyme reaction mixtures. Unlike nitroglycerin, which had a specific requirement for cysteine, nitrite activated guanylate cyclase in the presence of virtually any free sulfhydryl compound. As discussed above, nitroprusside and nitrosoguanidines activated guanylate cyclase to varying extents in the absence of added thiol but the addition of any thiol to enzyme reaction mixtures markedly enhanced guanylate cyclase activation. These observations led to the hypothesis and study that NO is responsible for the vascular smooth muscle relaxant effects of organic nitrate and nitrite esters and nitroprusside via the intermediate formation of *S*-nitrosothiols in vascular smooth muscle (Ignarro *et al.*, 1981a).

Studies in this laboratory were then focused on the mechanism of guanylate cyclase activation by NO and *S*-nitrosothiols. We found that in the presence of MgGTP substrate, low concentrations of free calcium inhibited guanylate cyclase activation by NO, nitroso compounds, and NO–heme complex (Gruetter *et al.*, 1980; Edwards *et al.*, 1981). During enzyme protein purification, thiols were found to enhance activation of crude hepatic fractions of guanylate cyclase by nitroso compounds, and an uncharacterized thioprotein present in the cytosolic fraction was found to enhance enzyme activation by the nitroso compounds (Ignarro *et al.*, 1981b). Partially purified cytosolic guanylate cyclase from liver was thought to be completely devoid of heme, but additional unpublished observations indicated that the spectrophotometric method used to measure heme was too insensitive and that trace amounts of heme were present in the enzyme fractions. Thus, the conclusion drawn that heme was not required for guanylate cyclase activation by added NO was erroneous. Indeed, in that study (Ignarro *et al.*, 1981b), the addition of heme or hemoglobin to enzyme reaction mixtures markedly enhanced enzyme activation by NO and *S*-nitrosocysteine but not NO–hemoglobin.

In order to develop a better understanding of the interaction between guanylate cyclase and NO, a study was conducted to ascertain whether enzyme protein —SH groups are involved in such an interaction. Partially

.lepatic soluble guanylate cyclase was found to be rapidly but reversibly inactivated upon exposure to 100% oxygen but retained complete catalytic activity in a nitrogen atmosphere (Ignarro *et al.*, 1981c). Enzyme inactivation by oxygen was prevented and reversed by addition of dithiothreitol to enzyme reaction mixtures. Various —SH oxidants, thiol alkylating agents, and disulfides also caused inhibition of enzymatic activity that was reversed by dithiothreitol and certain other thiols. These observations indicated that guanylate cyclase contains one or more —SH groups at its catalytic site. Further experiments revealed that guanylate cyclase contains two closely juxtaposed —SH groups located at the catalytic site. Similar observations had been made by Craven and DeRubertis (1978b), who studied relatively crude enzyme preparations, and these findings were consistent with those of others showing that enzyme bound thiols are involved in catalytic activity (White *et al.*, 1976; Goldberg and Haddox, 1977; DeRubertis and Craven, 1977; Braughler *et al.*, 1979). Passage of guanylate cyclase reaction mixtures, in which enzyme was maximally activated by NO, through a gel filtration column to separate low- from high-molecular-weight components resulted in elution of guanylate cyclase with only basal (unactivated) catalytic activity. The presence of cysteine or hematin in enzyme preincubates was mandatory in order to enable the activated form of guanylate cyclase to elute from the column (Ignarro *et al.*, 1981c). Guanylate cyclase activated by *S*-nitrosocysteine or NO–hemoglobin, however, was recovered in the maximally activated state by gel filtration, and this was prevented by preincubation of enzyme with —SH oxidants or excess MgGTP substrate. These data imply that cysteine, hematin, and their nitrosyl derivatives bind to —SH groups at the catalytic site of cytosolic guanylate cyclase.

C. Requirement of Heme for Guanylate Cyclase Activation by NO

We became temporarily sidetracked in our initial experiments using rat hepatic guanylate cyclase purified to apparent homogeneity because of the observation that such enzyme preparations did require the addition of heme or hemoglobin to allow the expression of enzyme activation by NO and nitroso compounds. Using more sensitive techniques for measuring heme than we had employed previously, we found that our purified hepatic soluble guanylate cyclase preparations did not contain heme. The enzyme, however, could be easily and readily reconstituted with heme by adding a 10-fold molar excess of hematin to guanylate cyclase in 5 m*M* dithiothreitol at 25°C followed 15 min later by gel filtration to remove all excess unbound heme (Ignarro *et al.*, 1982a). Spectral analysis of heme-reconstituted guanylate cyclase preparations indicated that 1 mol of heme was bound to 1 mol of holoenzyme dimer. Heme-deficient guanylate cyclase was activated only 1.5- to 2-fold by NO, whereas heme-reconstituted guanylate cyclase was activated nearly 100-fold by NO. While trying to understand the properties of hepatic guanylate cyclase, we purified and characterized cytosolic guanylate cyclase from bovine lung. It was often disconcerting that some purified enzyme preparations from lung contained bound heme, whereas other enzyme preparations were deficient in heme. Eventually, we found that the retention of heme is a function of the technique used to purify guanylate cyclase (Ignarro *et al.*, 1982a). Procedures such as isoelectric precipitation, ammonium sulfate precipitation alone or followed by isoelectric focusing, and DEAE-cellulose chromatography following either isoelectric or ammonium sulfate precipitation all result in the detachment of heme from guanylate cyclase during enzyme protein purification. We found that the elimination of isoelectric techniques and the substitution of DEAE-Sepharose CL-6B for DE-52 cellulose did not result in the removal of heme from guanylate cyclase (Ignarro *et al.*, 1982b). Other procedures that were found to cause detachment of heme from guanylate cyclase during enzyme purification were elevated temperatures (10–15°C), excessive agitation, and failure to store concentrated enzyme solutions in glycerol and dithiothreitol. Chromatography of purified enzyme preparations on columns of DE-52 cellulose resulted in the detachment of heme but not NO–heme complex or protoporphyrin IX from guanylate cyclase (Table I). These observations indicated that NO–heme complex and protoporphyrin IX bind with a greater affinity than does heme to guanylate cyclase.

During the conduct of our studies, Gerzer and co-workers (1981a) reported that cytosolic guanylate cyclase purified from bovine lung was markedly activated by nitroprusside in the absence of added heme. Further spectral analysis revealed that the guanylate cyclase preparations contained stoichiometric quantities of bound heme iron and copper (Gerzer *et al.*, 1981b). These investigators concluded also that endogenous soluble guanylate cyclase is a hemoprotein that is markedly activated by NO. We made similar observations with purified guanylate cyclase preparations from bovine lung (Ignarro *et al.*, 1982a,b) rat liver (Ohlstein *et al.*, 1982), human platelets (Mellion *et al.*, 1983), and bovine cerebellum (Ohlstein and Ignarro, unpublished observations).

Endogenous cytosolic guanylate cyclase is likely to exist in the hemoprotein form for numerous reasons. As will be addressed below, heme binds stoichiometrically to guanylate cyclase and the physical techniques that result in the detachment of heme suggest that the porphyrin ring and iron of heme bind to enzyme protein in a physiological and predictable

Table I

Binding of Heme, NO–Heme, and Protoporphyrin IX to Guanylate Cyclase

Guanylate cyclase preparation	Guanylate cyclase activity (μmol cyclic GMP/min/mg)		
	Control	+10 μM NO	+0.1 μM P-IX
Heme-containing	0.14	0.19	5.4
+1 μM heme added to reaction	0.10	6.2	5.0
Heme-reconstituted	0.16	0.18	5.7
+1 μM heme added to reaction	0.11	6.0	5.2
Heme-containing + NO	5.8	5.6	5.6
Heme-containing + P-IX	5.2	5.4	5.3
Heme-deficient + P-IX	5.6	5.3	5.4

Note. Heme-containing, heme-reconstituted, and heme-deficient preparations of purified bovine lung guanylate cyclase (1–1.4 μg protein), some of which were reacted with 10 μM NO or 0.1 μM protoporphyrin IX (P-IX) for 10 min at 25°C as indicated, were applied (0.1 ml) to columns (0.7 × 5 cm) of DE-52 cellulose preequilibrated with 40 mM TEA–HCl, pH 7.4, containing 2 mM dithiothreitol (Buffer). The column was washed with 5 ml of Buffer containing 0.1 M NaCl followed by 1 ml of Buffer containing 0.5 M NaCl. One milliliter of Buffer containing 0.5 M NaCl was added and the eluate was collected and used as the enzyme source. In some cases, as indicated, heme was added to enzyme reaction mixtures after chromatography. Enzyme assays were conducted for 10 min at 37°C in reaction mixtures containing Buffer, 1 mM GTP, 3 mM Mg^{2+}, 50–70 ng of guanylate cyclase, and NO or protoporphyrin IX as indicated.

manner. Heme-deficient preparations of guanylate cyclase can be readily reconstituted with heme simply by mixing the two together at pH 7.4 in a mild reducing environment. The principal endogenous and exogenous activator of cytosolic guanylate cyclase is NO, which requires enzyme-bound heme for enzyme activation. Heme-free guanylate cyclase does not bind NO and is not activated by NO, whereas heme-containing enzyme has a very high binding affinity for NO (Wolin *et al.*, 1982).

Up to this point in our research, we learned that mammalian cytosolic guanylate cyclase was likely a hemoprotein where bound heme served as a prosthetic group to facilitate enzyme activation by NO. Moreover, studies indicated clearly that —SH groups near or at the catalytic site were essential not only for catalytic activity but also for enzyme activation by NO and nitroso compounds. The next objective of our research was to unravel the precise mechanisms involved in the heme-dependent activation of guanylate cyclase by NO. The first study focused on the binding of heme to guanylate cyclase, and the first experiment in the series was to ascertain whether the iron of heme was necessary for binding. Preparations of heme-free guanylate cyclase purified from bovine lung were preincubated with excess heme at 25°C for 5 min and passed through a small gel filtration column. The starting enzyme preparations showed no appreciable absorbance in the Soret region (390–450 nm) and were not activated by NO. The enzyme preparation that was eluted from the gel filtration column after mixing with excess heme displayed significant absorbance in the Soret region and was markedly activated by NO. A similar experiment was performed with protoporphyrin IX instead of heme. The porphyrin became tightly bound to guanylate cyclase as did heme, thus indicating that the iron of heme was not obligatory for the binding of heme to guanylate cyclase.

III. Kinetic Mechanisms by Which Protoporphyrin IX Activates Guanylate Cyclase and Similarity to Nitric Oxide–Heme Complex

The most exciting observation of the above experiments was that the protoporphyrin IX-bound guanylate cyclase eluted from the column in the maximally activated state, and the addition of NO caused no further enzyme activation. Subsequent experiments with heme-containing and heme-deficient guanylate cyclase revealed that protoporphyrin IX activates guanylate cyclase in a heme-independent manner. Our first publication in the series (Ignarro *et al.*, 1982b) highlights the discovery that protoporphyrin IX activates purified cytosolic guanylate cyclase in a manner that is kinetically similar to enzyme activation by NO and NO—heme complex. With the exception of hematoporphyrin IX, close structural analogs of protoporphyrin IX and heme, including precursors and metabolites, did not activate guanylate cyclase. Heme was found to be a competitive inhibitor of protoporphyrin IX. The natural occurrence of protoporphyrin IX and heme in mammalian cells at concentrations similar to those used in this study suggests that these endogenous substances could play biological roles in modulating guanylate cyclase activity and cellular cyclic GMP levels. Thus, the finding that the incorporation of iron into protoporphyrin IX converts the latter from a potent activator (K_a = 15–25 nM) to an inhibitor (K_I = 3–4 μM) of guanylate cyclase could have physiological significance.

Additional experiments were focused on the mechanisms of guanylate cyclase activation by protoporphyrin IX and NO and whether or not a common mechanism is involved. A kinetic approach was taken to compare the properties of NO, NO—heme, protoporphyrin IX, and heme. Purified, heme-containing, cytosolic guanylate cyclase from bovine lung was used in the initial studies (Wolin *et al.*, 1982). Protoporphyrin IX, NO, and NO—heme increased the V_{max} up to 40- to 50-fold and decreased the K_m for GTP from 100 μM to 45–55 μM in the presence of excess Mg^{2+}. Protoporphyrin IX resembled NO and NO–heme also in lowering the K_m and K_i (apparent dissociation constant) for uncomplexed Mg^{2+}. Comparison with unactivated enzyme revealed that enzyme activation increases the V_{max} and eliminates the influence of GTP on the apparent K_m for free Mg^{2+}. Thus, the activated form of guanylate cyclase behaves in an identical manner regardless of whether the activator is protoporphyrin IX, NO, or NO—heme. This close similarity in the interactions of these activators with guanylate cyclase suggested that a common form of activated enzyme is generated.

Heme or hematin inhibited guanylate cyclase activity in two ways. Concentrations less than 1.5 μM were competitive with protoporphyrin IX, NO, and NO—heme (K_I = 0.35 μM), but higher concentrations were noncompetitive. The competitive interaction of the interaction indicates that protoporphyrin IX, NO–heme, and heme compete for a common binding site on guanylate cyclase (the porphyrin binding site). Higher concentrations of heme that interacted with protoporphyrin IX in a noncompetitive manner also inhibited basal guanylate cyclase activity in the absence of added protoporphyrin IX or NO.

The apparent K_a for protoporphyrin IX varied from 8 to 38 nM as a function of the guanylate cyclase concentration. The equilibrium dissociation constant of the guanylate cyclase–protoporphyrin IX complex was estimated by Scatchard analysis to be 1.4 nM (Wolin *et al.*, 1982). The stoichiometry of binding was found to be 0.92 to 1.0 mol/mol of holoenzyme dimer, which was the same as that found for enzyme-bound heme. Thus, protoporphyrin IX has a high binding affinity for guanylate cyclase and displaces enzyme-bound heme.

The precise mechanism of guanylate cyclase regulation by interaction with the porphyrin binding site is unknown. Lowering the affinity for free Mg^{2+} and an increase in the V_{max} are likely to be mechanisms associated with enzyme regulation. Stabilization of the configuration of guanylate cyclase in the transition state during catalysis, which is manifested as changes in the kinetic parameters, could represent the principal mechanism of enzyme regulation. These observations, together with the knowledge that protoporphyrin IX and heme occur naturally in mammalian tissues, indicate that both substances likely function biologically to regulate guanylate cyclase activity and, thus, tissue cyclic GMP levels.

Observations similar to those described above using heme-containing guanylate cyclase from bovine lung were subsequently made with heme-deficient guanylate cyclase purified from rat liver (Ohlstein *et al.*, 1982) and human platelets (Mellion *et al.*, 1983). Protoporphyrin IX and NO–heme activated hepatic guanylate cyclase by kinetically indistinguishable mechanisms, whereas NO was inactive. Heme-reconstituted guanylate cyclase, however, was activated by NO in a manner that was virtually identical to enzyme activation by protoporphyrin IX or NO–heme. A partially purified but uncharacterized, cytosolic, heat-stable factor (activation enhancing factor) was isolated from rat liver and found to enhance enzyme activation up to 35-fold without directly altering basal catalytic activity. This activation enhancing factor still enhanced enzyme activation in the presence of excess heme and thiol. Thus, endogenous factors in addition to heme and thiols may be required for the full expression of guanylate cyclase activation by NO.

IV. Role of Copper Bound to Guanylate Cyclase

Cytosolic guanylate cyclase purified from bovine lung was found to contain 1 mol of copper bound per mole of holoenzyme dimer (Gerzer *et al.*, 1981b). Following modification of these original observations for heme-containing and heme-deficient guanylate cyclase purified from bovine lung, we conducted several experiments designed to elucidate the role of enzyme-bound copper. Initial experiments with several copper chelating agents led to the finding that bathocuproine disulfonate reacts with guanylate cyclase to mask the detection of, or to detach, copper from the enzyme. Bathocuproine disulfonate increased basal guanylate cyclase activity and markedly enhanced the capacity of low concentrations of NO to activate the enzyme. Heme-containing guanylate cyclase was pretreated with bathocuproine disulfonate and passed through a gel filtration column to remove the copper chelator and any chelated copper. Bathocuproine disulfonate-pretreated enzyme displayed increased basal catalytic activity and was markedly more sensitive to activation by low concentrations of NO (Table II). Addition of superoxide dismutase to untreated control enzyme reaction mixtures yielded similar results.

Other experiments revealed that 0.01 to 0.1 mM concentrations of copper ($CuCl_2$) caused a marked decrease in basal catalytic activity and enzyme activation by protoporphyrin IX, and abolished enzyme activation by NO. Excess dithiothreitol prevented the effect of copper on basal activity and protoporphyrin IX-induced enzyme activation and partially protected against the loss of NO-induced enzyme activation. These observations suggest that enzyme-bound copper enhances superoxide anion-

Table II

Pretreatment of Guanylate Cyclase with Bathocuproine Disulfonate Enhances Enzyme Activity and Stimulation by NO

	Guanylate cyclase activity (μmol cyclic GMP/min/mg)		
Additions to reaction mixtures	Control	Control + SOD	BC-pretreated enzyme
None (basal activity)	0.10	0.21	0.41
1 m*M* Bathocuproine disulfonate	0.30	0.40	0.42
0.1 μM Protoporphyrin IX	4.4	8.1	12.2
100 μM NO	5.2	11.5	12.4
10 μM NO	4.1	11.3	12.7
1 μM NO	0.30	7.9	9.6
0.1. μM NO	0.10	3.1	4.5

Note. Heme-containing guanylate cyclase purified from bovine lung was used. Some enzyme preparations were pretreated with 1 m*M* bathocuproine disulfonate (BC) for 10 min at 25°C. Control and pretreated enzyme preparations (0.6–0.8 μg protein) were chromatographed on columns (0.7 × 4 cm) of Sephadex G-25 preequilibrated with 40 m*M* TEA–HCl, pH 7.4 (Buffer), to remove the dithiothreitol and glycerol used to stabilize enzyme preparations during storage, and to remove bathocuproine disulfonate and any chelated copper from pretreated enzyme mixtures. Enzyme assays were conducted for 5 min at 37°C in reaction mixtures containing Buffer, 1 m*M* GTP, 3 m*M* Mg^{2+}, 0.10–0.15 μg of guanylate cyclase and additional agents as indicated. Some reaction mixtures contained 50 units of superoxide dismutase (SOD) as indicated.

mediated oxidation of both enzyme —SH groups and NO. Thus, copper may serve to modulate basal catalytic activity as well as enzyme activation by NO.

V. Mechanism by Which Phenylhydrazine Activates Guanylate Cyclase

Hydralazine, a phthalazine derivative of hydrazine, is a well-known vasodilator that increases tissue cyclic GMP levels (Schultz *et al.*, 1977). During the course of testing various hydrazine analogs for relaxant activity on vascular smooth muscle preparations, we found that phenylhydrazine not only causes arterial relaxation (70% relaxation of bovine pulmonary artery at 100 μM) but also inhibits platelet aggregation (60% inhibition of ADP-induced human platelet aggregation at 100 μM). Consistent with these findings, we learned that phenylhydrazine had been shown to cause a slight activation of unpurified preparations of guanylate cyclase (Kimura *et al.*, 1975b). This prompted a study of the effects of hydrazines on purified preparations of guanylate cyclase (Ignarro *et al.*, 1984a). Phenylhydrazine failed to activate heme-deficient guanylate cyclase but activated heme-containing enzyme 3- to 4-fold. NO activated heme-containing enzyme nearly 100-fold. Addition of catalase or methemoglobin to enzyme reaction mixtures increased activation by phenylhydrazine to 12-fold. Reducing or anaerobic conditions inhibited, whereas oxidants enhanced, enzyme activation by phenylhydrazine plus catalase, and KCN had no effect. In contrast, guanylate cyclase activation by NO and azide (N_3) was inhibited by oxidants or KCN. Moreover, azide required native catalase, whereas phenylhydrazine also utilized heat-denatured catalase for enzyme activation. Therefore, the mechanism of guanylate cyclase activation by phenylhydrazine is distinctly different from that for NO or azide.

Further experimentation revealed that guanylate cyclase activation by phenylhydrazine in the presence of catalase or methemoglobin resulted from an oxygen-dependent reaction between phenylhydrazine and hemoproteins to generate stable iron–phenyl hemoprotein complexes. Mixtures of phenylhydrazine plus either catalase or methemoglobin were subjected to gel filtration chromatography, and guanylate cyclase-stimulating activity cochromatographed with catalase or methemoglobin. Guanylate cyclase-stimulating activity was completely retained after dialysis. Whereas prereacted mixtures of phenylhydrazine plus hemoproteins subjected to dialysis or to lyophilization and reconstitution activated heme-deficient guanylate cyclase similarly under anaerobic or aerobic conditions, the separate additions of phenylhydrazine and hemoproteins to enzyme reaction mixtures failed to cause enzyme activation under anaerobic conditions. These data indicated that oxygen is required for the initial reaction between phenylhydrazine and hemoprotein to form a new species that is capable of activating guanylate cyclase under anaerobic conditions. The new species was characterized as the iron–phenyl adduct of catalase or methemoglobin.

In order to characterize fully the reaction between phenylhydrazine and catalase or methemoglobin, [U-^{14}C]phenylhydrazine was reacted with hemoprotein and then subjected to gel filtration chromatography. Figure 1 illustrates that the radioactivity cochromatographed with the hemoproteins. The binding stoichiometry of the reactions between phenylhydrazine and hemoproteins was calculated to be 4 mol of phenyl per mole of methemoglobin. In view of the finding that phenylhydrazine reacts with the heme iron of methemoglobin to form the iron–phenyl complex (Augusto *et al.*, 1982; Kunze and Ortiz de Montellano, 1983), a binding stoichiometry of 1 mol of phenyl per mole of heme was likely. The identical binding

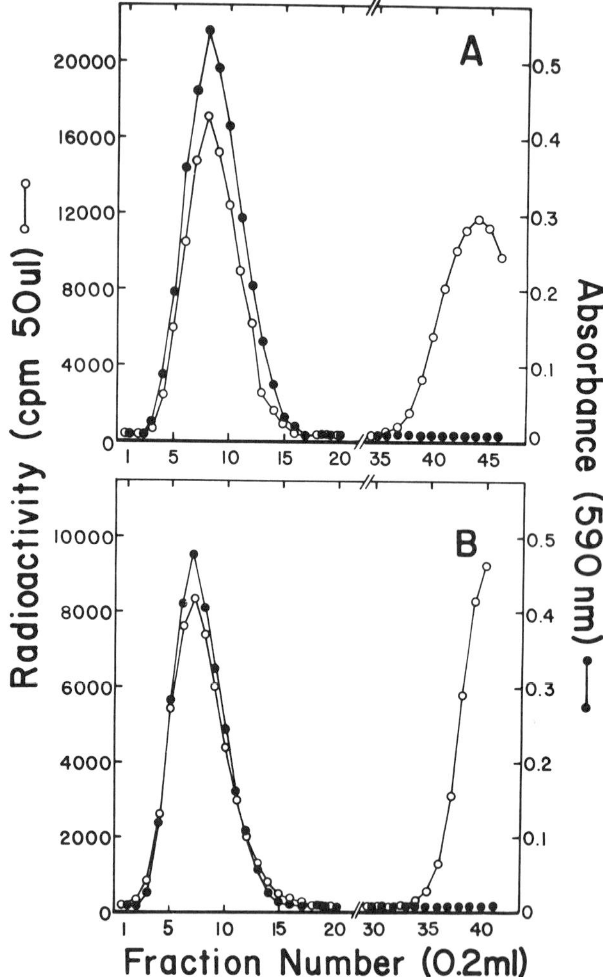

Fig. 1 Gel filtration of reaction mixtures of phenylhydrazine plus methemoglobin or catalase. One milliliter of 50 mM TEA–HCl, pH 7.4, containing 0.1 M NaCl, 1 mM [U-^{14}C]phenylhydrazine HCl (5 μCi), and either 30 μM methemoglobin (A) or 30 μM catalase (B) was incubated at 25°C for 30 min, centrifuged at 2000 × g for 10 min, and 0.5 ml of supernatant applied to a column (1.5 × 20 cm) of Sephadex G-25. Fractions (0.2 ml) were collected (12 ml/h) starting just prior to elution of protein in the void volume. Aliquots of 50 μl were added to 10 ml of Multisol cocktail and counted. Aliquots of 100 μl (methemoglobin reaction) or 25 μl (catalase reaction) were assayed for protein. See Ignarro *et al.* (1984a) for experimental details. Reproduced with permission from Ignarro *et al.* (1984a).

relationship was obtained for reactions between phenylhydrazine and catalase. These observations indicate that iron–phenyl hemoprotein complexes are responsible for the activation of guanylate cyclase by phenylhydrazine plus hemoprotein. This view was supported by the observations that phenylhydrazine plus freshly prepared apohemoglobin failed to activate heme-deficient guanylate cyclase, and radiolabeled phenylhydrazine failed to form a complex with apohemoglobin. Figure 2 is a schematic representation of the reaction between phenylhydrazine and hemoproteins to form iron–phenyl hemoprotein complexes.

Further experiments revealed that [^{14}C]phenylhydrazine reacts with heme-containing guanylate cyclase to form a stable complex with a binding stoichiometry of 1 mol of phenyl per mole of holoenzyme dimer. Moreover, [^{14}C]phenyl hemoprotein complexes reacted with guanylate cyclase to form a stable enzyme complex with a binding stoichiometry of 1 mol of phenyl per mole of holoenzyme dimer. Anaerobic conditions prevented the binding of phenyl to enzyme in reactions with phenylhydrazine but not iron–phenyl hemoprotein complexes. Reactions between guanylate cyclase and iron–[^{14}C]phenyl hemoprotein complexes resulted in the ex-

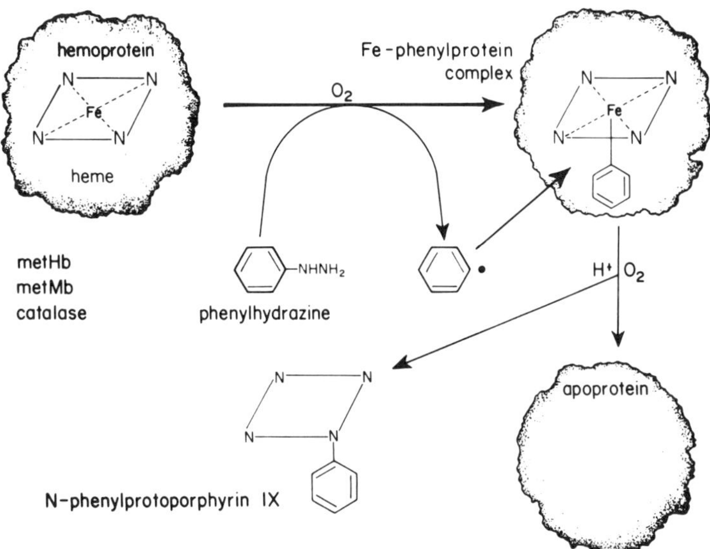

Fig. 2 Schematic illustration of the reaction between phenylhydrazine and hemoproteins to form iron–phenyl hemoprotein complexes. Abbreviations: metHb, methemoglobin; metMb, metmyoglobin. Complex formation occurs at neutral pH. Strongly acidic conditions in the presence of O_2 yield the green pigment *N*-phenylprotoporphyrin IX with metHb or metMb. Reproduced with permission from Ignarro *et al.* (1984c).

change or transfer of iron–phenyl heme to guanylate cyclase and this correlated with enzyme activation. The formation of the iron–phenyl heme adduct of guanylate cyclase was confirmed by spectral analysis (Ignarro *et al.*, 1984a).

A kinetic analysis revealed that phenylhydrazine, iron–phenyl hemoproteins, and iron–phenyl heme complex activate guanylate cyclase by identical mechanisms (Ignarro *et al.*, 1984a). Similarly, this mechanism of enzyme activation appears to be identical to that observed for protoporphyrin IX and NO–heme complex. These observations indicate that iron–phenyl heme, NO–heme, and protoporphyrin IX interact with guanylate cyclase to generate a similar activated form of enzyme.

VI. Structure–Activity Relationships Involving Porphyrins and Metalloporphyrins

The objective of this study was to develop a better understanding of the mechanisms by which protoporphyrin IX activates and ferro-protoporphyrin IX (heme) inhibits cytosolic guanylate cyclase. The approach taken was (*a*) to study the effects of alterations of the chemical structure of protoporphyrin IX on enzyme activation and (*b*) to determine the influence of several metalloporphyrins of protoporphyrin IX on guanylate cyclase activation by porphyrins, NO, and nitroso compounds (Ignarro *et al.*, 1984b).

Table III illustrates the porphyrins and metalloporphyrins studied, the major differences in chemical structure, and effects on guanylate cyclase activity. The structural modifications examined involve the vinyl side chains at positions 2 and 4, the vicinal propionic acid residues at positions 6 and 7, pyrrolic nitrogen substitutions, and the divalent metal in metalloporphyrins. The kinetic parameters are listed in Table IV. Substitution of ethyl for vinyl groups, forming mesoporphyrin IX, had no appreciable influence on K_a, but replacement of these hydrophobic groups with more polar hydroxyethyl groups, as in hematoporphyrin IX, resulted in a 10-fold increase in the K_a. Replacement with hydrogen atoms (deuteroporphyrin IX) resulted in a marked decrease in activity. An additional increase in K_a and decrease in V_{max} were obtained when propionic acid residues were introduced at positions 2 and 4 (coproporphyrin III). Substitution of highly polar groups (disulfonate or bisglycol deuteroporphyrin IX) resulted in inactive porphyrins. Methylation of the propionic acid residues resulted in a marked decrease in activity (protoporphyrin IX dimethyl ester). Moreover, methylation of hematoporphyrin IX and deuteroporphyrin IX converted these guanylate cyclase activators to enzyme inhibi-

Table III

Structural Modifications of Protoporphyrin IX and Their Effects on Guanylate Cyclase

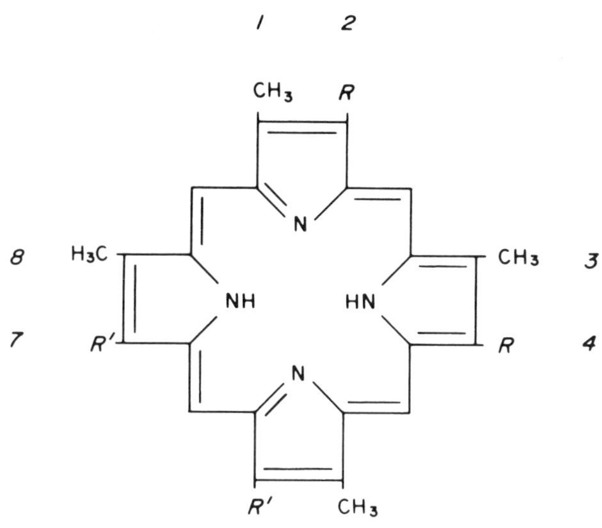

Porphyrin or metalloporphyrin	*R*	*R′*	Effect on enzyme
Protoporphyrin IX	$—CH{=}CH_2$	$—CH_2CH_2COOH$	Activate
Mesoporphyrin IX	$—CH_2CH_3$	$—CH_2CH_2COOH$	Activate
Hematoporphyrin IX	$—CHOHCH_3$	$—CH_2CH_2COOH$	Activate
Deuteroporphyrin IX	—H	$—CH_2CH_2COOH$	Activate
Deuteroporphyrin IX disulfonate	$—SO_3$	$—CH_2CH_2COOH$	Inactive
Deuteroporphyrin IX bisglycol	$—CHOHCH_2OH$	$—CH_2CH_2COOH$	Inactive
Coproporphyrin III	$—CH_2CH_2COOH$	$—CH_2CH_2COOH$	Activate
Coproporphyrin I	$—CH_2CH_2COOH$	$—CH_2CH_2COOH$[a]	Inactive
Protoporphyrin IX dimethyl ester	$—CH{=}CH_2$	$—CH_2CH_2COOCH_3$	Activate
Hematoporphyrin IX dimethyl ester	$—CHOHCH_3$	$—CH_2CH_2COOCH_3$	Inhibit
Deuteroporphyrin IX dimethyl ester	—H	$—CH_2CH_2COOCH_3$	Inhibit
N-Methylprotoporphyrin IX	$—CH{=}CH_2$	$—CH_2CH_2COOH$	Activate
N-Phenylprotoporphyrin IX	$—CH{=}CH_2$	$—CH_2CH_2COOH$	Inhibit
Ferro-protoporphyrin IX	$—CH{=}CH_2$	$—CH_2CH_2COOH$	Inhibit
Zinc-protoporphyrin IX	$—CH{=}CH_2$	$—CH_2CH_2COOH$	Inhibit
Manganese-protoporphyrin IX	$—CH{=}CH_2$	$—CH_2CH_2COOH$	Inhibit

[a] Positions 6 and 8.

Table IV

Effects of Structural Modifications of Protoporphyrin IX on Kinetic Parameters for Guanylate Cyclase[a]

		Kinetic parameters					
		MgGTP			Free Mg^{2+}		
Porphyrins	K_a[b]	K_m[c]	K_i[d]	V_{max}[e]	K_m[f]	K_i[d]	V_{max}[e]
None (basal activity)		122	194	0.13	91	190	0.14
Protoporphyrin IX, 1 μM	7.7	50	47	7.7	40	52	8.0
Mesoporphyrin IX, 1 μM	7.7	45	41	6.2	47	58	6.2
Hematoporphyrin IX, 10 μM	76	50	46	7.1	43	53	7.5
Deuteroporphyrin IX, 100 μM	2940	77	96	2.5	61	88	2.7
Protoporphyrin-IX, dimethyl ester, 100 μM[f]	1818	91	125	1.8	70	116	1.9
N-Methylprotoporphyrin-IX, 10 μM	115	48	52	3.0			

[a] Reproduced with permission from Ignarro *et al.* (1984b).
[b] Expressed as nanomolar.
[c] Obtained from intercept replots of primary plots; expressed as micromolar.
[d] Obtained from slope replots of primary plots; expressed as micromolar.
[e] Expressed as micromoles of cyclic GMP/min/mg.
[f] Preincubated with enzyme at 37°C for 10 min prior to initiation of reaction with MgGTP.

tors. The precise position of the propionate groups at positions 6 and 7 may be important for enzyme activation because coproporphyrin I, with propionate groups at positions 6 and 8, was completely inactive. Substitutions on the pyrrolic nitrogens had a marked inhibitory influence on activity, where *N*-methylprotoporphyrin IX was less active than protoporphyrin IX and *N*-phenylprotoporphyrin IX inhibited enzymatic activity and was a competitive inhibitor of protoporphyrin IX. Methylation of the propionic acid residues on the less active hematoporphyrin IX and deuteroporphyrin IX yielded porphyrin esters that were competitive inhibitors of protoporphyrin IX and, at higher concentrations, directly inhibited guanylate cyclase activity.

The hydrophobic vinyl side chains at positions 2 and 4 and the negatively charged carboxyl groups of the vicinal propionic acid residues at positions 6 and 7 in heme, as well as the coordination of heme iron to the apoprotein, contribute to the formation of stable hemoprotein complexes in hemoglobin, myoglobin, and other hemoproteins. Protoporphyrin IX, which contains no iron, binds tightly to guanylate cyclase with an apparent K_d of

1.4 n*M* (Wolin *et al.*, 1982). Substitution of the two vinyl groups with less hydrophobic or more polar groups results in only weakly active or inactive porphyrins. Deuteroporphyrin IX, which lacks substitutions at positions 2 and 4, caused only partial enzyme activation. Therefore, hydrophobic interactions between porphyrins and guanylate cyclase are essential for maximal enzyme activation. The hydrophobic binding sites in guanylate cyclase may lie buried in the interior of the molecule, and hydrophilic substitutions on porphyrins are not likely to interact with interiorized hydrophobic sites in hemoproteins.

Propionic acid residues at positions 6 and 7 of protoporphyrin IX and heme may form electrostatic bonds with basic groups such as arginine in the apoprotein of guanylate cyclase to form a stable hemoprotein complex. Thus, the dimethyl ester of protoporphyrin IX is only a weak enzyme activator and the dimethyl esters of hematoporphyrin IX and deuteroporphyrin IX are actually enzyme inhibitors. These propionic acid groups, which are ionized at pH 7.4, may form tight ion pairs with positively charged groups in guanylate cyclase and thereby contribute to the binding of porphyrins. This view is consistent with previous findings discussed above that isoelectric precipitation or isoelectric focusing of guanylate cyclase preparations results in the detachment of heme and renders the enzyme unresponsive to NO in the absence of added heme.

An open central core in the porphyrin ring is essential for maximal activation of guanylate cyclase. Addition of a methyl or more bulky phenyl group to a pyrrolic nitrogen results in decreased or no enzyme activation. Indeed *N*-phenylprotoporphyrin IX is a competitive inhibitor of protoporphyrin IX. Thus, although both porphyrins appear to bind at a common site on guanylate cyclase, they elicit opposing effects on enzymatic activity. One explanation of these observations is that enzyme-bound protoporphyrin IX may bind MgGTP and/or free Mg^{2+} and, thereby, facilitate their binding to guanylate cyclase. Consistent with the finding that protoporphyrin IX and structurally related porphyrins cause a two- to threefold increase in apparent affinities of guanylate cyclase for both MgGTP and uncomplexed Mg^{2+} was the observation that albumin-bound protoporphyrin IX binds MgGTP but not free GTP. Such an interaction with substrate or metal would be unlikely for *N*-phenylprotoporphyrin IX because of the projection of the phenyl moiety into the open central core of the porphyrin ring. Thus, *N*-phenylprotoporphyrin IX is a competitive inhibitor of protoporphyrin IX rather than an activator of guanylate cyclase.

Metallation drastically altered the effect of protoporphyrin IX on guanylate cyclase. Ferro-protoporphyrin IX inhibits guanylate cyclase activity and is a competitive inhibitor (K_I = 350 n*M*) of protoporphyrin IX. Zinc-protoporphyrin IX (K_I = 50 nM) and manganese-protoporphyrin IX (K_I = 9 nM) were much more potent than heme as competitive inhibitors of protoporphyrin IX. Metalloporphyrins inhibited also the activation of guanylate cyclase by NO and labile nitroso compounds. The affinities of zinc- and manganese-protoporphyrin IX for the porphyrin binding site are greater than that for heme, as assessed spectrally and by the observation that a mixture of heme-containing enzyme plus either metalloporphyrin subjected to gel filtration was unresponsive to added NO and showed only weak responsiveness to added protoporphyrin IX.

The competitive nature of the interaction between metalloporphyrins and porphyrins with guanylate cyclase indicates that all of these porphyrins interact at a common binding site on cytosolic guanylate cyclase. The coordinately bound metal (axial ligand) influences the binding of metalloporphyrins to the enzyme as indicated by the wide range of K_I values. Metal binding likely involves coordination to a nearby amino acid residue, and this somehow results in enzyme inhibition. One plausible explanation is that the axial ligand sterically hinders access of the catalytic site to MgGTP and uncomplexed Mg^{2+}, such that catalytic activity is markedly reduced. Removal of the axial ligand with continued binding of the porphyrin ring to guanylate cyclase may expose more of the catalytic site to MgGTP and free Mg^{2+} and thereby result in markedly increased enzymatic activity (enzyme activation). Removal of the axial ligand by NO to form NO–heme, which is structurally related to protoporphyrin IX, may account for the capacity of NO–heme to activate cytosolic guanylate cyclase (discussed below).

Additional studies revealed that stimulatory porphyrins, metalloporphyrins, and NO or NO–heme interacted with a common binding site on guanylate cyclase. Heme-deficient guanylate cyclase could be easily reconstituted with any porphyrin or metalloporphyrin to yield an activated or inhibited enzyme accordingly. These observations are consistent with previous findings that protoporphyrin IX, NO, nitroso compounds, and NO–heme activate cytosolic guanylate cyclase by kinetically identical mechanisms (Ignarro *et al.*, 1984c).

VII. Regulation of Guanylate Cyclase Activity by Porphyrins and Metalloporphyrins

Protoporphyrin IX binds tightly to a site on guanylate cyclase that also binds heme. Other porphyrins, metalloporphyrins, and nitrosyl-heme bind to the same site on the enzyme. We have termed this site the porphyrin binding site on cytosolic guanylate cyclase. The porphyrin binding site is distinct from the catalytic site, which binds MgGTP. Interaction with the porphyrin binding site influences catalytic activity. Porphyrins cause enzyme activation, whereas metalloporphyrins block enzyme activation and high concentrations inhibit basal enzymatic activity. Guanylate cyclase activation by NO requires heme and occurs via formation of NO–heme. Phenylhydrazine cayuses enzyme activation by a heme-dependent mechanism involving the formation of iron–phenyl heme complexes. Enzyme activation by NO–heme, iron–phenyl heme, and protoporphyrin IX occurs by a kinetically similar mechanism, suggesting that NO, phenylhydrazine, and related activators generate a modified porphyrin that resembles protoporphyrin IX in its interaction with guanylate cyclase.

A model was proposed to account for these observations in explaining the mechanism by which NO–heme activates guanylate cyclase (Ignarro *et al.*, 1984c). In binding to the iron of guanylate cyclase-bound heme, NO may weaken or break the axial ligand and thereby pull iron out from the plane of the porphyrin ring configuration (Fig. 3). Breakage of the axial ligand may expose the catalytic site to MgGTP and thereby constitute enzyme activation. Similarly, the binding of phenyl to heme iron may require displacement of the iron as a result of steric hindrance between the pyrrole nitrogens and *ortho* hydrogen atoms of the phenyl ring. Therefore, one surface of both NO–heme and iron–phenyl heme resembles the configuration of protoporphyrin IX and could account for the activation of guanylate cyclase.

A more recent study showed that both the 70- and 82-kDa subunits of cytosolic guanylate cyclase purified from rat lung must be present together and interactive with each other in order to observe catalytic activity, including enzyme activation by NO-generating compounds (Buechler *et al.*, 1991). On the basis of studies showing homology between certain portions of both subunits and with the catalytic region of a particulate isoform of guanylate cyclase (Nakane *et al.*, 1990), the authors suggested that a simplified model for the catalytic activity of the various isoforms of guanylate cyclase could involve a cooperative interaction of two catalytic domains located on different subunits. It is plausible that the porphyrin binding site spans both catalytic domains of the soluble isoform to influence the binding of enzyme substrate.

The interaction of NO with the heme bound to guanylate cyclase represents the signal transduction mechanism by which endogenous and exogenous NO leads to the intracellular biosynthesis of cyclic GMP. NO, a small, lipophilic, diffusible molecule, readily permeates target cells and interacts with intracellular soluble receptors on proteins. These receptors are the heme prosthetic groups bound to cytosolic guanylate cyclase.

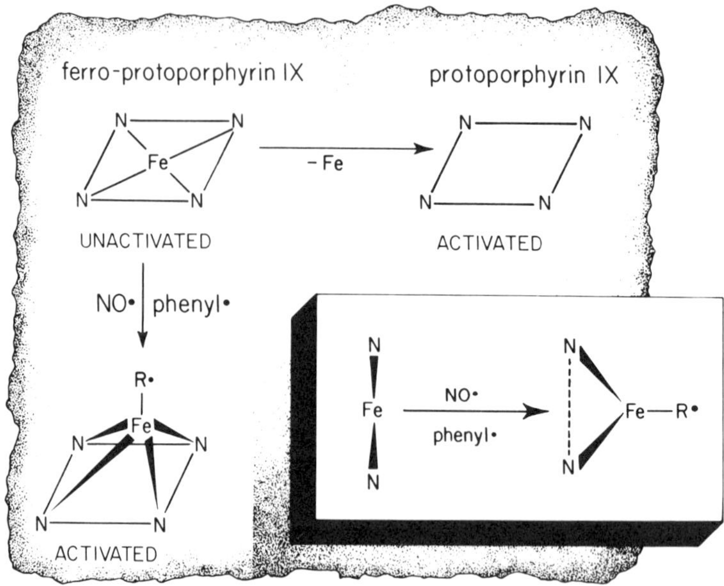

Fig. 3 Schematic illustration of the activation of cytosolic guanylate cyclase by NO or phenyl radical. The demetallation of ferroprotoporphyrin IX (heme) or its displacement by protoporphyrin IX yields the activated form of the enzyme. In addition, the reaction of ferroprotoporphyrin IX with NO radical or phenyl radical (both represented by R) results in the formation of a modified porphyrin and enzyme activation. Inset: side view of the displacement of iron from the plane of the planar porphyrin configuration. Reproduced with permission from Ignarro *et al.* (1984c).

The resulting NO–heme activates guanylate cyclase very rapidly, thereby causing the immediate and rapid formation of cyclic GMP, which then leads to the target cell response. Thus, the plasma membrane is not involved in processing the extracellular signal for intracellular expression. The onset time for guanylate cyclase activation in vascular smooth muscle cells, assessed by monitoring cyclic GMP formation, after addition of authentic NO or endothelium-dependent relaxants is less than 8 to 10 s (Ignarro *et al.*, 1981a, 1984d). The ensuing functional response of the cell occurs 3 to 5 s following measurable increases in intracellular cyclic GMP levels.

Although we now understand the role of nitrosyl-heme in stimulating the biosynthesis of cyclic GMP in mammalian cells, the physiological and/or pathophysiological significance of protoporphyrin IX and heme in modulating intracellular cyclic GMP levels is still unknown. Experimental alterations in heme metabolism do influence cyclic GMP levels. Chemicals such as 3,5-diethoxycarbonyl-1,4-dihydrocollidine (DDC) and 2-allyl-2-isopropylacetamide (AIA) are metabolized by cytochrome P450 to intermediates that, in turn, generate abnormal porphyrins. These abnormal porphyrins inhibit ferrochelatase, thereby resulting in increased tissue levels of protoporphyrin IX, decreased levels of heme, and increased activity of 5-aminolevulinate synthetase (De Matteis, 1967; Ortiz de Montellano *et al.*, 1979; Tephly *et al.*, 1979). These effects are exaggerated in animals pretreated with agents that induce the synthesis of hepatic cytochrome P450. Early experiments showed that the intraperitoneal or subcutaneous administration of 5–25 mg/kg of DDC or AIA to phenobarbital-pretreated mice produced a four- to sevenfold increase in hepatic cyclic GMP levels (Ignarro *et al.*, 1984c). This increase in hepatic cyclic GMP levels was comparable to the sixfold increase caused by the intravenous administration of 0.5 mg/kg of the labile *S*-nitrosothiol, *S*-nitroso-*N*-acetylpenicillamine. The reason for the effect elicited by DDC and AIA was attributed to an increase in the hepatic concentration ratio of protoporphyrin IX to heme, as assessed by HPLC techniques.

Phenobarbital induces marked heme synthesis, especially in the liver, and the protoporphyrin IX/heme concentration ratio declines after 48–72 h of treatment. Hepatic slices from phenobarbital-pretreated rats showed a decline in basal levels of cyclic GMP as well as an increase in the capacity of NO and nitrovasodilators to stimulate cyclic GMP formation in isolated hepatic slices (Wood and Ignarro, 1987). These observations could be attributed to elevated tissue levels of heme, which could inhibit basal guanylate cyclase activity but enhance heme-dependent activation of guanylate cyclase. Alternatively, phenobarbital could have induced the synthesis of new guanylate cyclase protein. The latter, however, should have resulted in a concomitant increase in basal cyclic GMP levels, which was not observed. Hepatic slices from rats pretreated with both phenobarbital and DDC in order to decrease the hepatic concentration of heme resulted in a marked increase in basal hepatic levels of cyclic GMP, which was associated with a marked decrease in the capacity of NO and nitrovasodilators to stimulate hepatic cyclic GMP formation. The increase in basal cyclic GMP levels could be attributed to decreased levels of heme and/or increased levels of protoporphyrin IX. The decreased responsiveness to NO could be attributed to decreased levels of heme. The influence of phenobarbital- and DDC-pretreatment on the effects of phenylhydrazine were even more marked than what was observed for NO. Phenylhydrazine did not stimulate hepatic cyclic GMP formation unless the hepatic slices were prepared from phenobarbital-pretreated rats, where a 10-fold stimulation was observed (Wood and Ignarro, 1987). These findings were consistent with previous observations that the further addition of heme or hemoglobin to heme-reconstituted guanylate cyclase preparations enhanced enzyme activation by phenylhydrazine (Ignarro *et al.*, 1984a).

These observations indicate that tissue levels of protoporphyrin IX and heme can markedly influence not only basal levels of cyclic GMP but also the capacity of heme-dependent guanylate cyclase activators to stimulate tissue cyclic GMP formation. Other tissue factors also have a marked influence on tissue cyclic GMP formation (Wood and Ignarro, 1987). Elevating the tissue cyanide concentration impairs cyclic GMP formation stimulated by NO, labile nitroso compounds, azide, and nitroglycerin. Decreasing tissue catalase activity by administration of aminotriazole impairs the capacity of azide but not NO or nitroglycerin to stimulate cyclic GMP formation. Decreasing tissue sulfhydryl levels by administration of high doses of acetaminophen does not affect basal cyclic GMP levels or the capacity of NO and azide to stimulate cyclic GMP formation, but markedly impairs the capacity of nitroglycerin to elevate tissue cyclic GMP levels. The effects of acetaminophen are reversed by the concomitant administration of *N*-acetylcysteine. The above observations made in hepatic slices after treatment of intact animals are consistent with the related findings made with crude and purified guanylate cyclase preparations.

VIII. Nitric Oxide–Heme Exchange between Hemoproteins and Guanylate Cyclase

High-molecular-weight nitrosyl-hemoproteins such as NO–hemoglobin, NO–myoglobin, and NO–catalase have been shown to activate crude preparations of cytosolic guanylate cyclase (Craven and DeRubertis, 1978a; Edwards *et al.*, 1981). Activation of heme-containing guanylate cyclase by NO–hemoprotein complexes could perhaps be explained by the rapid equilibrium of NO between both hemoproteins, thereby generating the active form of guanylate cyclase. However, a ready explanation of the activation of heme-deficient guanylate cyclase by high-molecular-weight NO–hemoproteins was more difficult to develop. One possible explanation was that the NO–heme complex is transferred from the NO–hemoprotein to guanylate cyclase to account for enzyme activation. In order to answer this question, we examined the interactions between nitrosyl-hemoproteins and heme-containing and heme-deficient forms of cytosolic guanylate cyclase purified from bovine lung (Ignarro *et al.*, 1986).

NO activated heme-containing and heme-reconstituted enzymes over 50-fold, with an accompanying shift in the Soret absorption peak from 431 to 398 nm (Fig. 4). NO did not activate or alter the spectral characteristics of heme-deficient enzyme. In contrast, preformed low-molecular-

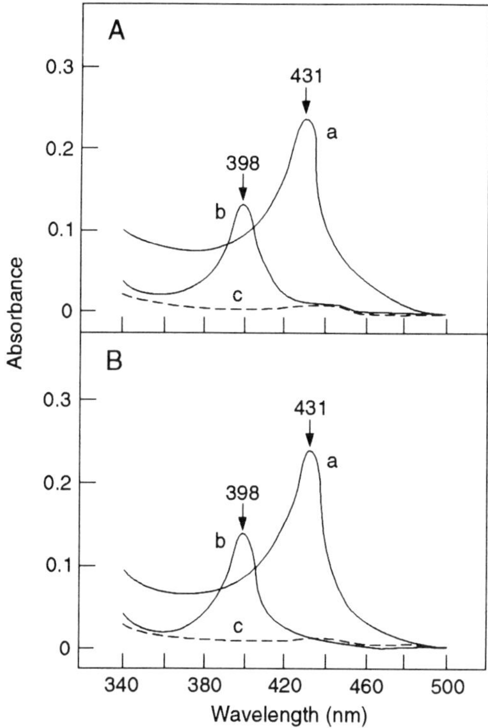

Fig. 4 Spectral properties of cytosolic guanylate cyclase. (A) Curve "a" represents 0.2 ml of N_2-purged 25 m*M* TEA–HCl, pH 7.4, containing 2 m*M* dithiothreitol, 0.1 *M* NaCl, and 88 μg of heme-containing guanylate cyclase. 431, Soret absorption maximum in nanometers. Curve "b" represents the same enzyme solution as above except that 5 ml of NO gas was injected as a fine stream of bubbles into the cuvette solution 30 s prior to scanning. 398, Soret absorption maximum in nanometers. Curve "c" represents 0.2 ml of the above buffer solution containing 88 μg of heme-deficient guanylate cyclase. (B) Curve "a" represents 0.2 ml of 25 m*M* TEA–HCl, pH 7.4, containing 2 m*M* dithiothreitol, 0.1 *M* NaCl, and 80 μg of heme-reconstituted guanylate cyclase. 431, Soret absorption maximum in nanometers. Curve "b" represents the same enzyme solution as above except that NO was introduced as described above. 398, Soret absorption maximum in nanometers. Curve "c" represents 0.2 ml of the above buffer solution containing 80 μg of heme-deficient guanylate cyclase. See Ignarro *et al.* (1986) for experimental details. Reproduced with permission from Ignarro *et al.* (1986).

weight NO–heme complex as well as the high-molecular-weight NO–hemoproteins activated all forms of guanylate cyclase. Heme-deficient guanylate cyclase was incubated with excess quantities of NO–hemoglobin, NO–myoglobin, or NO–catalase and then rapidly separated from the corresponding NO–hemoprotein by dye–ligand hydrophobic affinity column chromatography on Matrex Gel Blue-A. Spectral analysis revealed that the NO–heme moiety was transferred from each of the hemoproteins to heme-deficient guanylate cyclase. Figure 5 illustrates the spectral properties of heme-deficient guanylate cyclase after reaction with and complete separation from NO–hemoglobin or apohemoglobin. A distinct absorption maximum in the Soret region characteristic of NO–heme–guanylate cyclase was observed.

Calculation of the quantity of enzyme-bound NO–heme or heme, as the pyridine hemochrome, indicated binding of 1 mol of NO–heme or heme per mole of holoenzyme dimer. Thus, during the initial reaction between enzyme and NO–hemoglobin, the NO–heme complex was transferred from NO–hemoglobin to heme-deficient guanylate cyclase. Similar observations were made with NO–myoglobin and NO–catalase. Kinetic analysis revealed that NO activated heme-containing enzyme by kinetic mechanisms that were indistinguishable from those characteristic of activation of heme-deficient or heme-containing enzyme by NO–hemoproteins. Although NO–heme was readily transferred from NO–hemoglobin to heme-deficient guanylate cyclase to yield the NO–heme–enzyme complex, the reverse action was negligible. No detectable transfer of the

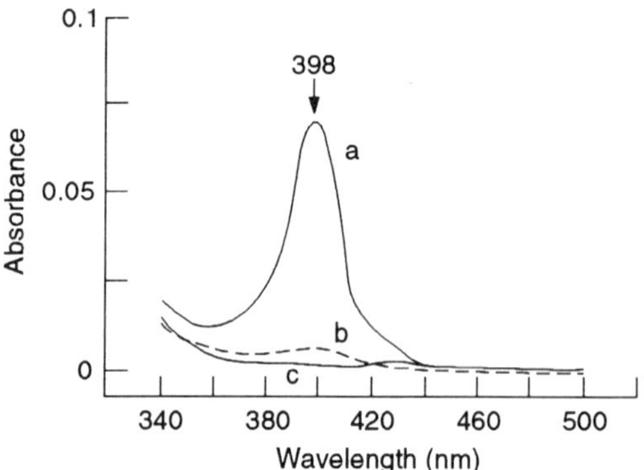

Fig. 5 Spectral properties of heme-deficient cytosolic guanylate cyclase after reaction with and separation from NO–hemoglobin and other proteins. Heme-deficient guanylate cyclase (88 μg) was reacted at 37°C with 300 μg of NO–hemoglobin for 5 min under N_2 (curve a), or with 300 μg of apohemoglobin for 5 min under N_2 and then bubbled with NO gas for 30 sec at 4°C (curve b), or with 300 μg of hemoglobin for 5 min under N_2 (curve c), and reaction mixtures were rapidly cooled and chromatographed on small columns of Matrex Gel Blue-A as described by Ignarro *et al.* (1986). Fractions 3 through 8 corresponding to the elution of guanylate cyclase were pooled and analyzed. All fractions were assayed for enzymatic activity as a check on the chromatographic separation of guanylate cyclase. 398, Soret absorption maximum in nanometers. Reproduced with permission from Ignarro *et al.* (1986).

NO–heme complex from nitrosyl-guanylate cyclase to apohemoglobin occurred upon reaction between the two proteins under various conditions. Similarly, although NO–heme was readily transferred from the NO–hemoprotein to the apo or heme-deficient form of guanylate cyclase, heme itself could not undergo a similar transfer or exchange reaction despite the fact the enzyme can be readily reconstituted with free heme. These observations are consistent with the view that NO–heme binds more tightly than does heme to guanylate cyclase (Wolin *et al.,* 1982).

The physiological implications of these observations are not well appreciated. The *in vivo* activation of guanylate cyclase by NO–hemoproteins may be important in cells containing appreciable quantities of hemoproteins, where such cells are also capable of synthesizing NO from L-arginine. Considering the chemical lability of NO, especially in the presence of oxygen and superoxide anion, and the much greater stability of nitrosyl-hemoproteins, resident hemoproteins could act as target binding sites for NO and transfer the NO as NO–heme to guanylate cyclase, thereby preserving the capacity of intracellularly generated NO to stimulate cyclic GMP formation.

IX. Summary and Conclusions

The experimental evidence is convincing that cytosolic guanylate cyclase is a hemoprotein containing stoichiometric amounts of heme, which functions as a prosthetic group for enzyme activation by NO. Nearly all of the studies described in this chapter were conducted before we began to appreciate in 1986 that mammalian vascular endothelial cells could synthesize their own NO. We know now that many different cell types synthesize NO, and that in most instances the NO interacts in a paracrine manner with adjacent target cells to activate cytosolic guanylate cyclase and elevate intracellular levels of cyclic GMP (Ignarro, 1990). The studies on endothelium-derived relaxing factor and authentic NO have shown clearly that heme and hemoproteins have a very high binding affinity for, and inhibit the actions of, these substances (Ignarro, 1989).

The interaction between NO and the heme prosthetic group of guanylate cyclase appears to constitute an important signal transduction mechanism whereby NO raises intracellular cyclic GMP levels. This novel signal transduction mechanism is highly conducive to the efficient functioning of NO as a paracrine mediator of cellular function. As a small, lipophilic, and chemically labile molecule, NO diffuses out of its cells of origin and into nearby target cells. The very high binding affinity of enzyme-bound heme for NO ensures interaction of the two to cause guanylate cyclase activation and cyclic GMP formation. Thus, relatively uncomplicated mechanisms can account for the paracrine function of endogenous NO in transcellular communication.

References

Arnold, W. P., Mittal, C. K., Katsuki, S., and Murad, F. (1977). Nitric oxide activates guanylate cyclase and increases guanosine 3′, 5′-cyclic monophosphate levels in various tissue preparations. *Proc. Natl. Acad. Sci. U.S.A.* **74,** 3203–3207.

Augusto, O., Kunze, K. L., and Ortiz de Montellano, P. R. (1982). N-Phenylprotoporphyrin IX formation in the hemoglobin-phenylhydrazine reaction. Evidence for a protein-stabilized iron-phenyl intermediate. *J. Biol. Chem.* **257,** 6231–6241.

Braughler, J. M., Mittal, C. K., and Murad, F. (1979). Purification of soluble guanylate cyclase from rat liver. *Proc. Natl. Acad. Sci. U.S.A.* **76,** 219–222.

Buechler, W. A., Nakane, M., and Murad, F. (1991). Expression of soluble guanylate cyclase activity requires both enzyme subunits. *Biochem. Biophys. Res. Commun.* **174,** 351–357.

Craven, P. A., and DeRubertis, F. R. (1978a). Restoration of the responsiveness of purified guanylate cyclase to nitrosoguanidine, nitric oxide, and related activators by heme and heme proteins. Evidence for the involvement of the paramagnetic nitrosyl-heme complex in enzyme activation. *J. Biol. Chem.* **253,** 8433–8443.

Craven, P. A., and DeRubertis, F. R. (1978b). Effects of thiol inhibitors on hepatic guanylate cyclase activity. Evidence for the involvement of vicinal dithiols in the expression of basal and agonist-stimulated activity. *Biochim. Biophys. Acta* **524,** 231–244.

Craven, P. A., DeRubertis, F. R., and Pratt, D. W. (1979). Electron spin resonance study of the role of NO-catalase in the activation of guanylate cyclase by NaN_3 and NH_2OH. Modulation of enzyme responses by heme proteins and their nitrosyl derivatives. *J. Biol. Chem.* **254,** 8213–8222.

De Matteis, F. (1967). Disturbances of liver porphyrin metabolism caused by drugs. *Pharmacol. Rev.* **19,** 523–557.

DeRubertis, F. R., and Craven, P. A. (1976). Calcium-independent modulation of cyclic GMP and activation of guanylate cyclase by nitrosoamines. *Science* **193,** 897–899.

DeRubertis, F. R., and Craven, P. A. (1977). Activation of hepatic guanylate cyclase by N-methyl-N′-nitro-N-nitrosoguanidine. Effects of thiols, N-ethylmaleimide, and divalent cations. *J. Biol. Chem.* **252,** 5804–5814.

Diamond, J., and Blisard, K. S. (1976). Effects of stimulant and relaxant drugs on tension and cyclic nucleotide levels in canine femoral artery. *Mol. Pharmacol.* **12,** 688–692.

Diamond, J., and Holmes, T. G. (1975). Effects of potassium chloride and smooth muscle relaxants on tension and cyclic nucleotide levels in rat myometrium. *Can. J. Physiol. Pharmacol.* **53,** 1099–1107.

Edwards, J. C., Barry, B. K., Gruetter, D. Y., Ohlstein, E. H., Baricos, W. H., and Ignarro, L. J. (1981). Activation of hepatic guanylate cyclase by nitrosyl-heme complexes. Comparison of unpurified and partially purified enzyme. *Biochem. Pharmacol.* **30,** 2531–2538.

Gerzer, R., Bohme, E., Hofmann, F., and Schultz, G. (1981a). Soluble guanylate cyclase purified from bovine lung contains heme and copper. *FEBS Lett.* **132,** 71–74.

Gerzer, R., Hofmann, F., and Schultz, G. (1981b). Purification of a soluble sodium nitroprusside-stimulated guanylate cyclase from bovine lung. *Eur. J. Biochem.* **116,** 479–488.

Goldberg, N. D., and Haddox, M. K. (1977). Cyclic GMP metabolism and involvement in biological regulation. *Annu. Rev. Biochem.* **46,** 823–896.

Gruetter, C. A., Barry, B. K., McNamara, D. B., Gruetter, D. Y., Kadowitz, P. J., and Ignarro, L. J. (1979). Relaxation of bovine coronary artery and activation of coronary arterial guanylate cyclase by nitric oxide, nitroprusside and a carcinogenic nitrosoamine. *J. Cyclic Nucleotide Res.* **5,** 211–224.

Gruetter, D. Y., Gruetter, C. A., Barry, B. K., Baricos, W. H., Hyman, A. L., Kadowitz, P. J., and Ignarro, L. J. (1980). Activation of coronary arterial guanylate cyclase by nitric oxide, nitroprusside, and nitrosoguanidine. Inhibition by calcium, lanthanum, and other cations, enhancement by thiols. *Biochem. Pharmacol.* **29,** 2943–2950.

Ignarro, L. J. (1989). Biological actions and properties of endothelium-derived nitric oxide formed and released from artery and vein. *Circ. Res.* **65,** 1–21.

Ignarro, L. J. (1990). Nitric oxide: A novel signal transduction mechanism for transcellular communication. *Hypertension (Dallas)* **16,** 477–483.

Ignarro, L. J., and Gruetter, C. A. (1980). Requirement of thiols for activation of coronary arterial guanylate cyclase by glyceryl trinitrate and sodium nitrite. Possible involvement of S-nitrosothiols. *Biochim. Biophys. Acta* **631,** 221–231.

Ignarro, L. J., Edwards, J. C., Gruetter D. Y., Barry, B. K., and Gruetter, C. A. (1980a). Possible involvement of S-nitrosothiols in the activation of guanylate cyclase by nitroso compounds. *FEBS Lett.* **110,** 275–278.

Ignarro, L. J., Barry, B. K., Gruetter, D. Y., Edwards, J. V., Ohlstein, O. H., Gruetter, C. A., and Baricos, W. H. (1980b). Guanylate cyclase activation by nitroprusside and nitrosoguanidine is related to formation of S-nitrosothiol intermediates. *Biochem. Biophys. Res. Commun.* **94,** 93–100.

Ignarro, L. J., Lippton, H., Edwards, J. C., Baricos, W. H., Hyman, A. L., Kadowitz, P. J., and Gruetter, C. A. (1981a). Mechanism of vascular smooth muscle relaxation by organic nitrates, nitrites, nitroprusside and nitric oxide: Evidence for the involvement of S-nitrosothiols as active intermediates. *J. Pharmacol. Exp. Ther.* **218,** 739–749.

Ignarro, L. J., Barry, B. K., Gruetter, D. Y., Ohlstein, E. H., Gruetter, C. A., Kadowitz, P. J., and Baricos, W. H. (1981b). Selective alterations in responsiveness of guanylate cyclase to activation by nitroso compounds during enzyme purification. *Biochim. Biophys. Acta* **673,** 394–407.

Ignarro, L. J., Kadowitz, P. J., and Baricos, W. H. (1981c). Evidence that regulation of hepatic guanylate cyclase activity involves interactions between catalytic site —SH groups and both substrate and activator. *Arch. Biochem. Biophys.* **208,** 75–86.

Ignarro, L. J., Degnan, J. N., Baricos, W. H., Kadowitz, P. J., and Wolin, M. S. (1982a). Activation of purified guanylate cyclase by nitric oxide requires heme. Comparison of heme-deficient, heme-reconstituted and heme-containing forms of soluble enzyme from bovine lung. *Biochim. Biophys. Acta* **718,** 49–59.

Ignarro, L. ., Wood, K. S., and Wolin, M. S. (1982b). Activation of purified soluble guanylate cyclase by protoporphyrin IX. *Proc. Natl. Acad. Sci. U.S.A.* **79,** 2870–2873.

Ignarro, L. J., Wood, K. S., Ballot, B., and Wolin, M. S. (1984a). Guanylate cyclase from bovine lung. Evidence that enzyme activation by phenylhydrazine is mediated by iron-phenyl hemoprotein complexes. *J. Biol. Chem.* **259,** 5923–5931.

Ignarro, L. J., Ballot, B., and Wood, K. S. (1984b). Regulation of soluble guanylate cyclase activity by porphyrins and metalloporphyrins. *J. Biol. Chem.* **259,** 6201–6207.

Ignarro, L. J., Wood, K. S., and Wolin, M. S. (1984c). Regulation of purified soluble guanylate cyclase by porphyrins and metalloporphyrins: A unifying concept. *Adv. Cyclic Nucleotide Protein Phosphorylation Res.* **17,** 267–274.

Ignarro, L. J., Burke, T. M., Wood, K. S., Wolin, M. S., and Kadowitz, P. J. (1984d). Association between cyclic GMP accumulation and acetylcholine-elicited relaxation of bovine intrapulmonary artery. *J. Pharmacol. Exp. Ther.* **228,** 682–690.

Ignarro, L. J., Adams, J. B., Horwitz, P. M., and Wood, K. S. (1986). Activation of soluble guanylate cyclase by NO-hemoproteins involves NO-heme exchange. Comparison of heme-containing and heme-deficient enzyme forms. *J. Biol. Chem.* **261,** 4997–5002.

Katsuki, S., and Murad, F. (1977). Regulation of adenosine cyclic 3′, 5′-monophosphate and guanosine cyclic 3′, 5′-monophosphate levels and contractility in bovine tracheal smooth muscle. *Mol. Pharmacol.* **13,** 330–341.

Katsuki, S., Arnold, W., Mittal, C., and Murad, F. (1977). Stimulation of guanylate cyclase by sodium nitroprusside, nitroglycerine and nitric oxide in various tissue preparations and comparison to the effects of NaN_3 and NH_2OH. *J. Cyclic Nucleotide Res.* **3,** 23–35.

Kimura, H., Mittal, C. K., and Murad, F. (1975a). Increases in cyclic GMP levels in brain and liver with NaN_3 an activator of guanylate cyclase. *Nature (London)* **257,** 700–702.

Kimura, H., Mittal, C. K., and Murad, F. (1975b). Activation of guanylate cyclase from rat liver and other tissues by NaN_3. *J. Biol. Chem.* **250,** 8016–8022.

Kunze, K. L., and Ortiz de Montellano, P. R. (1983). Formation of a 6-bonded aryliron complex in the reaction of arylhydrazines with hemoglobin and myoglobin. *J. Am. Chem. Soc.* **105,** 1380–1381.

Lawley, P. D., and Thatcher, C. J. (1970). Methylation of deoxyribonucleic acid in cultured mammalian cells by N-methyl-N′-nitro-N-nitrosoguanidine. *Biochem. J.* **116,** 693–707.

McCalla, D. R., Reuvers, A., and Kitai, R. (1968). Inactivation of biologically active N-methyl-N-nitroso compounds in aqueous solution. Effect of various conditions of pH and illumination. *Can. J. Biochem.* **46,** 808–811.

Mellion, B. T., Ignarro, L. J., Myers, C. B., Ohlstein, E. H., Ballot, B. A., Hyman, A. L., and Kadowitz, P. J. (1983). Inhibition of human platelet aggregation by S-nitrosothiols. Heme-dependent activation of soluble guanylate cyclase and stimulation of cyclic GMP accumulation. *Mol. Pharmacol.* **23,** 653–664.

Murad, F., Mittal, C. K., Arnold, W. P., Katsuki, S., and Kimura, H. (1978). Guanylate cyclase: Activation by azide, nitro compounds, nitric oxide, and hydroxyl radical and inhibition by hemoglobin and myoglobin. *Adv. Cyclic Nucleotide Res.* **9,** 145–158.

Nakane, M., Arai, K., Saheki, S., Kuno, T., Buechler, W., and Murad, F. (1990). Molecular cloning and expression of cDNAs coding for soluble guanylate cyclase from rat lung. *J. Biol. Chem.* **265,** 16841–16845.

Ohlstein, E. H., Barry, B. K., Gruetter, D. Y., and Ignarro, L. J. (1979). Methemoglobin blockade of coronary arterial soluble guanylate cyclase activation by nitroso compounds and its reversal with dithiothreitol. *FEBS Lett.* **102,** 316–320.

Ohlstein, E. H., Wood, K. S., and Ignarro, L. J. (1982). Purification and properties of heme-deficient hepatic soluble guanylate cyclase: Effects of heme and other factors on enzyme activation by NO, NO-heme, and protoporphyrin IX. *Arch. Biochem. Biophys.* **218,** 187–198.

Ortiz de Montellano, P. R., Yost, G. S., Mico, B. A., Dinizo, S. E., Correia, M. A., and Kumbara, H. (1979). Destruction of cytochrome P-450 by 2-isopropyl-4-pentenamide and methyl 2-isopropyl-4-pentenoate: Mass spectrometric characterization of prosthetic heme adducts and nonparticipation of epoxide metabolites. *Arch. Biochem. Biophys.* **197,** 524–533.

Schoental, R., and Rive, D. J. (1965). Interaction of N-alkyl-N-nitrosourethanes with thiols. *Biochem. J.* **97,** 466–474.

Schultz, K. D., Schultz, K., and Schultz, G. (1977). Sodium nitroprusside and other smooth muscle relaxants increase cyclic GMP levels in rat ductus deferens. *Nature* (*London*) **265,** 750–751.

Schulz, U., and McCalla, D. R. (1969). Reactions of cysteine with N-methyl-N-nitroso-p-toluenesulfonamide and N-methyl-N′-nitro-N-nitrosoguanidine. *Can. J. Chem.* **47,** 2021–2027.

Tephly, T. R., Gibbs, A. H., and De Matteis, F. (1979). Studies on the mechanism of experimental porphyria produced by 3,5-diethoxycarbonyl-1,4-dihydrocollidine. Role of a porphyrin-like inhibitor of protohaem ferro-lyase. *Biochem. J.* **180,** 241–244.

White, A. A., Crawford, K. M., Patt, C. S., and Lad, P. J. (1976). Activation of soluble guanylate cyclase from rat lung by incubation or by hydrogen peroxide. *J. Biol. Chem.* **251,** 7304–7312.

Wolin, M. S., Wood, K. S., and Ignarro, L. J. (1982). Guanylate cyclase from bovine lung. A kinetic analysis of the regulation of the purified soluble enzyme by protoporphyrin IX, heme, and nitrosyl-heme. *J. Biol. Chem.* **257,** 13312–13320.

Wood, K. S., and Ignarro, L. J. (1987). Hepatic cyclic GMP formation is regulated by similar factors that modulate activation of purified hepatic soluble guanylate cyclase. *J. Biol. Chem.* **262,** 5020–5027.

Regulation of Particulate Guanylate Cyclase by Natriuretic Peptides and Escherichia coli Heat-Stable Enterotoxin

Dale C. Leitman,* Scott A. Waldman,† and Ferid Murad‡

* *Metabolic Research Unit*
University of California, San Francisco
School of Medicine
San Francisco, California 94143

† *Division of Clinical Pharmacology*
Departments of Medicine and Pharmacology
Thomas Jefferson University
Philadephia, Pennsylvania 19107

‡ *Molecular Geriatrics Corporation*
Lake Bluff, Illinois 60044

I. Introduction

The signal transducing enzyme guanylate cyclase is activated in response to specific extracellular signals, leading to the formation of the intracellular second messenger, cyclic GMP (Murad, 1986; Waldman and Murad, 1987; Leitman and Murad, 1987). Two isoenzyme forms of guanylate cyclase exist in cells, which are physically and biochemically distinct and regulated by different agents. The soluble isoenzyme of guanylate cyclase, located in the cytosol, is a heterodimer composed of 70- and 82-kDa subunits and is activated by nitrate containing compounds, endothelial-derived relaxing factors, and oxytocin (Murad, 1986; Waldman and Murad, 1987; Leitman *et al.*, 1988b). A second isoenzyme, particulate guanylate cyclase is located in the cell membrane. Initial interest in this isoenzyme was prompted by studies demonstrating that the heat-stable enterotoxin (ST) derived from *Escherichia coli*-activated particulate guanylate cyclase in intestinal mucosa (Hughes *et al.*, 1978; Field *et al.*, 1978). This isoenzyme has become the focus of intense investigation after it was discovered that atrial natriuretic peptide (ANP) activated particulate guanylate cyclase in cell-free preparations (Waldman *et al.*, 1984, 1985a; Leitman *et al.*, 1987, 1988a). This was the first demonstration that a circulating peptide hormone exerted its biological effects by activating particulate guanylate cyclase (Waldman *et al.*, 1984). Currently, ST and natriuretic peptides are the only known agents that exert their biological effects by activating particulate guanylate cyclase in mammalian cells.

Enterotoxigenic *E. coli* elaborate ST, a plasmid-encoded low-molecular-weight peptide exotoxin that produces profuse watery diarrhea by increasing fluid and electrolyte secretion in the intestinal mucosa (Field *et al.*, 1978). Interestingly, a mammalian homologue of this peptide, termed guanylin, has been identified in small intestinal cells and presumably mediates local fluid and electrolyte balance in that organ (Currie *et al.*, 1992). Our previous studies with ST prompted us to investigate the effects of ANP on particulate guanylate cyclase because both are small, heat-stable peptides that regulate ion transport. ANP belongs to a unique family of peptides, known as natriuretic peptides (NP) that produce diuresis, natriuresis, and hypotension (Atlas, 1986; Brenner *et al.*, 1990; Rosenzwieg and Seidman, 1991). NP are sythesized as a larger precursor protein that are characterized by the presence of a 17 amino acid ring structure that is joined by a disulfide bond. NP have been classified into three groups. The major circulating natriuretic peptide, ANP (Type A), is a 28 amino acid peptide synthesized and secreted from atrial cardiocytes in response to expanded extracellular volume (Flynn *et al.*, 1983; Rosenzwieg and Seidman, 1991). Also, ANP has been shown to promote vascular smooth muscle relaxation, inhibit the secretion of renin, vasopressin and aldosterone, and increase the secretion of testosterone and progesterone (Atlas, 1986; Brenner *et al.*, 1990; Rosenzwieg and Seidman, 1991). A 32 amino acid natriuretic peptide derived from the brain and heart (Sudoh *et al.*, 1988), brain natriuretic peptide (BNP), is classified as a Type B natriuretic peptide (Rosenzwieg and Seidman, 1991). The most recently discovered natriuretic peptide, designated Type C natriuretic peptide (CNP), is a 22 amino acid peptide located in the brain (Sudoh *et al.*, 1990; Rosenzwieg and Seidman, 1991). Although these NP have a primary structure different from that of ANP, they also activate particulate guanylate cyclase and produce hypotension, diuresis, and natriuresis.

It is clear that many of the cellular effects of NP and ST are mediated by the activation of particulate guanylate cyclase and increase in intracellular concentrations of cyclic GMP. The signaling pathways mediating the increased formation of cyclic GMP induced by these peptides are similar and involve several key steps. First, ANP and ST bind to specific receptors in the cell membrane that are coupled to the activation of particulate guanylate cyclase. The activated form of particulate guanylate cyclase converts GTP into cyclic GMP, which regulates specific biochemical pathways that mediate selected biological effects. In this paper, we describe the membrane receptors for ANP and ST and particulate guanylate cyclase to which they are coupled.

II. Atrial Natriuretic Peptide (ANP) Receptors

A. Pharmacological Heterogeneity of ANP Receptors

The pharmacological properties of ANP receptors in cultured cells and tissues have been investigated with radioligand binding studies and biological assays. In most binding studies a single class of receptors have been identified by Scatchard analysis with equilibrium binding constants (K_d) that range from 0.025 to 2 nM (Martin and Ballermann, 1989; Leitman and Murad, 1990). Surprisingly, high-affinity ANP receptors have been identified in many different cell types, including some not directly involved in blood pressure, electrolyte and intravascular volume homeostasis (Martin and Ballerman, 1989; Leitman and Murad, 1990). The adrenal cortex and glomeruli have the highest density of ANP receptors (Martin and Ballermann, 1989; Leitman and Murad, 1990). The cultured cell type with the most ANP receptors is aortic smooth muscle with 200,000–500,000 receptors per cell (Martin and Ballermann, 1989; Leitman and Murad, 1990). The widespread distribution of ANP receptors suggests that this peptide hormone may regulate general cellular processes, such as metabolism, differentiation, and proliferation.

Initially, only a single class of ANP receptors was apparent in cells, since Scatchard analyses of equilibrium binding experiments revealed a single linear isotherm. However, evidence for multiple ANP receptors was derived from studies comparing the ability of ANP analogues to compete for ^{125}I-labeled ANP binding and stimulate cyclic GMP production in cultured aortic endothelial cells (Leitman and Murad, 1986). These studies demonstrated that C-terminal truncated analogues, such as atriopeptin I, were nearly as effective as native ANP at binding to receptors. However, atriopeptin I was 1000-fold weaker at stimulating cyclic GMP production compared to ANP, which contains the native C-terminal amino acids. Furthermore, atriopeptin I was unable to antagonize the increase in cyclic GMP levels produced by ANP. These studies suggested that two functionally distinct classes of ANP receptors exist in cells (Leitman and Murad, 1986). The vast majority of ANP receptors in cells have high affinity for C-terminal truncated analogues but are not coupled to the activation of particulate guanylate cyclase. A less abundant second class of ANP receptors with low affinity for C-terminal truncated analogues is apparently coupled to particulate guanylate cyclase and responsible for mediating many of the biological effects of ANP.

B. Heterogeneity of ANP Receptor Subunit Structure

Affinity crosslinking studies revealed that cells contain two physically distinct ANP receptors, designated ANP-R1 and ANP-R2 (Leitman *et al.*, 1986; Martin and Ballermann, 1989; Leitman and Murad, 1987, 1990). Crosslinking studies demonstrated that ANP-R1, which accounted for 1–5% of ANP binding sites in cultured cells, consisted of a single, nonreducible protein with a molecular mass of approximately 130 kDa (Leitman *et al.*, 1986). Other studies established that ANP-R1 was linked to the activation of particulate guanylate cyclase and that this receptor subtype was responsible for promoting the biological effects mediated by cyclic GMP (Leitman *et al.*, 1988a). In contrast, ANP-R2 is a protein of 66 kDa that exists in the cell membrane as either a monomer or a 130-kDa dimer that is joined by disulfide bridges (Schenk *et al.*, 1987; Leitman *et al.*, 1988a). Although ANP-R2 receptors are highly abundant (95–99% of ANP binding sites; Leitman *et al.*, 1988a) the second messenger system(s) coupled to this subtype and its physiological role remain unclear. Whereas ANP-R2 are not coupled to particulate guanylate cyclase, it has been reported that ANP binding to this receptor subtype stimulates phosphoinositide breakdown (Hirata *et al.*, 1989) and inhibits adenylate cyclase activity (Anand-Srivastava *et al.*, 1990). ANP-R2 also has been termed the C-receptor because of its suggested role in the uptake, internalization, and clearance of ANP from the circulation (Maack *et al.*, 1987; Maack, 1992).

C. ANP-R1 and Particulate Guanylate Cyclase Reside on the Same Transmembrane Protein

After it was discovered that ANP activated particulate guanylate cyclase in cell-free systems, studies were initiated to purify the ANP receptor and particulate guanylate cyclase to establish an *in vitro* system to explore how this receptor and enzyme are coupled. Unexpectedly, the ANP-R1 receptor and particulate guanylate cyclase solubilized from rat lung membranes could not be separated by a variety of chromatographic techniques (Kuno *et al.*, 1986). Indeed, these activities copurified in homogeneous preparations containing a single protein of 120 to 130 kDa (Kuno *et al.*, 1986). The copurification of these activities suggested that ANP-R1 is a single transmembrane protein containing ANP binding and guanylate cyclase catalytic sites (Kuno *et al.*, 1986). The extracellular domain possesses the ligand-binding region, whereas the catalytic site is located in the cytoplasm, where it converts GTP into cyclic GMP in the presence of intracellular cofactors. ANP-R1 and particulate guanylate cyclase also copurify from the adrenal gland (Takayanagi *et al.*, 1987; Meloche *et al.*, 1988). In contrast to the ANP-R1 receptor, purified ANP-R2 from smooth muscle cells are devoid of guanylate cyclase activity (Schenk *et al.*, 1987). The purification of the ANP receptor–particulate guanylate cyclase protein and ANP-R2 receptor confirmed the existence of at least two functionally and physically distinct ANP receptors.

D. Cloning of ANP Receptors

Oligonucleotide probes derived from the amino acid sequence of sea urchin sperm particulate guanylate cyclase were used to isolate a cDNA for particulate guanylate cyclase from a rat brain cDNA library (Singh *et al.*, 1988). The transfection of this cDNA into COS-7 cells resulted in a marked increase in both ^{125}I-labeled ANP binding and guanylate cyclase activity (Chinkers *et al.*, 1989). These studies confirmed that the ANP-R1 receptor and particulate guanylate cyclase reside on a single transmembrane protein. By screening a human cDNA library, two distinct cDNA were isolated that encode ANP receptor/particulate guanylate cyclase proteins, designated NPR-A (or GC-A) and NPR-B (or GC-B; Lowe *et al.*, 1989; Garbers, 1989; Chinkers and Garbers, 1991; Koller and Goeddel, 1992). The GC-A cDNA encodes a protein consisting of four functional domains: a 441 amino acid extracellular segment containing the ligand binding site, a single 21 amino acid transmembrane domain, and a 568 residue intracellular segment containing a protein tyrosine kinase-like and guanylate cyclase catalytic domains (Lowe *et al.*, 1989). The GC-B is 91, 72, and 43% identical to GC-A in the guanylate cyclase, protein kinase-like, and extracellular domains, respectively (Garbers, 1989; Chinkers and Garbers, 1991). The isolation of two forms of ANP-R1 receptors raised the intriguing possibility that different natriuretic peptides may interact more selectively with the different receptors. Indeed, it has been shown that ANP and BNP bind to the GC-A receptor with higher affinity than CNP (Chang *et al.* 1989), whereas CNP binds with much greater affinity to GC-B (Koller *et al.*, 1991). A cDNA for the mouse GC-A has been cloned that has 97% homology to the rat cDNA and 94% homology to the human cDNA, demonstrating that this protein is highly conserved among vertebrates (Pandey and Singh, 1990). Recently, the gene for the GC-A receptor has been isolated from the rat (Yamaguchi *et al.*, 1990). The gene comprises 22 exons and 21 introns and is 17.5 kb in length. Analysis of the sequence of the 5′ promoter region found no TATA box, but an initiator sequence (Smale and Baltimore, 1989) was present at −3 to +5. The promoter region contains an inverted CCAAT box that is potentially activated by CTF/NF1 and three GC boxes that potentially interact with the non-tissue-specific transcription factor, Sp-1 (Yamaguchi *et al.*, 1990).

A bovine and human cDNA for the non-guanylate cyclase coupled ANP-R2 (C-receptor) has been isolated (Fuller *et al.*, 1988; Lowe *et al.*, 1990; Porter *et al.*, 1990). The cDNA encodes for a 496 amino acid protein that comprises a 436 amino acid extracellular domain, a single 23 amino acid transmembrane segment, and a very short intracellular region of 37 amino acids (Fuller *et al.*, 1988). The extracellular domain, which contains the binding site for ANP, exhibits 30% homology to the extracellular domain of GC-A and GC-B. Furthermore, the expression of the cDNA in mammalian cells results in high-affinity ANP binding, but no ANP-induced stimulation of cyclic GMP (Porter *et al.*, 1988). These results confirm that ANP-R2 does not have guanylate cyclase activity and is not coupled to the activation of guanylate cyclase.

E. Coupling of ANP-R1 Receptor to the Activation of Particulate Guanylate Cyclase

ANP has been shown to activate particulate guanylate cyclase in several rat tissues and multiple cultured cells (Waldman *et al.*, 1984, 1985a; Leitman *et al.*, 1987, 1988a). The mechanism by which ANP activates the catalytic domain of guanylate cyclase is not known, but is of key importance to understanding the actions of ANP and the development of possible therapeutic agents that mimic the effects of ANP. Interestingly, purified particulate guanylate cyclase binds ANP ith high affinity but is not activated by this hormone, suggesting that accessory factors may participate in the signaling process, since these putative factors may be removed or inactivated during purification (Kuno *et al.*, 1986). One possible accessory factor may be ATP, since it markedly potentiates the ANP-induced activation of particulate guanylate cyclase (Kurose *et al.*, 1987; Chang *et al.*, 1990; Gazzano *et al.*, 1991b; Leitman *et al.*, 1991). Several possible mechanisms may account for ATP potentiating the activation of guanylate cyclase by ANP. First, ATP may serve as a substrate for autophosphorylation of the receptor–enzyme or phosphorylation of an accessory protein involved in signal transduction. This possibility is unlikely since nonhydrolyzable adenine nucleotides, such as AMP-PNP and ATP-δS, also potentiate the effect of ANP on guanylate cyclase activity, although they cannot serve as a substrate for protein phosphorylation.

Another possible mechanism to explain the potentiation of ANP activation of guanylate cyclase by ATP is that the adenine nucleotide binds to the ANP receptor–enzyme complex and activates the catalytic domain by an allosteric action. Thus, when the kinase-like domain was removed from the ANP receptor–guanylate cyclase protein by deletion mutagenesis, the resulting mutant receptor retained guanylate cyclase activity, but is not activated adenine nucleotides (Chinkers and Garbers, 1989). These data were interpreted as suggesting that the kinase-like domain directly mediated adenine nucleotide-dependent regulation of guanylate cyclase activation by ANP (Chinkers and Garbers, 1989). However, these data need to be interpreted cautiously since deletion of the kinase-like domain also results in the loss of ANP-dependent activation of guanylate cyclase (Koller *et al.*, 1992), which was absolutely required to observe adenine nucleotide regulation of this enzyme in that system (Chinkers and Garbers, 1989). Also, guanylate cyclase (GC-A) expressed from a baculovirus in Sf9 insect cells was activated by ATP in the presence of ANP (Chinkers *et al.*, 1991). These results suggest that GC-A is activated directly by ATP, since it is unlikely that an accessory coupling protein is expressed in insect cells. Direct binding of ATP to the ANP receptor–particulate guanylate cyclase protein purified from rat testicular membranes has been demonstrated previously (Marala *et al.*, 1991). Adenine nucleotide binding to purified receptor–cyclase appears to be mediated by the kinase-like region since point mutations of this domain by site-directed mutagenesis results in the loss of ATP binding to the mutant protein (Marala *et al.*, 1992). These studies suggest that ATP regulates the signal transducing functions of the ANP receptor–guanylate cyclase protein, in part, by binding to the protein kinase-like domain, which allosterically activates the guanylate cyclase catalytic domain.

Finally, ATP may regulate the activity of accessory proteins that participate in this signal transduction pathway. Indeed, washing lung membranes decreased the ATP activation of particulate guanylate cyclase in the absence and presence of ANP (Marala *et al.*, 1992). Also, adenine nucleotide regulation could be separated from particulate guanylate cyclase activity by partial purification of this enzyme using affinity chromatography (Gazzano *et al.*, 1991b). These observations suggest that an accessory regulatory factor, presumably a membrane protein, was removed during membrane washing or partial purification of the enzyme. Thus, an adenine nucleotide binding protein (A-protein) may participate in the activation of guanylate cyclase, in a fashion similar to the role of the guanine nucleotide-binding protein (G-protein) in the activation of adenylate cyclase. Taken together, data from various studies suggest that basal and ANP activation of particulate guanylate cyclase may be regulated by ATP directly through allosteric interactions and may require the interaction of ATP with an accessory coupling protein. Clearly, the mechanisms of ANP and ATP activation of particulate guanylate cyclase require further examination.

III. Heat-Stable Enterotoxin (ST) Receptor

A. Pharmacological Heterogeneity of ST Receptors

In studies utilizing ^{125}I-labeled ST with low (< 100 Ci/mmol) specific activity, membrane-bound and detergent-solubilized receptors exhibited linear Scatchard plots with a K_d of 10^{-9} M, suggesting a single class of receptors (Gianella *et al.*, 1980, 1983; Thomas and Knoop, 1983; Frantz *et al.*, 1984; Dreyfus *et al.*, 1984; Dreyfus and Robertson, 1984; Kuno *et al.*, 1985; Thompson, 1987). However, a novel class of high-affinity, low-capacity receptors was identified in these membranes when ^{125}I-labeled ST with high (1000–2000 Ci/mmol) specific activity and NaCl were utilized in binding assays (Hugues *et al.*, 1991). Scatchard analyses of equilibrium binding in the absence of NaCl demonstrated a single class of binding sites with a K_d of 1.9×10^{-9} M and a B_{max} of 0.75 pmol/mg of protein. Identical experiments performed in the presence of physiological concentrations of NaCl yielded curvilinear Scatchard analyses, suggesting an additional high-affinity binding site for ST, with a K_d of 2.1×10^{-11} M and B_{max} of 75 fmol/mg of protein. These sites were confirmed in studies of competitive and dynamic binding of ST to intestinal membranes. Thus, the K_d for high- and low-affinity sites calculated from kinetic parameters, 1.3×10^{-11} and 6×10^{-10} M, respectively, agreed closely with those values determined by equilibrium binding. The mechanism by which NaCl "unmasks" high-affinity ST binding sites is unclear. The role of these novel high affinity sites in guanylin and toxin-mediated regulation of fluid and electrolyte balance in intestine remains to be defined.

Previous studies suggested that ST-induced guanylate cyclase activation occurred at concentrations (EC_{50} = 10^{-7} M) that were higher than those required for receptor occupancy (K_d = 10^{-11}–10^{-9} M) or induction of intestinal secretion (about 10^{-10} M; Gianella *et al.*, 1980, 1983; Thomas and Knoop, 1983; Frantz *et al.*, 1984; Dreyfus *et al.*, 1983, 1984; Dreyfus and Robertson, 1984; Kuno *et al.*, 1985; Thompson, 1987; Greenberg and Saeed, 1988; Carr *et al.*, 1989; Hugues *et al.*, 1991). However, studies of the activation of guanylate cyclase by ST are conducted over a time course (5 min) that is shorter than that to achieve equilibrium binding (100 min) or induce secretion (3 hr). Therefore, the time course of ST–receptor interaction was examined in order to correlate receptor occupancy with enzyme activation (Crane *et al.*, 1992). High-affinity receptors for ST saturated almost instantaneously at concentrations of ST that were lower than those necessary to stimulate guanylate cyclase, confirming that these sites were not coupled to guanylate cyclase activation (Hugues *et al.*, 1991; Crane *et al.*, 1992). The signaling pathways to which high-affinity ST receptors are coupled remain unclear. However, ST has been demonstrated to induce phosphatidylinositol metabolism and increase calcium concentrations in intestinal cells (Banik and Ganguly, 1988; Knoop *et al.*, 1991). Occupation of high-affinity receptors by ST may activate these pathways, which could be important in the pathophysiological actions of this toxin (Hugues *et al.*, 1991).

In contrast to high-affinity receptors, low-affinity receptors exhibited curvilinear association kinetics, demonstrating at least two different rates of association with ST (Crane *et al.*, 1992). Only a single rate of dissociation of ST from these receptors was observed. Although two affinities for ST were observed in kinetic studies, equilibrium binding studies revealed a single class of low-affinity receptors. Biphasic association kinetics is most consistent with a model in which binding is a two-step process involving a separate regulatory component (Crane *et al.*, 1992). Thus, ST binding to the basal, higher affinity state of this receptor may induce an interaction between the ligand–receptor complex and a separate regulatory protein. This interaction could result in the production of a new lower affinity state of the liganded receptor, observed later in the time course of association once the initial ligand–receptor complex had dissociated, whereas at equilibrium, a single population of receptors possessing the lowest affinity for ST would be observed. This model is particularly attractive since a potential candidate for such a regulatory component has been identified (Gazzano *et al.*, 1991a). Indeed, an adenine nucleotide-binding protein may couple ST–receptor interaction to guanylate cyclase as suggested for ANP receptors and in a fashion analogous to the coupling of ligand–receptor interaction and adenylate cyclase by guanine nucleotide-binding proteins (Kurose *et al.*, 1987; Chang *et al.*, 1990; Gazzano *et al.*, 1991b). In the adenylate and guanylate cyclase systems, purine nucleotides mediate through a coupling protein ligand-induced regulation of the enzyme and alterations in receptor affinity. Thus, an adenine nucleotide-dependent regulatory component may mediate the alterations in affinity of low-affinity ST receptors.

As indicated above, high-affinity receptors are not coupled to activation of particulate guanylate cyclase (Dreyfus *et al.*, 1983; Crane *et al.*, 1992). Similarly, the highest affinity state of low-affinity receptors is saturated at concentrations of ST that are lower than those required to activate particulate guanylate cyclase at 5 min (Crane *et al.*, 1992). However, occupation of the lowest affinity state of these receptors correlates with activation of guanylate cyclase (Crane *et al.*, 1992). These data suggest that the lowest affinity state of low-affinity receptors is directly coupled to activation of particulate guanylate cyclase and increases in intracellular cyclic GMP in intestinal membranes. The role of alterations in the affinity of ST receptors and its importance for toxin activation of particulate guanylate cyclase and regulation of intestinal secretion remains to be defined.

B. Heterogeneity of ST Receptor Subunit Structure

Radiolabeled ST was covalently crosslinked to receptors and analyzed by sodium dodecyl sulfate–polyacrylamide gel electrophoresis (SDS–PAGE; Kuno *et al.*, 1985; Ivens *et al.*, 1990). Experiments performed under nonreducing conditions yielded proteins specifically labeled with molecular masses of 160, 136, 78, 71, and 56 kDa. Under these conditions the label was distributed approximately uniformly in these proteins. In the presence of a reducing agent, β-mercaptoethanol, the same proteins were specifically labeled. However, labeling of the proteins was not uniform. The majority of radioactivity was associated with the 78-, 71-, and 56-kDa proteins and only a minor fraction associated with 160- and 136-kDa proteins. When the 160-kDa protein was excised from nonreducing denaturing gels and then subjected to electrophoresis under reducing conditions, most of the protein was reduced to a 78-kDa subunit while a portion was resistant to reduction. Subjecting the 136-kDa protein to the identical protocol yielded subunits of 136 and 71 kDa. The 78-, 71-, and 56-kDa subunits were not further reducible. These data suggest that there are several structural forms of the ST receptor: nonreducible 160-, 136-, 78-, 71-, and 56-kDa binding subunits and 160- and 136-kDa proteins, which are reducible to 78- and 71-kDa binding subunits, respectively. The presence of multiple binding subunits for ST in native intestinal cells from rats, pigs, and humans has been confirmed in other laboratories (Thompson and Giannella, 1990; Katwa *et al.*, 1991). The structural and functional relationships of these toxin binding subunits to each other and particulate guanylate cyclase remain to be defined. However, it is notable that this observed pattern of high-molecular-weight monomeric receptors and oligomeric receptors with subunit molecular masses of about 70 kDa is reminiscent of the different structural receptors for ANP (Leitman *et al.*, 1986, 1988a; Schenk *et al.*, 1987; Martin and Ballermann, 1989; Leitman and Murad, 1990).

C. Heterogeneity of ST Receptor Subcellular Distribution

In addition to the structural and functional heterogeneity outlined above, ST receptors demonstrate heterogeneity of membrane localization in intestinal brush borders. Only about 40% of the total ST receptors in these membranes can be extracted with detergents and appear to be associated with the lipid bilayer (Waldman *et al.*, 1985b). The remaining receptors resistant to solubilization appear to be associated with the cytoskeleton of brush border membranes. Although ST receptors and particulate guanylate cyclase are present in detergent extracts, these activities are not coupled (Kuno *et al.*, 1985; Waldman *et al.*, 1985b). However, addition of ST to the cytoskeletal residue remaining after detergent extraction results in guanylate cyclase activation (Waldman *et al.*, 1985b). Sequential extraction of intestinal membranes with detergent followed by a combination of detergent and KCl results in the extraction of 80% of the total ST binding and guanylate cyclase activities in these membranes (Hakki *et al.*, 1992). This technique permitted selective extraction and characterization of receptors and enzyme associated with the lipid bilayer and cytoskeleton. Interestingly, ST receptor binding was preferentially coupled to activation of guanylate cyclase in cytoskeleton compared to lipid bilayer-associated fractions. Thus, coupled ST receptor and particulate guanylate cyclase appear to be selectively associated with the cytoskeleton in intestinal membranes (Waldman *et al.*, 1985b).

D. Cloning of ST Receptors

To determine if ST receptors in intestinal cells are members of the guanylate cyclase–peptide receptor family of proteins, degenerate oligonucleotide primers based on conserved sequences in both soluble and particulate guanylate cyclases were employed to amplify cDNA from rat small intestinal cells by the method of PCR (Schultz *et al.*, 1990). Unique PCR-generated sequences were used to probe intestinal cell cDNA libraries, which yielded a novel member of this family of cyclase receptors, termed GC-C. This receptor is derived from an open reading frame of 3225 nucleotides that code for an unprocessed protein of 1053 amino acids with a molecular mass of 121 kDa. The cloned protein exhibits the same structural motif as the ANP receptor–particulate guanylate cyclases: an extracellular domain encoding a peptide-binding region, a single, short transmembrane domain, and a cytoplasmic region containing protein tyrosine kinase-like and guanylate cyclase catalytic domains. The extracellular ligand-binding domain of GC-C exhibits only 10% sequence homology with the natriuretic receptors GC-A and GC-B, congruent with the different ligand-binding specificities of these receptors. The protein tyrosine kinase-like domain of GC-C is 39 and 35% identical with GC-A and GC-B, respectively. Interestingly, this domain in GC-C lacks the consensus sequence for protein kinases found in the natriuretic receptors and important for adenine nucleotide binding (Schulz *et al.*, 1990). The region of greatest homology of primary structure is in the guanylate cyclase catalytic domain, wherein GC-C is 55% identical to GC-A and GC-B. This novel peptide receptor–cyclase appears to be unique in possessing an extended carboxy terminal enriched in uncharged amino acids (Schulz *et al.*, 1990). It was suggested that this tail may form the structural basis for tight association with the cytoskeleton of intestinal cells (Schulz *et al.*, 1990).

Thus, cloned GC-C encodes a novel member of the particulate guanylate cyclase–peptide receptor family presumably associated with the cytoskeleton of intestinal cells and possessing ligand-binding characteristics that are different from those of GC-A and GC-B. Indeed, when cloned ST receptors were transiently expressed in COS-7 cells, these cells exhibited increased particulate guanylate cyclase activity and acquired the ability to specifically bind ^{125}I-labeled ST. ST binding in transfected cells was of the low-affinity type, with a K_d of about 10^{-9} M (Schulz *et al.*, 1990). Also, the addition of ST resulted in an increased accumulation of cyclic GMP in COS-7 cells expressing this receptor in culture. Thus, GC-C appears to possess ST binding activity, which is coupled to particulate guanylate cyclase activity and accumulation of intracellular cyclic GMP. Similar results have been obtained with the human homologue of GC-C expressed in COS-7 and other mammalian cells (de Sauvage *et al.*, 1991, 1992; Singh *et al.*, 1991).

E. Purification of ST Receptors from Intestinal Mucosa

Active receptors for ST were purified by ligand affinity chromatography from membranes isolated from the lipid-associated compartment of rat intestinal mucosa (Hugues *et al.*, 1992). The novel affinity column was prepared by coupling ST to biotin derivatized with an extended *N*-hydroxy-succinylated spacer arm prior to binding to monomeric avidin immobilized on agarose. Detergent extracts of rat intestinal mucosa membranes were quantitatively depleted of ST binding activity when chromatographed on this affinity matrix. Biotinylated ST–receptor complexes were specifically eluted from affinity columns with 2 mM biotin and these complexes dissociated with bile salts. Using this technique, functional ST receptors were purified maximally about 2000-fold, with about 3% of the total activity in crude extracts recovered in these purified preparations. Analysis of affinity-purified preparations by SDS–PAGE and silver staining demonstrated a major protein subunit of 74 kDa. Affinity crosslinking of these preparations to ^{125}I-labeled ST demonstrated specific labeling predominantly of the 74-kDa subunit. In addition, lower amounts of labeled ST were incorporated into higher and lower molecular weight subunits, confirming the structural heterogeneity of ST receptors. This technique of affinity chromatography also was applied to ST receptors in the cytoskeleton-associated compartment of intestinal membranes (S. Hakki and S. A. Waldman, unpublished observations). Again, these preparations were composed of a major ST-binding protein of 71 kDa, with minor contributions from higher and lower molecular weight toxin-binding proteins. Preparations purified from the lipid bilayer and cytoskeleton demonstrated both high- and low-affinity binding with curvilinear Scatchard plots. Of significance, purified ST receptors were not associated with particulate guanylate cyclase activity, supporting previous studies in detergent extracts of intestinal membranes suggesting that particulate guanylate cyclase and ST receptor activities were contained on separate proteins (Kuno *et al.*, 1985; Waldman *et al.*, 1985b).

F. Relationship of ST Receptors and Particulate Guanylate Cyclase in Intestinal Cells

Studies of ST receptors in crude and purified preparations suggest that multiple structural and functional forms of these receptors, which are separate from particulate guanylate cyclase, exist in intestinal cells (Kuno *et al.*, 1985; Waldman *et al.*, 1985a,b; Banik and Ganguly, 1988; Ivens *et al.*, 1990; Thompson and Giannella, 1990; Hugues *et al.*, 1991, 1992; Katawa *et al.*, 1991; Crane *et al.*, 1992; Hakki *et al.*, 1992; Knoop *et al.*, 1991). In contrast, studies of cloned ST receptors suggest that they are members of the family of proteins possessing peptide ligand-binding and particulate guanylate cyclase activities on a single transmembrane protein (Schulz *et al.*, 1990; de Sauvage *et al.*, 1991, 1992; Singh *et al.*, 1991). Resolution of the apparent discrepancy in these data will require further detailed studies of purified and cloned receptor proteins. It is notable that receptors identified by molecular cloning exhibit a single affinity for toxin but that receptors identified in crude membranes and purified preparations exhibit complex binding characteristics suggesting multiple isoreceptor forms of this protein. Thus, ST receptors may resemble natriuretic peptide receptors in their structural and functional heterogeneity in native membranes, with high- and low-molecular-weight forms of the receptor coupled to particulate guanylate cyclase and other signaling cascades, respectively. Alternatively, diverse ST receptors may be the translation product of a single RNA transcript that undergoes post-translational processing into proteins that differ in subunit molecular weights, ligand affinities, and associated enzyme activities. Indeed, this hypothesis is supported by recent studies demonstrating that the cloned human ST receptor undergoes post-translational processing, presumably proteolysis, to yield toxin-binding subunits with high and low molecular weights identified previously in native intestinal cell membranes (de Sauvage *et al.*, 1992). The relationship of the different structural and functional forms of ST receptors in intestinal membranes and the cloned GC-C will be elucidated once quantities of purified receptors are obtained to permit comparison of native and cloned amino acid sequences.

G. Coupling of ST Receptors and Activation of Particulate Guanylate Cyclase

As discussed above, ANP binding appears to be coupled to activation of guanylate cyclase by adenine nucleotides, possibly involving an allosteric mechanism and a separate regulatory protein (Kurose *et al.*, 1987; Chang *et al.*, 1990; Chinkers and Garbers, 1989; Chinkers *et al.*, 1991; Marala *et al.*, 1991). Similarly, particulate guanylate cyclase is regulated by adenine nucleotides in membranes of intestinal mucosa cells (Gazzano *et al.*, 1991a). Basal guanylate cyclase is activated about two-fold by adenine nucleotides. Activation is specific for adenine nucleotides, of which the most potent is the nonhydrolyzable analogue of ATP, adenosine 5′-*O*-(3-thiotriphosphate). Also, adenine nucleotides potentiate ST activation of guanylate cyclase by increasing the maximum velocity of the enzyme without altering its affinity for substrate or cooperativity. In addition to stimulating guanylate cyclase, adenine nucleotides decrease specific binding of the heat-stable enterotoxin to receptors in intestinal membranes. The coordinated regulation of toxin–receptor interaction and guanylate cyclase activity by a process utilizing nonhydrolyzable analogues of a purine nucleotide is similar to mechanisms underlying hormone regulation of adenylate cyclase by guanine nucleotide-binding proteins. These data support the suggestion that an adenine nucleotide-dependent protein may couple toxin–receptor interaction to regulation of particulate guanylate cyclase in intestinal membranes. This mechanism may be mediating the biphasic association kinetics of ST binding to low-affinity receptors important in coupling ligand–receptor interaction to particulate guanylate cyclase activation (Crane *et al.*, 1992). Whether the adenine nucleotide-dependent protein is GC-C, another form of ST receptor, or a separate coupling protein remains to be defined. The observation that the consensus sequence postulated to be important for ATP binding in protein kinases is completely absent from GC-C suggests that a separate coupling protein mediates the effect of adenine nucleotides on ST activation of particulate guanylate cyclase (Schulz *et al.*, 1990; de Sauvage *et al.*, 1991, 1992; Singh *et al.*, 1991).

IV. Particulate Guanylate Cyclase–Cyclic GMP Second Messenger System

After the discovery of guanylate cyclase and cyclic GMP, this second messenger system has often been compared to the previously characterized adenylate cyclase–cyclic AMP pathway. The adenylate cyclase–cyclic AMP pathway is composed of three separate components: a cell membrane receptor that binds hormone; adenylate cyclase, which serves as the catalytic component that converts ATP to cyclic AMP; and a guanine nucleotide binding protein that functions to couple the hormone–receptor complex to the activation of adenylate cyclase. The purification and cloning of a cDNA for particulate cyclase has unexpectedly revealed that the ligand-binding domain, the catalytic domain, and possibly an adenine nucleotide-binding domain reside on a single transmembrane protein. These observations suggest that particulate cyclase may represent a more primitive signal transduction pathway than the adenylate cyclase system. Perhaps the adenylate cyclase system originally existed as a single protein with multiple functional domains that later diverged into three separate components. A major advantage of the three component system is that multiple hormones interacting with different membrane receptors can converge to regulate the activity of adenylate cyclase. In this case, only a single transduction system is needed to mediate the biological effects of a variety of different hormones, which is consistent with the fact that numerous distinct hormones and neurotransmitters activate adenylate cyclase. One of the most intriguing aspects of particulate guanylate cyclase is whether multiple forms exist that have ligand-binding domains different than the NP or ST/guanylin binding sites. Interestingly, particulate guanylate cyclase from LLC-PK_1 cells does not bind ANP with high affinity, but exhibits physical and kinetic properties similar to those of the enzyme from lung membranes that binds ANP with high affinity (Waldman *et al.*, 1989). These findings suggest that particulate guanylate cyclase in LLC-PK_1 cells may have a binding site for a ligand other than ANP or guanylin/ST. Recently, a cDNA for membrane guanylate cyclase (GC_a) was cloned and expressed from a rat adrenal library (Duda *et al.*, 1991). This particulate guanylate cyclase is identical to GC-A with the exception of two amino acids at positions 338 and 364 within the extracellular domain. GC_a has basal guanylate cyclase activity, but does not have ANP binding, suggesting that GC_a may be the form of particulate guanylate cyclase in LLC-PK_1 cells. Furthermore, a cDNA encoding another potential member of the peptide receptor–particulate guanylate cyclase family of proteins, with unknown binding specificity, has been isolated from a small intestinal cell cDNA library, but remains to be characterized. The identification of other forms of particulate guanylate cyclase will provide additional systems for studying the regulation of this enzyme and may permit the identification of known or undiscovered hormones that use cyclic GMP as a second messenger.

References

Anand-Srivastava, M. B., Sairam, M. R., and Cantin, M. (1990). Ring-deleted analogs of atrial natriuretic factor inhibit adenylate cyclase/cAMP system. Possible coupling of clearance atrial natriuretic factor receptors to adenylate cyclase/cAMP signal transduction system. *J. Biol. Chem.* **265,** 8566–8572.

Atlas, S. A. (1986). Atrial natriuretic factor: A new hormone of cardiac origin. *Recent Prog. Horm. Res.* **42,** 207–242.

Banik, N., and Ganguly U. (1988). Stimulation of phosphoinositides breakdown by heat stable E. Coli enterotoxin in rat intestinal epithelial cells. *FEBS Lett.* **236,** 489–492.

Brenner, B. M., Ballermann, B. J., Gunning, M. E., and Zeidel, M. L. (1990). Diverse biological actions of atrial natriuretic peptide. *Physiol. Rev.* **70,** 665–669.

Carr, S., Gazzano, H., Chang, L. Y., and Waldman, S. A. (1989). Regulation of particulate guanylate cyclase by *Escherichia coli* heat-stable enterotoxin: receptor binding and enzyme kinetics. *Int. J. Biochem.* **21,** 1211–1215.

Chang, C.-H., Kohse, K. P., Chang, B., Hirata, M., Jiang, B., Douglas, J. E., and Murad, F. (1990). Characterization of ATP-stimulated guanylate cyclase activation in rat lung membranes. *Biochim. Biophys. Acta* **1052,** 159–165.

Chang, M.-S., Lowe, D. G., Lewis, M., Hellmiss, R., Chen, E., and Goeddel, D. V., (1989). Differential activation by atrial and brain natriuretic peptides of two different receptor guanylate cyclases. *Nature (London)* **341,** 68–72.

Chinkers, M., and Garbers, D. L. (1989). The protein kinase domain of the ANP receptor is required for signaling. *Science* **245,** 1392–1394.

Chinkers, M., and Garbers, D. L. (1991). Signal transduction by guanylyl cyclases. *Annu. Rev. Biochem.* **60,** 553–575.

Chinkers, M., Garbers, D. L., Chang, M.-S., Lowe, D. G., Chin, H., Goeddel, D. V., and Schulz, S. (1989). A membrane form of guanylate cyclase is an atrial natriuretic peptide receptor. *Nature (London)* **338,** 78–83.

Chinkers, M., Singh, S., and Garbers, D. L. (1991). Adenine nucleotides are required for activation of rat atrial natriuretic peptide receptor/guanylyl cyclase expressed in a baculovirus system. *J. Biol. Chem.* **266,** 4088–4093.

Crane, M. R., Hugues, M., O'Hanley, P., and Waldman, S. A. (1992). Identification of two affinity states of low affinity receptors for *Escherichia coli* heat-stable enterotoxin: Correlation of occupation of lower affinity state with guanylate cyclase activation. *Mol. Pharmacol.* **41,** 1073–1080.

Currie, M. G., Fok, K. F., Kato, J., Moore, R. J., Hamra, F. K., Duffin, K. L., and Smith, C. E. (1992). Guanylin: An endogenous activator of intestinal guanylate cyclase. *Proc. Natl. Acad. Sci. U.S.A.* **89,** 947–951.

de Sauvage, F. J., Camerato, T. R., and Goeddel, D. V. (1991). Primary structure and functional expression of the human receptor for *Escherichia coli* heat-stable enterotoxin. *J. Biol. Chem.* **266,** 17912–17918.

de Sauvage, F. J., Horuk, R., Bennet, G., Quan, C., Burnier, J. P., and Goeddel, D. V.

(1992). Characterization of the recombinant human receptor for *Escherichia coli* heat-stable enterotoxin. *J. Biol. Chem.* **267,** 6479–6482.

Dreyfus, L. A., and Robertson, D. C. (1984). Solubilization and partial characterization of the intestinal receptor for *Escherichia coli* heat-stable enterotoxin. *Infect. Immun.* **46,** 537–543.

Dreyfus, L. A., Frantz, J. C., and Robertson, D. C. (1983). Chemical properties of heat-stable enterotoxin produced by enterotoxigenic *Escherichia coli* of different host origins. *Infect. Immun.* **42,** 539–548.

Dreyfus, L. A., Jaso-Friedman, L., and Robertson, D. C. (1984). Characterization of the mechanism of the intestinal receptor for *Escherichia coli* heat-stable enterotoxin. *Infect. Immun.* **44,** 493–501.

Duda, T., Goraczniak, R. M., and Sharma, R. K. (1991). Site-directed mutational analysis of a membrane guanylate cyclase cDNA reveals the atrial natriuretic factor signaling site. *Proc. Natl. Acad. Sci. U.S.A.* **88,** 7882–7886.

Field, M., Graf, L. H., Jr., Laird, W. J., and Smith, P. L. (1978). Heat stable enterotoxin *Escherichia coli:* In vitro effects on guanylate cyclase activity, cyclic GMP accumulation, and ion transport in small intestine. *Proc. Natl. Acad. Sci. U.S.A.* **75,** 2800–2804.

Flynn, T. G., de Bold, M. L., and de Bold, A. J. (1983). The amino acid sequence of an atrial peptide with potent diuretic and natriuretic properties. *Biochem. Biophys. Res. Commun.* **117,** 859–865.

Frantz, J. C., Jaso-Friedman, L., and Robertson, D. C. (1984). Binding of *Escherichia coli* heat stable enterotoxin to rat intestinal cells and brush border membranes. *Infect. Immun.* **43,** 622–630.

Fuller, F., Porter, J. G., Arfsen, A. E., Miller, J., Schilling, J. W. Scarborough, R. M., Lewicki, J. A., and Shenk D. B. (1988). Atrial natriuretic peptide clearance receptor: Complete sequence and functional expression of cDNA clones. *J. Biol. Chem.* **19,** 9395–9401.

Garbers, D. L. (1989). Guanylate cyclase, a cell surface receptor. *J. Biol. Chem.* **264,** 9103–9106.

Gazzano, H., Wu, H. I., and Waldman S. A. (1991a). Activation of particulate guanylate cyclase by *Escherichia coli* heat-stable enterotoxin is regulated by adenine nucleotides. *Infect. Immun.* **59,** 1552–1557.

Gazzano, H., Wu, H. I., and Waldman S. A. (1991b). Adenine nucleotide regulation of particulate guanylate cyclase from rat lung. *Biochim. Biophys. Acta* **1077,** 99–106.

Giannella, R. A., Luttrell, M., and Drake, K. (1980). Binding of pure *E. coli* heat-stable enterotoxin to isolated rat intestinal vilous cells. *Clin. Res.* **28,** 764.

Giannella, R. A., Luttrell, M., and Thompson M. (1983). Binding of *Escherichia coli* heat-stable enterotoxin to receptors on rat intestinal cells. *Am. J. Physiol.* **245,** G492–G498.

Greenberg, R. N., and Saeed, A. M. K., (1988). Purification of *Escherichia coli* heat-stable enterotoxin. *In* "Methods in Enzymology" (S. Harshman, ed.), Vol. 165, pp. 126–137. Academic Press, San Diego, CA.

Hakki, S., Crane, M. R., Hugues, M., O'Hanley, P., and Waldman, S. A. (1993). Solubilization and characterization of functionally coupled *Escherichia coli* heat-stable toxin receptors and particulate guanylate cyclase associated with the cytoskeleton compartment of intestinal membranes. *Int. J. Biochem.* **25,** 557–566.

Hirata, M., Chang, C.-H., and Murad, F. (1989). Stimulatory effects of atrial natriuretic factor on phosphoinositide hydrolysis in cultured bovine aortic smooth muscle cells. *Biochim. Biophys. Acta* **1010,** 346–351.

Hughes, J. M., Murad, F., Chang, B., and Guerrant, R. L. (1978). The role of cyclic

GMP in the action of heat-stable enterotoxin of *Escherichia coli. Nature* (*London*) **271,** 755–756.

Hugues, M., Crane, M. R., Hakki, S., O'Hanley, P., and Waldman, S. A. (1991). Identification and characterization of a new family of high-affinity receptors for *Escherichia coli* heat-stable enterotoxin in rat intestinal membranes. *Biochemistry* **30,** 10738–10745.

Hugues, M., Crane, M. R., Thomas, B. R., Robertson, D. C., Gazzano, H., O'Hanley, P., and Waldman, S. A. (1992). Affinity purification of functional receptors for *Escherichia coli* heat-stable enterotoxin from rat intestine. *Biochemistry* **31,** 12–16.

Ivens, K., Gazzano, H., and Waldman, S. A. (1990). Heterogeneity of intestinal receptors for *Escherichia coli* heat-stable enterotoxin. *Infect. Immun.* **58,** 1817–1820.

Katwa, L. C., Parker, C. D., and White A. A. (1991). Age-dependent changes in affinity-labeled receptors for *Escherichia coli* heat-stable enterotoxin in the swine intestine. *Infect. Immun.* **59,** 4318–4323.

Knoop, F., Owens, M., Marcus, J. N., and Murphy, B. (1991). Elevation of calcium in enterocytes by *Escherichia coli* heat-stable (STa) enterotoxin. *Curr. Microbiol.* **23,** 291–296.

Koller, K. J., and Goeddel, D. V. (1992). Molecular biology of natriuretic peptides and their receptors. *Circulation* **86,** 1081–1088.

Koller, K. J., Lowe, D. G., Bennett, G. L., Minamino, N., Kangawa, K., Matsuo, H., and Goeddel, D. V. (1991). Selective activation of the B natriuretic peptide receptor by C-type natriuretic peptide (CNP). *Science* **252,** 120–123.

Koller, K. J., de Sauvage, F. J., Lowe, D. G., and Goeddel, D. V. (1992). Conservation of the kinaselike regulatory domain is essential for activation of the natriuretic peptide receptor guanylyl cyclases. *Mol. Cell. Biol.* **12,** 2581–2590.

Kuno, T., Kamisaki, Y., Waldman, S. A., Gariepy, J., Schoolnik, G., and Murad, F. (1985). Characterization of the receptor for heat-stable enterotoxin from *Escherichia coli* in rat intestine. *J. Biol. Chem.* **261,** 1470–1476.

Kuno, T., Andresen, J. W., Kamisaki, Y., Waldman, S. A., Chang, L. Y., Saheki, S., Leitman, D. C., Nakane, M. and Murad, F. (1986). Co-purification of an atrial natriuretic factor receptor and particulate guanylate cyclase from rat lung. *J. Biol. Chem.* **261,** 5817–5823.

Kurose, H., Inagami, T., and Ui, M. (1987). Participation of adenosine 5′ triphosphate in the activation of membrane-bound guanylate cyclase by atrial natriuretic factor. *FEBS Lett.* **219,** 375–379.

Larose, L., McNicoll, N., Ong, H., and De Lean, A. (1991). Allosteric modulation by ATP of the bovine adrenal natriuretic factor R1 receptor functions. *Biochemistry* **30,** 8990–8995.

Leitman, D. C., and Murad, F. (1986). Comparison of binding and cyclic GMP accumulation by atrial natriuretic peptides in endothelial cells. *Biochim. Biophys. Acta* **885,** 74–79.

Leitman, D. C., and Murad, F. (1987). Atrial natriuretic factor receptor heterogeneity and stimulation of particulate guanylate cyclase and cyclic GMP accumulation. *Endocrinol. Metab. Clin. North Am.* **16,** 79–105.

Leitman, D. C., and Murad, F. (1990). Stucture and function of atrial natriuretic receptor subtypes. *In* "Atrial Natriuretic Peptides," pp. 77–93. CRC Press, Boca Raton, FL.

Leitman, D. C., Andresen, J. W., Kuno, T., Kamasaki, Y., Chang, J.-K. and Murad, F. (1986). Identification of multiple binding sites for atrial natriuretic factor by affinity cross-linking cultured endothelial cells. *J. Biol. Chem.* **261,** 11650–11655.

Leitman, D. C., Agnost, V. L., Tuan, J. J., Andresen, J. W., and Murad, F. (1987). Atrial natriuretic factor and sodium nitroprusside increase cyclic GMP in cultured rat lung fibroblasts by activating different forms of guanylate cyclase. *Biochem. J.* **244,** 69–74.

Leitman, D. C., Andresen, J. W., Catalano, R. M., Waldman, S. A., Tuan, J. J., and Murad,

F. (1988a). Atrial natriuretic peptide binding, cross-linking, and stimulation of cyclic GMP accumulation and particulate guanylate cyclase activity in cultured cells. *J. Biol. Chem.* **263,** 3720–3728.

Leitman, D. C., Agnost, V. L., Catalano, R. M., Schroder, H., Waldman, S. A., Bennett, B. M., Tuan, J. J., and Murad, F. (1988b). Atrial natriuretic peptide, oxytocin, and vasopressin increase guanosine 3′,5′-monophosphate in LLC-PK1 kidney epithelial cells. *Endocrinology (Baltimore)* **122,** 1478–1485.

Leitman, D. C., Chang, C.-H., Kohse, K. P., Hirata, M., Song, D. L., Waldman, S. A., and Murad, F. (1991). Signal transduction pathways of atrial natriuretic peptide receptor subtypes. *In* "Atrial and Brain Natriuretic Peptides" (H. Imura and H. Matsuo, eds.), pp. 149–164.

Lowe, D. G., Chang, M.-S., Hellmiss, R., Chen, E., Singh, S., Garbers, D. L., and Goeddel, D. V. (1989). Human atrial natriuretic peptide receptor defines a new paradigm for second messenger signal transduction. *EMBO J.* **8,** 1377–1384.

Lowe, D. G., Camerato, T. R., and Goeddel, D. V. (1990). cDNA sequence of the human atrial natriuretic peptide clearance receptor. *Nucleic Acids Res.* **18,** 3412–3418.

Maack, T. (1992). Receptors of atrial natriuretic factor. *Annu. Rev. Physiol.* **54,** 11–27.

Maack, T., Suzuki, M., Almeida, F. A., Nussenzveig, D., Scarborough, R. M., McEnroe, G. A., and Lewicki, J. A. (1987). Physiological role of silent receptors of atrial natriuretic factor. *Science* **238,** 675–678.

Marala, R. B., Sitaramayya, A., and Sharma, R. K. (1991). Dual regulation of atrial natriuretic factor-dependent guanylate cyclase activity by ATP. *FEBS. Lett.* **281,** 73–76.

Marala, R. B., Duda, T., Goraczniak, R. M., and Sharma, R. K. (1992). Genetically tailored atrial natriuretic factor-dependent guanylate cyclase. Immunological and functional identity with 180 kDa membrane guanylate cyclase and ATP signaling site. *FEBS Lett.* **296,** 254–258.

Martin, E. R., and Ballermann, B. J. (1989). Atrial natriuretic peptides receptors. *Contemp. Issues Nephrol.* **21,** 105–136.

Meloche, S., McNicoll, N., Liu, B., Ong, H., and De Lean, A. (1988). Atrial natriuretic factor R_1 receptor from bovine adrenal zona glomerulosa: Purification, characterization, and modulation by amiloride. *Biochemistry* **27,** 8151–8158.

Murad, F. (1986). Cyclic guanosine monophosphate as a mediator of vasodilation. *J. Clin. Invest.* **78,** 1–5.

Pandey, K., and Singh, S. (1990). Molecular cloning and expression of murine guanylate cyclase/atrial natriuretic factor receptor cDNA. *J. Biol. Chem.* **265,** 12342–12348.

Porter, J. G., Wang, Y., Schwartz, K., Arfsen, A., Loffredo, A., Spratt, K., Schenk, D., Fuller, F., Scarborough, R. M., and Lewicki, J. A. (1988). Characterization of the atrial natriuretic peptide clearance receptor using a vaccinia virus expression vector. *J. Biol. Chem.* **263,** 18827–18833.

Porter, J. G., Arfsen, Y., Fuller, A. F., Miller, J. A., Gregory, L. C., and Lewicki, J. A. (1990). Isolation and functional expression of the rat human atrial natriuretic peptide clearance receptor cDNA. *Biochem. Biophys. Res. Commun.* **171,** 796–803.

Rosenzwieg, A., and Seidman, C. E. (1991). Atrial natriuretic factor and related peptide hormones. *Annu. Rev. Biochem.* **60,** 229–255.

Schulz, S., Green, K. C., Green, P. S. T., and Garbers, D. L. (1990). Guanylyl cyclase is a heat-stable enterotoxin receptor. *Cell* **63,** 941–948.

Schenk, D., Phelps, M. N., Porter, J. G., Scarborough, R. M. McEnroe, G. A., and Lewicki, J. A. (1987). Purification and subunit composition of atrial natriuretic peptide receptor. *Proc. Natl. Acad. Sci. U.S.A.* **84,** 1521–1525.

Singh, S., Lowe, D. G., Thorpe, D. S., Rodriguez, H., Kuang, W.-J., Dangott, L. J.,

Chinkers, M., Goeddel, D. V., and Garbers, D. L. (1988). Membrane guanylate cyclase is a cell-surface receptor with homology to protein kinases. *Nature (London)* **334,** 708–712.

Singh, S., Singh, G., Heim, J. M., and Gerzer, R. (1991). Isolation and expression of a guanylate cyclase-coupled heat stable enterotoxin receptor cDNA from a human colonic cell line. *Biochem. Biophys. Res. Commun.* **179,** 1455–1463.

Smale, S. T., and Baltimore, D. (1989). The "initiator" as a transcription control element. *Cell (Cambridge, Mass.)* **57,** 103–113.

Sudoh, T., Kangawa, K., Minamino, N., and Matsuo, H. (1988). A new natriuretic peptide in porcine brain. *Nature (London)* **332,** 78–80.

Sudoh, T., Minamino, N., Kangawa, K., and Matsuo, H. (1990). C-type natriuretic peptide (CNP): A new member of natriuretic peptide family identified in porcine brain. *Biochem. Biophys. Res. Commun.* **168,** 863–870.

Takayanagi, R., Inagami, T., Snajdar, R. M., Imada, T., Tamura, M., and Misono, K. S. (1987). Two distinct forms of receptors for atrial natriuretic factor in bovine adrenocortical cells: Purification, ligand binding, and peptide mapping. *J. Biol. Chem.* **262,** 12104–12113.

Thomas, D. D., and Knoop, F. C. (1983). Effect of heat-stable enterotoxin of *Escherichia coli* on cultured mammalian cells. *J. Infect. Dis.* **147,** 450–459.

Thompson, M. R. (1987). *Escherichia coli* heat-stable enterotoxin and their receptors. *Pathol. Immunopathol. Res.* **6,** 103–116.

Thompson, M. R., and Giannella, R. A. (1990). Different crosslinking agents identify distinctly different putative *Escherichia coli* heat-stable enterotoxin rat intestinal cell receptor proteins. *J. Recep. Res.* **10,** 97–117.

Waldman, S. A., and Murad, F. (1987). Cyclic GMP synthesis and function. *Pharmacol. Rev.* **39,** 163–196.

Waldman, S. A., Rapoport, R. M., and Murad, F. (1984). Atrial natriuretic factor selectively activates particulate guanylate cyclase and elevates cyclic GMP in rat tissues. *J. Biol. Chem.* **259,** 14332–14334.

Waldman, S. A., Rapoport, R. M., Fiscus, R. R., and Murad, F. (1985a). Effects of atriopeptin on particulate guanylate cyclase from rat adrenal. *Biochim. Biophys. Acta* **845,** 298–302.

Waldman, S. A., Kuno, T., Kamisaki, Y., Chang, L. Y., Gariepy, J., O'Hanley, P. O., Schoolnik, G. K., and Murad, F. (1985b). Intestinal receptor for heat-stable enterotoxin of *Escherichia coli* is tightly coupled to a novel form of particulate guanylate cells. *Infect. Immun.* **51,** 320–326.

Waldman, S. A., Leitman, D. C., Chang, L. Y., and Murad, F. (1989). Comparison of particulate guanylate cyclase in cells with and without atrial natriuretic peptide receptor binding activity. *Mol. Cell. Biochem.* **90,** 19–25.

Yamaguchi, M., Rutledge, L. J., and Garbers, D. L. (1990). The primary structure of the rat guanylyl cyclase A/atrial natriuretic peptide receptor gene. *J. Biol. Chem.* **265,** 20414–20420.

Cyclic GMP and Regulation of Cyclic Nucleotide Hydrolysis

William K. Sonnenburg and Joseph A. Beavo

Department of Pharmacology
University of Washington
Seattle, Washington 98195

I. Introduction

A. Scope of Chapter

It is increasingly clear that the physiological responses of a cell to cGMP are determined in large part by which isozymes of cyclic nucleotide phosphodiesterase (PDE) are expressed by that cell. For example, the length and magnitude of the cGMP response, the interaction of cGMP with the cAMP and Ca^{2+} signaling pathways, and the modes of feedback regulation are all affected by which PDE or combination of PDEs are expressed. This chapter gives a brief review of the control of cyclic nucleotide degradation with particular emphasis on how cGMP hydrolysis is regulated and on how cGMP regulates cAMP hydrolysis. In order to do this, the PDE families, which are thought to be primarily responsible for degrading cGMP and thereby turning off the cGMP signal, are discussed in some detail. These include the CaM-dependent and cGMP-specific PDE families as well as the cGMP-stimulated PDEs. Other sections discuss those PDEs for which cGMP is thought to regulate cAMP hydrolysis. These include the cGMP-stimulated and cGMP-inhibited PDEs. These cGMP-regulated PDEs provide a mechanism by which cGMP can alter the concentrations of cAMP in the cell in either a positive or a negative manner and thereby indirectly affect all of the processes mediated by cAMP. The CaM-

Advances in Pharmacology, Volume 26

dependent PDEs provide a mechanism by which signal transducing pathways generating Ca^{2+} can alter cGMP (and cAMP) metabolism. Another isozyme family, the cGMP-specific photoreceptor PDEs, also are important mediators of cGMP function. These very interesting and specialized PDEs are unique to photoreceptors where they are pivotal to the mechanism of light-mediated visual transduction. Because of their specialized role they are not discussed in this chapter. Interested readers are referred to recent reviews on their function and regulation (Gillespie, 1990; Pittler and Baehr, 1991).

B. General Control of Cyclic Nucleotide Steady-State Levels

As with any metabolite, the level of cGMP in a cell is controlled by its relative rate of synthesis and degradation. It has been known for many years that in most tissues the maximal catalytic capacity for synthesis of either cAMP or cGMP is much less than for degradation. Therefore, most of the phosphodiesterase activity in intact cells is operating at less than maximal capacity. In other words, these PDE activities are regulated. More recently, it has become appreciated that the process of cyclic nucleotide degradation can be catalyzed by any one or combination of a large number of different phosphodiesterase isoenzymes. Despite this fact, it is still common for many authors to use the term cGMP phosphodiesterase or cAMP phosphodiesterase. In most cases this is not appropriate since, although enzymes with such selectivity do exist in a few tissues, most do not show this specificity and in many cells cGMP degradation is due to the action a PDE isoenzyme(s) that can hydrolyze both cyclic nucleotides.

C. Multiple PDEs Control Cyclic GMP Hydrolysis

Current evidence suggests that at least seven different families of PDE isoenzymes exist in higher eukaryotes. Each family is encoded by a different gene or series of very closely related genes. Additional diversity is created by alternative splicing of many of the genes. A list of these families is shown in Table I.

The number of different families recognized has expanded rapidly in the last few years as the tools of molecular biology and molecular genetics have begun to be applied to the question of how many different PDEs exist in animal cells. It is to be expected that both the number of new isozymes within a family and the number of families will continue to increase in the coming years.

Table I

PDE isozyme family	Number of unique gene products identified to date	Number of alternative splice products	Evidence for additional members
I. CaM-dependent	2	2	Yes
II. cGMP-stimulated	1	2	Yes
III. cGMP-inhibited	1	2	Yes
IV. cAMP-specific	4	3	Yes
V. cGMP-specific	1	2	Possible
VI. Photoreceptor	3	2	Possible
VII. HPC1	1	-	Possible

D. Basis for PDE Family Designation, General Domain Organization, and Conserved Motifs

To date all mammalian PDEs for which sequence data are available have been shown to contain a region of approximately 270 amino acids that is highly conserved. This region is located near the carboxy terminus of the enzyme and is thought to contain the catalytic domain of the enzymes (Charbonneau *et al.*, 1986). Several lines of direct experimental evidence support this conclusion including photoaffinity labeling (Stroop *et al.*, 1989), proteolysis (Charbonneau *et al.*, 1986; Kincaid *et al.*, 1985; Stroop *et al.*, 1989), and chemical modification (Ahn *et al.*, 1991) experiments. In general, alignments within these conserved regions show greater than 60% amino acid identity within an isozyme family but less than 30% identity between families (Charbonneau, 1990; Charbonneau *et al.*, 1986). The conserved catalytic domain is made up of two smaller regions that are themselves weakly homologous presumably due to a very ancient gene duplication event.

Sequence alignment of the catalytic and cGMP-binding domains from all known PDEs identifies a number of amino acids that are invariant or very highly conserved. Several of these contain sufficient information to allow recognition of signature sequences or motifs that appear to be specific for cyclic nucleotide PDEs. For example, all mammalian cyclic nucleotide PDEs for which data are available contain the general sequence HDxxHxxxxN near the N-terminal part of the conserved catalytic domain (Charbonneau, 1990; Charbonneau *et al.*, 1986). In fact, this motif is more restrictive in that the third amino acid is hydrophobic (L,I,V,F, or Y) in all sequences determined to date and the second residue after the asparagine has the same restriction. Also, the second residue after the middle two histidines is always A or G. Computer searches[1] for the motif HD(L,I,V,M,F,Y)HxHx(A,G)xxNx(L,I,V,M,F,Y) will uniquely define a cyclic nucleotide PDE (Bairoch, 1991). Other motifs within the conserved domain, such as EF(F,W)xxQGD(R,K,L)E, also are present and will uniquely identify PDEs. No protein sequences other than PDEs from the 21,000 present in the SwissProt database (Release 19) are recognized by these motifs. Several other residues and small clusters of residues are shared by nearly all PDEs (Charbonneau, 1990). The mechanistic function(s) for these conserved residues within the catalytic domain remains to be determined but presumably includes determinants for cyclic nucleotide and metal ion binding and for phosphodiester hydrolysis.

Since this C-terminal conserved region of the protein comprises less than half of the total sequence for any of the PDEs (less than one-third for most), it is assumed that the other regions are important for functions such as regulation, subunit interaction, and localization of the isozyme. Functionally, the best characterized of these other regions is located just N-terminal to the conserved catalytic domain of the enzymes. This region is conserved in all of those PDEs known to have high affinity, noncatalytic cGMP binding sites (Charbonneau *et al.*, 1990). Since this region is present only in those PDEs that have high affinity sites for cGMP, it is presumed to contain the so-called high affinity, noncatalytic site(s) of the enzyme. A substantial amount of direct evidence for this hypothesis has been obtained (Stroop and Beavo, 1991; Stroop *et al.*, 1989).

Like the catalytic domains, the cGMP binding domains are formed from two regions of recognizable internal homology, presumably due to an ancient gene duplication event. Since for some of the PDEs at least, a weak homology can be seen between the catalytic and the noncatalytic domains, they may have arisen from a common ancestral sequence. The long general sequence motif, LxxPIxNxxxxxxGVAxxxNxxxG, or the more specific sequence, L(C,S)(F,L,M)PI(K,V)NXX(E,Q)(E,D)(I,V)(I,V)-GVAX(A,F,L)(I,V,Y)N(K,R)(I,K)XG, will uniquely identify the cGMP binding domains of all PDEs know to contain them. These include the cGMP-stimulated, cGMP-specific, and photoreceptor a, a′, and b subunit sequences. In fact either half of this sequence is sufficient to form a unique signature. Although the generality of this motif remains to be determined, it is not found in any other proteins in the SwissProt data bank (Release 19).[2]

[1] This more restricted motif is currently the only one of these patterns defined in the *PROSITE Dictionary of Protein Sites and Patterns* distributed by EMBL Data Library, Postfach 10.2209, Meyerhofstrasse 1, 6900 Heidelberg, Germany.

[2] All searches were done using the FINDpatterns routine of the University of Wisconsin Genetics Computer Group (GCG) or PROSITE (Motif) sequence analysis program for the VAX. The database searched was SwissProt (Release 19). PROSITE was release 7.1.

The function(s) for the amino acids outside of these two conserved domains is less clear. The N-terminal region appears to be important for CaM binding to the CaM-dependent PDEs (Charbonneau *et al.*, 1991; Novack *et al.*, 1991); however, recent evidence suggests that other regions are also important for this function. In addition, since all mammalian PDEs studied to date are dimers,[3] it seems likely that some of these regions will provide subunit association domains. Some suggestive evidence for sequence-dependent specific localization signals has also been reported (Baehr *et al.*, 1991; Sonnenburg *et al.*, 1991). Since several of the PDEs exist in states that can be activated by allosteric, protein binding, or phosphorylation mechanisms, some regions of the sequence may function to inhibit activity of the catalytic site until the enzyme can be activated.

II. Ca^{2+}/CaM-Dependent PDE Family

A. Multiple Isoforms

The CaM-stimulated cyclic nucleotide phosphodiesterases (CaM-dependent PDEs) constitute a genetically diverse and expanding family of enzymes that catalyze both cAMP and cGMP hydrolysis (Beavo, 1990). Calmodulin in the presence of Ca^{2+} stimulates the activity of these isozymes several-fold. At present, at least six different CaM-dependent PDE isoforms are thought to exist (Table II).

Current nomenclature for the CaM-PDE isozymes is based on the apparent molecular weight of each isozyme as estimated by SDS–PAGE (Beavo, 1990). This is somewhat unsatisfactory since the real molecular weights

Table II

Name	SDS–PAGE (kDa)	Sequence (MW)	Reference
58-kDa lung CaM-PDE	58	ND[a]	Sharma and Wang (1986b)
59-kDa heart CaM-PDE	59–60	59,200 Da	Novack *et al.* (1991)
61-kDa brain CaM-PDE	60–61	60,800 Da	Charbonneau *et al.* (1991)
63-kDa heart CaM-PDE	63	61,000 Da	Bentley *et al.* (1992); Repaske *et al.* (1992)
67-kDa low K_m CaM-PDE	67–68	ND	Purvis and Rui (1988); Rossi *et al.* (1988)
75-kDa brain CaM-PDE	75	ND	Shenolikar *et al.* (1985)

[a] ND, not determined.

[3] However, some, like the CaM-dependent and the photoreceptor PDEs, also have smaller subunits associated with them under certain physiological conditions.

determined by direct sequencing or cDNA cloning do not correspond directly to mobility in SDS gels; however, it will probably continue to be used until all family members are identified and a better nomenclature system based on function or regulation can be applied. The primary structure of the 61- and 63-kDa isozymes purified from bovine brain, and the 59-kDa isozyme isolated from bovine heart has been reported (Bentley *et al.,* 1992; Charbonneau *et al.,* 1991; Novack *et al.,* 1991; Repaske *et al.,* 1992). At this time, little structural information about the 58-, 67-, and 75-kDa isozymes is known.

B. Kinetic Properties

Substantial variation in the substrate specificity of the different isozymes has been reported. For example, the 61-, 59-, and 58-kDa CaM-dependent PDEs exhibit similar specificity for cAMP and cGMP (Grewal *et al.,* 1989; Sharma and Wang, 1986c). On the other hand, the 63-kDa isoform catalyzes cGMP hydrolysis several times better than cAMP (Sharma *et al.,* 1984; Sharma and Wang, 1986a). Finally, the 75-kDa CaM-dependent PDE expressed in brain appears to hydrolyze specifically cGMP (Shenolikar *et al.,* 1985).

Kinetically, one of the more distinctive CaM-dependent PDEs is the 67-kDa isoform expressed in testis (Purvis and Rui, 1988; Rossi *et al.,* 1988) and spermatozoa (Chaudhry and Casillas, 1988; Geremia *et al.,* 1984). This isozyme hydrolyzes both cAMP and cGMP. However, the apparent K_m values for these substrates are 10-fold lower than those for all of the other CaM-dependent isoforms. Moreover, unlike the other CaM-dependent PDEs, cGMP inhibits cAMP hydrolysis (and vice versa) in a noncompetitive manner, suggesting that this "low K_m" 67-kDa CaM-dependent PDE possesses two discrete nucleotide binding sites (Geremia *et al.,* 1984; Rossi *et al.,* 1988).

In general, the kinetic parameters of Ca^{2+}/CaM activation are similar for all of the isozymes. Calmodulin in the presence of Ca^{2+} increases the maximum velocity (V_{max}) 5- to 10-fold, with little effect on the apparent K_m for either cAMP or cGMP. However, several of these isozymes exhibit notable differences in CaM-activation properties. Both the 61- and the 63-kDa CaM-dependent PDEs have CaM activation constants of about 1–2 nM (Hansen and Beavo, 1986; Sharma and Wang, 1986a). However, CaM is a more potent activator of the heart 59-kDa CaM-dependent PDE, attaining half-maximal stimulation at a concentration of about 0.1 nM (Hansen and Beavo, 1986; Sharma, 1991). Additionally, CaM activation of each isozyme is attenuated by phosphorylation *in vitro*. The 63-kDa isozyme serves as a substrate for CaM-dependent protein kinase II and the 61- and 59-kDa CaM-dependent PDE isozymes are substrates for the cAMP-dependent protein kinase. In each case phosphorylation blunts stimulation of enzyme activity by Ca^{2+}/CaM (Hashimoto *et al.*, 1989; Sharma, 1991; Sharma and Wang, 1985, 1986a).

C. Structure and Domain Organization

The recent reports of the amino acid sequence for several of the CaM-dependent PDEs have advanced our understanding of the mechanism by which Ca^{2+}/CaM activates the enzyme. The complete primary structure of the 61-kDa (Charbonneau *et al.*, 1991) and 63-kDa CaM-dependent PDEs (Bentley *et al.*, 1992) and partial sequence of the 59-kDa CaM-dependent PDE (Novack *et al.*, 1991) have been determined. Comparison of the 61- and 59-kDa sequence shows that these isozymes are identical except for a very short segment of the amino terminus (Novack *et al.*, 1991). Moreover, the nucleotide sequence of the cDNAs corresponding to the regions of amino acid identity are also the same, suggesting that these isozymes are products of an alternately spliced gene (Sonnenberg et al., 1993). However, the 63-kDa CaM-dependent PDE is the product of a different, albeit homologous, gene. Each PDE isozyme exists as a homodimer of two large (~60 kDa) subunits. In the presence of Ca^{2+}, 2 mol of CaM associate with the native enzyme (Charbonneau *et al.*, 1991; Wang *et al.*, 1990). It is not yet clear whether heterodimers formed from the ~60-kDa monomers of different PDEs can form *in vivo* or what their kinetic properties might be if they do exist.

Current evidence suggests that at least part of the CaM binding domain spans the amino-terminal segment of the isozyme (Charbonneau *et al.*, 1991; Novack *et al.*, 1991). However, recent mutagenesis studies by the authors suggest that additional regions also must be involved. The mechanism by which the CaM-dependent PDE is activated by CaM binding to a domain so distant from the catalytic site is not known. Early models proposed that CaM relieves suppression of CaM-dependent PDE activity by binding to a domain at or near the catalytic site (Kincaid *et al.*, 1985; Krinks *et al.*, 1984). One proposed mechanism involves displacement of an inhibitory segment, as may occur in other CaM-regulated enzymes (Falchetto *et al.*, 1991; Jarrett and Madhavan, 1991; Kemp *et al.*, 1987). However, evidence for such a discrete domain in the CaM-dependent PDE molecule is currently lacking. Mechanisms involving a conformational alteration in the catalytic site upon CaM binding that convert it from an inactive to an active form are also possible (Klee, 1980). Approaches using deletion and site-directed mutagenesis of the CaM-dependent PDE cDNAs should provide answers to these questions.

D. Tissue and Cellular Distribution

As mentioned above, CaM-stimulated PDEs have been identified in a variety of different tissues. In addition to brain, heart, lung, and testis, CaM-dependent PDE activities have been detected in aorta (Hagiwara *et al.*, 1984), thyroid gland (Miot *et al.*, 1983), lymphocytes (Hurwitz *et al.*, 1990), kidney (Marcus and Grant, 1983), pancreas (Sugden and Ashcroft, 1981; Vandermeers *et al.*, 1983), and adrenal gland (Uzunov *et al.*, 1976). Until recently, however, relatively little was known about the tissue distribution of individual CaM-dependent PDE isozymes.

The 61-kDa CaM-dependent PDE is most concentrated in the brain. The mRNA encoding this isozyme has been detected only in this tissue, suggesting that it may be specifically expressed in neurons (Sonnenburg *et al.*, 1993). In brain this isozyme is widely expressed since nearly equal concentrations of mRNA have been detected in cerebral cortex, basal ganglia, hippocampus, cerebellum, and spinal cord/medulla. This observation is consistent with previous immunohistochemical analyses using antisera reactive with a brain CaM-dependent PDE preparation (Kincaid *et al.*, 1987). These latter studies revealed that one or more CaM-dependent PDE isozymes are expressed in specific neuronal populations. CaM PDE-specific staining was detected in the dendrites and cytoplasm of hippocampal and cortical pyramidal cells, and cerebellar Purkinje cells. However, it is not certain how many different CaM-dependent PDE isoforms were recognized by the sera. Interestingly, the pattern of CaM-dependent PDE expression changes during development (Billingsley *et al.*, 1990).

The 63-kDa CaM-dependent PDE is also highly concentrated in the brain. Unlike the 61-kDa CaM-dependent PDE, Northern and RNase protection analyses suggest that this isozyme is concentrated in the basal ganglia (Bentley *et al.*, 1992). A CaM-dependent cGMP phosphodiesterase activity has been detected in specific neuronal populations of the rat caudate-putamen using histochemical methods (Ariano, 1983). This activity appears to be concentrated in postsynaptic structures near the plasma membrane. The identity of the specific isoform was not determined in these experiments; however, this activity may represent the 63-kDa enzyme since mRNA encoding it is also concentrated in this brain region. RNA species that are identical, or closely related, to 63-kDa CaM-dependent PDE mRNA have also been detected in lymphocytes and kidney papillae.

The heart appears to express the highest levels of the 59-kDa CaM-dependent PDE isoform. However, one recent report indicates that isolated rat cardiomyocytes contain no CaM-dependent PDE activity, even though this isozyme can be purified from extracts of heart tissue (Bode *et al.*, 1991). This intriguing observation suggests that very high concentrations of this PDE may be expressed in a minor cell population in cardiac tissue.

Relatively little is known about the distribution of the 75-kDa CaM-dependent PDE, which has been reported only in brain (Shenolikar *et al.*, 1985). Similarly, the distribution of the 58- and 67-kDa CaM-dependent PDEs in tissues other than lung, testis, or spermatozoa, respectively, is currently not known. Some of these isozymes may be structurally related, or indeed identical, to other known CaM-dependent PDE isoforms. For example, the 58-kDa CaM-dependent PDE detected in lung may in fact be structurally identical to the 59-kDa isoform.

E. Calcium/CaM-Dependent PDE Regulation of Cyclic GMP

In contrast to the large number of studies implicating CaM-dependent PDEs in the Ca^{2+} mediated control of cAMP (Wang *et al.*, 1990), much less is known about their effect on cGMP. The kinetic characteristics of many CaM-dependent PDE isozymes suggest that they should serve to attenuate cGMP accumulation *in vivo*. Indirect pharmacological evidence suggests that a CaM-dependent PDE may modulate cGMP in several tissues. For example, the xanthine derivative 1-methyl-3-isobutyl-8-methoxymethylxanthine selectively inhibits bovine coronary artery CaM-dependent PDE activity and potentiated cGMP formation in response to activators of guanylate cyclase (Lorenz and Wells, 1983). Similar results have been observed by other investigators (Ahn *et al.*, 1989; Chiu *et al.*, 1988; Souness *et al.*, 1989). These results suggest that a CaM-dependent PDE is involved in regulating cGMP concentrations in vascular smooth muscle. One might predict that many of the agonists causing smooth muscle vasoconstriction by increasing Ca^{2+} might also activate the CaM-dependent PDE. The actions of endothelin-1 to lower cGMP levels in isolated blood vessels supports this hypothesis (Godfraind *et al.*, 1989).

It also seems likely that CaM-dependent PDEs are important modulators of cGMP degradation in isolated trachealis. This tissue expresses a number of phosphodiesterases, including a CaM-dependent PDE. Using a battery of selective and nonselective inhibitors, Torphy *et al.* presented evidence suggesting that a CaM-dependent PDE, in addition to a cGMP-specific phosphodiesterase, preferentially catalyzes cGMP hydrolysis in this tissue (Torphy *et al.*, 1991).

Although brain is a rich source of the 61-, 63-, and 75-kDa CaM-dependent PDE activities, specific examples demonstrating physiologically relevant activation of these isozymes are lacking. Experimentally, the situation is complicated by the Ca^{2+}/CaM dependence of nitric oxide formation (Snyder and Bredt, 1991), which will increase cGMP and potentially could increase cAMP hydrolysis by the cGMP-stimulated PDEs. As a result, we can only speculate about how CaM-dependent PDEs may mediate physiological responses governed by cGMP and Ca^{2+} metabolism in the brain. In an intriguing study utilizing histochemical and immunohistochemical techniques, several components of the cGMP signal transduction pathway, including a CaM-dependent cGMP phosphodiesterase activity, were all found in the rat caudate-putamen (Ariano, 1983). These enzymes were all concentrated around the inner membrane of the postsynaptic dendrite, suggesting that cGMP (and CaM PDE) may regulate postsynaptic neurotransmission. It was not clear from this particular study which CaM-dependent PDE isozyme is represented. However, the 63-kDa CaM-dependent PDE is concentrated in the bovine basal ganglia (Bentley *et al.*, 1992), which includes the caudate putamen. In another recent study, a CaM-dependent PDE has been implicated in angiotensin II regulation of cGMP accumulation in hypothalamic and brain stem neurons (Sumners and Myers, 1991). Angiotensin II decreases cGMP levels in these neurons, and this effect is blocked by certain Ca^{2+} channel blockers, or by pretreatment with 3-isobutyl-1-methylxanthine. These results suggest that angiotensin II may stimulate CaM-dependent PDE activity and subsequent cGMP hydrolysis by increasing intracellular Ca^{2+}. Finally, another possible role for CaM PDE is suggested by the recent studies of Nawy and Jahr, who reported that glutamate induces hyperpolarization in depolarizing retinal bipolar cells (DBCs) by decreasing a cGMP mediated conductance (Nawy and Jahr, 1990, 1991). Since DBC hyperpolariazation is thought to be mediated by a metabotropic glutamate receptor that is coupled to inositol trisphosphate formation (Shoepp and Johnson, 1988), it seems possible that activation of a CaM-dependent PDE is involved in this response.

III. Cyclic GMP-Specific PDE Family

This family of PDEs has also been firmly implicated as a regulator of cGMP levels in some tissues. Until recently, this isozyme family name has been used to describe all PDEs with substrate specificity for cGMP (Beavo and Reifsnyder, 1990). These included the cGMP binding PDE(s) from lung and platelets as well as each of the photoreceptor PDEs. In addition to their common substrate specificity, all of these isoenzymes share the common property of having a noncatalytic, high affinity binding site for cGMP and also of being selectively inhibited by the drugs dipyridamole or zaprinast. Therefore, it came as somewhat of a surprise to the authors when the first sequence of a cDNA encoding a member of this isozyme family present in bovine lung showed a sequence greatly divergent from the photoreceptor isoenzymes (McAllister-Lucas *et al.*, 1993). Since the overall homology between the photoreceptor and the lung isoenzymes is less than 25%, it now seems most appropriate to list them as different families. The following section describes the nonphotoreceptor cGMP binding, cGMP specific PDE(s).

A. General Properties

Much less is known about this isozyme family than about the CaM-dependent PDEs. Many of the features about the kinetic properties, structural features, and regulation of the enzyme have been reviewed recently (Francis *et al.*, 1990). The enzyme was first identified and partially purified from bovine lung (Francis *et al.*, 1980) and platelets (Coquil *et al.*, 1983). More recently, a procedure for obtaining pure protein has been described (Francis and Corbin, 1988). The most striking kinetic characteristics of this PDE family are a strong preference for cGMP as substrate (over 100-fold) and the presence of a high affinity, noncatalytic binding site(s) for cGMP (Francis *et al.*, 1990).

B. Cellular Distribution and Functions

In addition to lung and platelets where the activity was originally isolated, more recent indirect data indicate that this isozyme is also present in vascular smooth muscle (Malta, 1989) and kidney (Wilkins *et al.*, 1990a). This isozyme is present in lung at concentrations at least as high as those of cGMP-dependent protein kinase (Mumby *et al.*, 1982). The cell type(s) in the lung in which the enzyme is localized has not been determined.

The function of this family is presumably to regulate the steady-state levels of cGMP since it hydrolyzes cAMP poorly. Although little direct *in vivo* evidence has been reported, the effects of isozyme-selective inhibitors support this hypothesis. In two recent examples, zaprinast (M&B 22948) augmented the effect of either endogenous or exogenous atrial natriuretic peptide on sodium excretion (Wilkins *et al.*, 1990a,b). This was particularly striking since it occurred even in the face of decreased blood pressure, which would normally be expected to decrease sodium output. These data suggest that the cGMP binding, cGMP-specific PDE can play a key role in kidney function. In platelets selective inhibition of this PDE potentiates the effect of agents like NO that stimulate guanylate cyclase (Radomski *et al.*, 1990a,b; Salvemini *et al.*, 1990), therefore suggesting that the cGMP binding PDE may play an important role in modulating the effects of EDRF on platelet aggregation.

The cGMP binding PDE also has been implicated in cGMP control of vascular resistance. Zaprinast is reported to potentiate a decrease in mean arterial pressure in response to sodium nitroprusside but not atrial natriuretic peptide, implying that this PDE modulates the response to EDRF (Dundore *et al.*, 1990). Similarly, the enzyme also has been implicated as playing a role in endothelium-dependent relaxation (Rapoport *et al.*, 1989). One report suggests that it is most important in aortic smooth muscle under conditions of low intracellular Ca^{2+} (Ahn *et al.*, 1989). Relaxant effects of inhibitors of this PDE also have been seen in airway smooth muscle (Langlands *et al.*, 1989; Malta, 1989; Rimele *et al.*, 1988; Torphy *et al.*, 1991) where cGMP and zaprinast appear to modulate airway resistance. Obviously much work remains to be done to define clearly the role(s) for the cGMP-specific cGMP binding PDE in cellular regulation.

C. Regulation of Activity

Only a little is known about regulation of activity for this family of PDEs. Since many of its properties are so similar to those of the photoreceptor PDEs, it has been suggested that it could be regulated by a GTP binding protein cascade analogous to transducin activation of the photoreceptor PDEs. However, to date no such evidence has been reported. The differences in sequence mentioned above may provide an explanation for the lack of such regulation. The enzyme contains a high affinity binding site that is distinct from the catalytic site (Charbonneau *et al.*, 1990; Francis *et al.*, 1980). However, no changes in activity or function have been seen in response to cGMP binding at this site. Substantial differences exist between the specificity of binding of different cyclic GMP analogs to the catalytic and the noncatalytic binding sites and, curiously, inhibitors of the catalytic activity of the enzyme are reported to stimulate cGMP binding (Francis *et al.*, 1990), implying communication between the two sites. Finally, the enzyme is phosphorylated by both cAMP- and cGMP-dependent protein kinase (Thomas *et al.*, 1990). Phosphorylation occurs best when cGMP is bound to the noncatalytic cGMP binding site and in fact the enzyme is the best substrate for cGMP-dependent protein kinase yet described. However, since no functional change has yet been determined either for cGMP binding or for the phosphorylation, the physiological significance of these observations is not clear.

D. Structural Features

The enzyme from bovine lung has been purified to homogeneity (Francis and Corbin, 1988). It shows an apparent size of 93 kDa on SDS–PAGE, a value that is close to the subunit molecular weight of 97,600 predicted by the cDNA sequence. As expected from the cGMP binding data, this enzyme has an easily identifiable cGMP binding domain that is distinct from the catalytic domain (McAllister-Lucas *et al.*, 1993). As with most other PDEs, hydrodynamic measurements suggest that it is a dimer in its native state, probably composed of two identical monomers. No evidence for smaller subunits analogous to the inhibitory subunits of the photoreceptor PDEs has been reported.

IV. Cyclic GMP-Stimulated Phosphodiesterase Family

A. Multiple Isoforms

At least two different cGS PDE isozymes have been identified. These include a membrane-associated isoform and a cytosolic isoform. The soluble form is present in bovine heart and adrenal gland (Martins *et al.*, 1982), liver (Yamamoto *et al.*, 1983a), and platelets (Grant *et al.*, 1990) and is composed of two identical subunits of M_r ~103,200 (Trong *et al.*, 1990). The membrane-associated isozyme has been purified from rabbit (Whalin *et al.*, 1988) and bovine brain (Murashima *et al.*, 1990) and, like its cytosolic counterpart, is also composed of two identical subunits of M_r ~105,600.[4] Cytosolic and membrane-associated cGS PDE activities have also been isolated from rat liver (Pyne *et al.*, 1986). Unlike the bovine and rabbit enzymes, the rat liver cGS PDE is composed of two smaller subunits of M_r ~67,000. The authors suggest that the smaller subunit size is not due to proteolysis during purification since protease inhibitors do not affect the mobility of the subunits on SDS–PAGE. Therefore, it is possible that the rat liver cGS PDE isozymes are different than those isolated from rabbit and bovine sources. Clearly, resolution of the question will require sequencing the rat liver cGS PDE isozymes or cloning the corresponding cDNAs.

B. Kinetic Properties

The kinetic characteristics of the cGS PDE have been reviewed extensively (Manganiello *et al.*, 1990a). In general, the kinetic properties of

[4] W. K. Sonnenburg, D. Seger, and J. A. Beavo, unpublished data.

both cytosolic and membrane-associated cGS PDE isozymes are very similar (Grant *et al.*, 1990; Martins *et al.*, 1982; Murashima *et al.*, 1990; Pyne *et al.*, 1986; Whalin *et al.*, 1988; Yamamoto *et al.*, 1983a). The kinetics for hydrolysis of both cAMP and cGMP are positively cooperative. The apparent K_m value for cAMP is ~30 μM and the Hill coefficient is ~1.9. For cGMP, the apparent K_m is ~10 μM and the Hill coefficient ~1.5. The V_{max} values for either cGMP or cAMP are ~120 μmol/min/mg. cGMP stimulation of cAMP hydrolytic activity results in a less cooperative kinetic behavior, and a lowering of the apparent K_m for cAMP with little or no effect on V_{max} (Pyne *et al.*, 1986; Whalin *et al.*, 1988; Yamamoto *et al.*, 1983b). The apparent activation constant for cGMP (Kact) is 0.3 μM (Whalin *et al.*, 1988; Yamamoto *et al.*, 1983b). The degree of stimulation by cGMP ranges from 13- to 30-fold, depending on the substrate levels (Murashima *et al.*, 1990; Yamamoto *et al.*, 1983a). An illustration of the effect of cGMP on cAMP hydrolysis by cGS PDE is shown in Fig. 1.

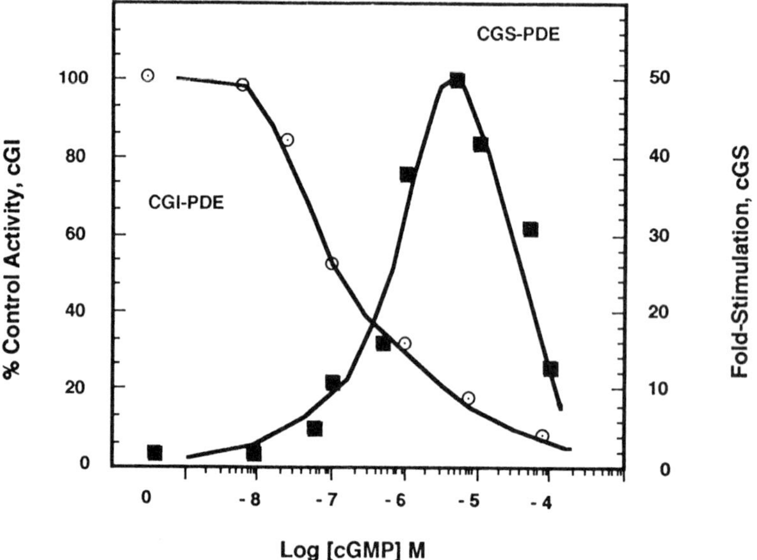

Fig. 1 The effects of cGMP on the hydrolysis of cAMP by cGMP-stimulated (cGS) or cGMP-inhibited (cGI) PDEs isolated from bovine heart. Substrate was 1 μM [^{3}H]cAMP. The biphasic nature of the stimulatory effect of cGMP with the cGS PDE is due to the fact that cGMP is also a substrate for this enzyme and at higher concentrations competes for cAMP at the catalytic site. Redrawn from data obtained in the authors' laboratory by T. Martins and D. Reifsnyder.

C. Domain Organization

The nature of the kinetics and the differential effects of various cGMP analogs on activity (Erneux *et al.*, 1981; Miot *et al.*, 1985) suggested that cGMP stimulates cGS PDE activity by interacting with an allosteric domain separate from the catalytic site. Several lines of direct evidence for a noncatalytic cGMP binding site within the cGS PDE have been obtained (Erneux *et al.*, 1981; Martins *et al.*, 1982; Stroop and Beavo, 1991). Half-maximal binding was observed at 0.2 μM cGMP, and binding sites were saturated at about 1 mol of cGMP bound per mole of enzyme (Martins *et al.*, 1982). A more recent analysis using different binding conditions revealed a second class of cGMP binding sites with apparent K_d values of 1 and 30 μM for the native enzyme (Stroop *et al.*, 1989). All of these observations, plus many others, suggest that the cGS PDE is composed of at least two physically and functionally distinct domains: (*a*) a catalytic domain that catalyzes both cAMP and cGMP hydrolysis and (*b*) an allosteric domain that preferentially binds cGMP with high affinity and in some way stimulates cyclic nucleotide hydrolysis. Some direct evidence for this hypothesis is now available (Stroop and Beavo, 1991; Stroop *et al.*, 1989).

It is now evident that the kinetic similarities between the soluble and the membrane-associated isozymes result from the fact that they are nearly identical structurally through most of their sequence. However, some evidence for structural diversity between these two cGS PDE isoforms was obtained by peptide mapping (Murashima *et al.*, 1990). One of the peptides generated from the membrane-associated isoform was unique. The simplest explanation of these results was that either the carboxy- or the amino-terminal ends of these two cGS PDE isoforms are different. More recently, the authors have reported a mRNA species encoding a variant form of the isozyme (Sonnenburg *et al.*, 1991). These results indicated that the amino-terminal domain encoded by the membrane-associated cGS PDE in the brain is structurally different from that of the cGS PDE in adrenal gland. Moreover, the data suggest that these isozymes are products of an alternately spliced gene. Since these experiments were conducted, we have isolated a cDNA encoding a 943 residue cGS PDE isoform from bovine brain confirming this supposition.[4] The data indicate that a short segment of amino-terminal residues somehow facilitates association of the brain cGS PDE with the membrane.

D. Tissue Distribution

As might be expected from its mode of regulation, the cGS PDE usually is found in cells where the physiological effects of cAMP and cGMP are opposite. The cytosolic isoform of cGS PDE has been isolated from adrenal gland, heart, liver and platelets. The membrane-associated activity has also been purified from brain and liver. Using a specific monoclonal antibody, Martins found that the soluble cGS PDE was most concentrated in adrenal gland (Martins, 1984). More recently, it has been shown that most of this activity is found in the glomerulosa region of the adrenal cortex. Roughly one-fifth of the adrenal gland activity was detected in brain, and even lower levels in heart, spleen, liver, lung, and testis. However, since these values only take into account the cytosolic cGS PDE activity, the total quantity of brain and heart cGS PDE activity is probably underestimated, as these tissues express large amounts of the membrane-associated isoform (Murashima *et al.*, 1990; Simmons and Hartzell, 1988; Sonnenburg *et al.*, 1991; Whalin *et al.*, 1988).

We have recently reported the distribution of mRNAs encoding two structurally related cGS PDE isoforms among total RNAs extracted from a variety of bovine tissues using probes derived from the adrenal cGS PDE cDNA (Sonnenburg *et al.*, 1991). As expected the adrenal cortex expresses the greatest concentration of cGS PDE mRNA. However, cGS PDE transcript was also abundant in heart, adrenal medulla, brain, and kidney. RNase protection assays further revealed that the major isoform expressed in brain was the membrane-associated isoform, whereas the heart appeared to express an equal amount of both the cytosolic and the membrane-associated cGS PDE mRNAs. Considerably less cytosolic cGS PDE mRNA was detected in liver, trachea, lung, spleen, and T-lymphocytes. No cGS PDE transcript was detected in RNA isolated from aorta. This observation is consistent with the apparent lack of this isozyme activity in vascular smooth muscle cells (Lugnier *et al.*, 1986; Schoeffter *et al.*, 1987). However, a cGS PDE activity has been detected in aortic endothelial cells (Lugnier and Schini, 1990; Souness *et al.*, 1990). These analyses further revealed that the cGS PDE was concentrated in anatomically distinct brain and kidney regions. In the brain, cGS PDE mRNA was detected in cerebral cortex, basal ganglia, and hippocampus. Little transcript was detected in cerebellum and spinal cord. Similarly in kidney, greater concentrations of cGS PDE mRNA were detected in the outer red medulla and papilla; however, some mRNA was detected in the cortex.

E. Regulation

Several examples now indicate that hormones that stimulate cGMP formation can reduce cAMP content in cells that express cGS PDE activity. For example, it has been shown that atrial natriuretic peptide (ANP) inhibits isoproterenol and PGE1-induced cAMP accumulation in human fibroblasts (Lee *et al.*, 1988) by a mechanism that involved activation of a phosphodiesterase, presumably the cGS PDE. Similar observations have been made recently using a PC12 pheochromocytoma cell line that predominantly expresses the cGS PDE isozyme (Whalin *et al.*, 1991). In these cells, ANP and sodium nitroprusside blocked adenosine-induced cAMP formation and increased the rate of elimination of cAMP. Both of these studies provide indirect evidence suggesting that cGMP may antagonize hormone-stimulated cAMP accumulation by activating the cGS PDE.

More direct evidence suggesting that activation of the cGS PDE is a primary mechanism of cGMP-mediated effects has been obtained in heart and adrenal gland. Simmons and Hartzell demonstrated that the cGS PDE may play a pivotal role in regulating the slow, inward Ca^{2+} current in frog cardiomyoctyes (Simmons and Hartzell, 1988). It is well established that cAMP increases the trans-sarcolemmal Ca^{2+} current in frog cardiomyocytes, and acetylcholine can blunt this response by increasing intracellular concentrations of cGMP (Hartzell and Fischmeister, 1986). These authors demonstrated that addition of the phosphodiesterase inhibitor IBMX reverses the inhibitory action of cGMP on cAMP-stimulated Ca^{2+} current. More importantly, 8-bromo cGMP, an analog that does not stimulate the PDE but will activate cGMP-dependent protein kinase, did not mimic cGMP. These data infer that cGMP may, at least in part, blunt the trans-sarcolemmal Ca^{2+} current by activating the cGS PDE in frog cardiomyocytes. However, very recent studies in mammalian heart myofibrils by the same authors suggest that cGMP-dependent protein kinase may be more important than cGS PDE in this effect (Mery *et al.*, 1991). In a similar study, Doerner and Alger have presented evidence suggesting that cGMP-mediated suppression of inward Ca^{2+} conductance in hippocampal neurons involves stimulation of a cGS PDE (Doerner and Alger, 1988). Clearly more work needs to be done to resolve the role for cGS PDE in modulating Ca^{2+} conductance.

More recently, a physiological role for the cGS PDE in ANP-mediated inhibition of aldosterone production in the adrenal cortex has been reported (MacFarland *et al.*, 1991). These investigators reported several lines of evidence supporting the idea that inhibition of ACTH-induced aldosterone production by ANP is a cGS PDE-mediated event. Most importantly, they found that ANP could effectively block aldosterone production in response to either 8-bromo-cAMP or *p*-chlorophenylthio-cAMP, agents that stimulate cAMP-dependent protein kinase and also are substrates for the cGS PDE. However, dibutyryl-cAMP-induced aldosterone synthesis was not blocked by ANP as expected since it was a very poor substrate for the CGS PDE. Figure 2 illustrates how this mechanism is thought to function in these cells.

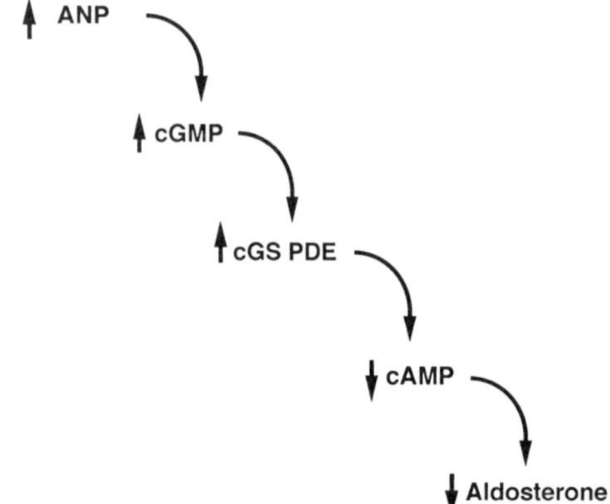

Fig. 2 Model for the mechanism by which atrial natriuretic peptide (ANP) inhibits steroidogenesis in adrenal cortex glomerulosa cells. Data supporting this model have been previously presented (MacFarland *et al.*, 1991).

V. Cyclic GMP-Inhibited PDE Family

A. General Properties and Multiple Isoforms within Family

The cGMP-inhibited PDE(s) hydrolyzes both cAMP and cGMP. Currently, it is thought that this isozyme does not play a major role in modulating cGMP levels. However, it probably does play a role in mediating the effect of cGMP on cAMP levels. The family is named for the property that relatively low concentrations of cGMP act as an inhibitor of cyclic AMP hydrolysis (Fig. 1). Although the mechanism by which this inhibition occurs is still not entirely clear, it is probably due to a direct competition by cGMP at the catalytic site since the K_m for cyclic GMP hydrolysis and the IC_{50} for inhibition are both approximately 1×10^{-7} *M*. (Harrison *et al.*, 1986). At first thought, it seems unusual for an inhibitor of an enzyme to also be a substrate. However, because the V_{max} for cGMP is 10-fold lower than that for cAMP, the energy consumption for such a control mechanism would not be too great. This mechanism also would allow a cell to rapidly alter both the level of the inhibitor and the substrate.

It is not clear how many different isoforms there are within this family although a substantial amount of kinetic data would suggest that more than one must exist. These include the observations that different size variants and different drug inhibition profiles are seen in enzyme preparations made from different tissues and species. Substantial variation is seen for the size(s) of the cGI PDE(s). Molecular masses ranging from 53 to 135 kDa have been reported for "purified" preparations of the enzyme. It is known that the enzyme is very labile to proteolysis and it is likely that this underlies at least part of the variation observed (Macphee *et al.*, 1986; Manganiello *et al.*, 1990b). The great heterogeneity of size, regardless of whether it is intrinsic to the native enzymes or is artifactual due proteolysis, has made purification and characterization very difficult. Nevertheless, several highly purified preparations have been described (Degerman *et al.*, 1987; Harrison *et al.*, 1988). Very recently two apparently different cDNAs encoding variants of this isozyme have been reported (Meacci *et al.*, 1991; Taira *et al.*, 1991). Whether there are other forms remains to be determined.

B. Tissue and Cellular Distribution

Direct evidence for the cellular and subcellular distribution for this isozyme family is largely unavailable since no isozyme-specific antibodies or nucleotide sequences suitable for localization have been available. As might be expected from its mode of regulation by cGMP, this isozyme generally is present in cells that also contain high concentrations of cGMP-dependent protein kinase and/or in cells where the effects of cGMP are similar to cAMP. The enzyme has been purified from heart and adipose tissue (Degerman *et al.*, 1987; Harrison *et al.*, 1988). In addition, PDE activity analysis and drug inhibition studies suggest that cardiocytes, vascular (Lindgren *et al.*, 1991) and trachael smooth muscle (Torphy *et al.*, 1991), platelets (Macphee *et al.*, 1988), adipocytes (Degerman *et al.*, 1990), and liver (Pyne *et al.*, 1987b) are all good sources for the enzyme. Within the cardiocyte, at least some of the activity is associated with the sarcoplasmic reticulum (Kauffman *et al.*, 1987). It is not yet clear whether this activity is the same isozymic form as that found in the soluble fractions.

C. Regulation

Despite the fact that little is known about the physical properties of the enzyme, a large amount is known about its regulation. As the name implies, cGMP can regulate the activity of the enzyme *in vitro* at concentration ranges likely to be achieved in intact tissue in response to agents like ANP and endothelial-derived relaxation factor. The effects of cGMP on cAMP hydrolysis are illustrated in Fig. 1. Finally, the enzyme can be phosphorylated and activated by cAMP-dependent protein kinase *in vitro* (Macphee *et al.*, 1988).

Perhaps more importantly, an even larger set of observations shows that the activity of this isozyme can be modulated in intact tissue. One study indicates that cGMP will inhibit the activity and raise cAMP in intact platelets (Maurice and Haslam, 1990). A large number of studies suggest that it can be activated both by insulin and by hormones that increase cAMP (Loten and Snyed, 1970; Manganiello *et al.*, 1990b; Pyne *et al.*, 1987a). At least part of this activation seems to be post-translational. Tissues where activation has been demonstrated include platelets (Macphee *et al.*, 1988), where prostacyclin will increase the degree of phosphorylation and activity, and adipose tissue (Degerman *et al.*, 1990; Smith *et al.*, 1991), where both β-receptor stimulation and insulin receptor activation will increase the state of phosphorylation and activity. Other reports suggest that the stimulation of PDE activity seen in liver after insulin or epinephrine treatment is due at least in part to activation of this isozyme (Pyne *et al.*, 1987a). In nearly all cases studied, the phosphorylation is reported to be on serine or threonine residues. In most studies, the enzyme apparently is not a direct substrate of the tyrosine kinase activity of the insulin receptor (Manganiello *et al.*, 1990b), although one report suggests that it may be (Pyne *et al.*, 1989). At the present time it is not yet clear why insulin and epinephrine both should increase the activity of the enzyme since they usually act as physiological antagonists. Perhaps different isozymes or subcellular compartments are involved. In the case of insulin, the fact that many of the metabolic effects of insulin are most conspicuous only after cAMP has been elevated in the cell would suggest that activation of this PDE is a very important component of many metabolic effects of insulin in the body. In the case of epinephrine and prostacyclin, activation of this PDE is thought to mediate at least part of the rapid fall in cAMP that commonly occurs after the initial stimulatory response to an activator of adenylate cyclase.

D. Structural Features

Although a large amount of kinetic and regulatory information is available for the cGMP-inhibited phosphodiesterase isozyme family, very little structural information has been published. As mentioned above, this isoenzyme has been extremely recalcitrant toward purification efforts, at least in part because it is extremely sensitive to proteolysis. Very recently, Manganiello and colleagues reported the predicted primary sequence for at least one member of this isozyme family (Meacci *et al.*, 1991; Taira *et al.*, 1991). As with all other mammalian cyclic nucleotide PDEs, the predicted sequence contains a conserved region thought to be the catalytic domain. However, the rest of the sequence bears little homology to any other PDE. The open reading frame of the human heart cDNA predicts a protein product of approximately 135 kDa.

VI. Summary

Several of the different PDE isozyme families have the ability *in vitro* to hydrolyze cGMP. In particular they include the CaM-dependent PDEs, the cGMP-stimulated PDEs, and the cGMP binding, cGMP-specific PDEs. Existing evidence suggests or demonstrates that in different cell types, each of these can be important determinants for the control of cGMP steady-state levels. Each of these enzymes is differentially expressed and regulated; moreover, the amount of the enzyme expressed and the mode of regulation determine to a large extent the rate of rise, maximal level, rate of fall, and duration of the cGMP signal in the cell. In addition to enzymes that function to degrade cGMP at least two also are regulated by cGMP both *in vitro* and in the intact cell. The cGMP-stimulated PDE has the ability to decrease cAMP levels in response to cGMP and the cGMP-inhibited PDE can increase cAMP levels in response to cGMP. We are just beginning to define how many different isozymes of PDE exist in mammalian tissues, where they are located, and how they are regulated. Selective inhibitors to each are being developed and studies designed to define structural features that determine the mechanisms of action and regulation of the PDEs have been initiated. It is expected that in the next few years more PDEs will be discovered and the functions of the new and existing ones will be more clearly defined.

Acknowledgments

Much of the work reported here was funded by NIH Grants DK21723 and EY08197.

References

Ahn, H. S., Crim, W., Romano, M., Sybertz, E., and Pitts, B. (1989). Effects of selective inhibitors on cyclic nucleotide phosphodiesterases of rabbit aorta. *Biochem. Pharmacol.* **38,** 3331–3339.

Ahn, H. S., Foster, M., Foster, C., Sybertz, E., and Wells, J. N. (1991). Evidence for essential histidine and cysteine residues in calcium/calmodulin-sensitive cyclic nucleotide phosphodiesterase. *Biochemistry* **30,** 6754–6760.

Ariano, M. A. (1983). Distribution of components of the guanosine 3′-5′-phosphate system in rat cuadate-putamen. *Neuroscience* **10,** 707–723.

Baehr, W., Champagne, M. S., Lee, A. K., and Pittler, S. J. (1991). Complete cDNA sequences of mouse rod photoreceptor cGMP phosphodiesterase alpha- and beta-subunits, and identification of beta′-, a putative beta-subunit isozyme produced by alternative splicing of the beta-subunit gene. *FEBS Lett.* **278,** 107–114.

Bairoch, A. (1991). Prosite: A dictionary of sites and patterns in proteins. *Nucleic Acids Res.* **19,** Suppl., 2241–2245.

Beavo, J. (1990). Multiple phosphodiesterase isoenzymes: Background, nomenclature, and implications. *In* "Cyclic Nucleotide Phosphodiesterases: Structure, Regulation and Drug Action" (J. Beavo and M. D. Houslay , eds.), Vol. 2, pp. 3–19. Wiley, Chichester.

Beavo, J. A., and Reifsnyder, D. H. (1990). Primary sequence of cyclic nucleotide phosphodiesterase isozymes and the design of selective inhibitors. *Trends Pharmacol. Sci.* **11,** 150–155.

Bentley, J. K., Kadlecek, A., Sonnenburg, W. K., Charbonneau, H., Novack, J. P., and Beavo, J. A. (1992). Molecular cloning of cDNA encoding a "63" kDa calmodulin-stimulated phosphodiesterase from bovine brain. *J Biol Chem.* **267,** 18676–18682.

Billingsley, M. L., Polli, J. W., Balaban, C. D., and Kincaid, R. L. (1990). Developmental expression of calmodulin-dependent cyclic nucleotide phosphodiesterase in rat brain. *Brain Res. Dev. Brain Res.* **53,** 253–263.

Bode, D. C., Kanter, J. R., and Brunton, L. L. (1991). Cellular distribution of phosphodiesterase isoforms in rat cardiac tissue. *Circ. Res.* **68,** 1070–1079.

Charbonneau, H. (1990). Structure-function relationships among cyclic nucleotide phosphodiesterases. *In* "Cyclic Nucleotide Phosphodiesterases: Structure, Function, Regulation and Drug Action" (J. Beavo and M. D. Houslay, eds.), Vol. 2, pp. 267–298. Wiley, Chichester.

Charbonneau, H., Beier, N., Walsh, K. A., and Beavo, J. A. (1986). Identification of a conserved domain among cyclic nucleotide phosphodiesterases from diverse species. *Proc. Natl. Acad. Sci. U.S.A.* **83,** 9308–9312.

Charbonneau, H., Prusti, R. K., LeTrong, H., Sonnenburg, W. K., Mullaney, P. J., Walsh, K. A., and Beavo, J. A. (1990). Identification of a noncatalytic cGMP-binding domain conserved in both the cGMP-stimulated and photoreceptor cyclic nucleotide phosphodiesterases. *Proc. Natl. Acad. Sci. U.S.A.* **87,** 288–292.

Charbonneau, H., Kumar, S., Novack, J. P., Blumenthal, D. K., Griffin, P. R., Shabanowitz, J., Hunt, D. F., Beavo, J. A., and Walsh, K. A. (1991). Evidence for domain organization within the 61-kDa calmodulin-dependent cyclic nucleotide phosphodiesterase from bovine brain. *Biochemistry* **30,** 7931–7940.

Chaudhry, P. S., and Casillas, E. R. (1988). Calmodulin-stimulated cyclic nucleotide phosphodiesterases in plasma membranes of bovine epididymal spermatozoa. *Arch. Biochem. Biophys.* **262,** 439–444.

Chiu, P. J., Tetzloff, G., Ahn, H. S., and Sybertz, E. J. (1988). Comparative effects of vinpocetine and 8-Br-cyclic GMP on the contraction and 45Ca-fluxes in the rabbit aorta. *Am. J. Hypertens.* **1,** 262–268.

Coquil, J. F., Franks, D. J., Wells, J. N., Dupuis, M., and Hamet, P. (1983). Characteristics of a new binding protein distinct from the kinase for guanosine 3′,5′-monophosphate in rat platelets. *Biochim. Biophys. Acta* **743,** 359–369.

Degerman, E., Belfrage, P., Newman, A. H., Rice, K. C., and Manganiello, V. C. (1987). Purification of the putative hormone-sensitive cyclic AMP phosphodiesterase from rat adipose tissue using a derivative of cilostamide as a novel affinity ligand. *J. Biol. Chem.* **262,** 5797–5807.

Degerman, E., Smith, C. J., Tornqvist, H., Vasta, V., Belfrage, P., and Manganiello, V. C. (1990). Evidence that insulin and isoprenaline activate the cGMP-inhibited low-Km cAMP phosphodiesterase in rat fat cells by phosphorylation. *Proc. Natl. Acad. Sci. U.S.A.* **87,** 533–537.

Doerner, D., and Alger, B. E. (1988). Cyclic GMP depresses hippocampal Ca2+ current through a mechanism independent of cGMP-dependent protein kinase. *Neuron* **1,** 693–699.

Dundore, R. L., Pratt, P. F., Hallenbeck, W. D., Wassey, M. L., Silver, P. J., and Buchholz, R. A. (1990). Sodium nitroprusside potentiates the depressor response to the phosphodiesterase inhibitor zaprinast in rats. *Eur. J. Pharmacol.* **185,** 91–97.

Erneux, C., Miot, F., Van, H. P. J. M., and Jastorff, B. (1981). The binding of cyclic nucleotide analogs to a purified cyclic GMP-stimulated phosphodiesterase from bovine adrenal tissue. *Eur. J. Biochem.* **115,** 503–510.

Falchetto, R., Vorherr, T., Brunner, J., and Carafoli, E. (1991). The plasma membrane Ca2+ pump contains a site that interacts with its calmodulin-binding domain. *J. Biol. Chem.* **266,** 2930–2936.

Francis, S. H., and Corbin, J. D. (1988). Purification of cGMP-binding protein phosphodiesterase from rat lung. *In* "Methods in Enzymology" (J. Corbin and R. Johnson, eds.), Vol. 159, pp. 722–729. Academic Press, San Diego, CA.

Francis, S. H., Lincoln, T. M., and Corbin, J. D. (1980). Characterization of a novel cGMP binding protein from rat lung. *J. Biol. Chem.* **255,** 620–626.

Francis, S. H., Thomas, M. K., and Corbin, J. D. (1990). Cyclic GMP-binding cyclic GMP-specific phosphodiesterase from lung. *In* "Cyclic Nucleotide Phosphodiesterases: Structure, Regulation and Drug Action" (J. Beavo and M. D. Houslay, eds.), Vol. 2, pp. 117–140. Wiley, Chichester.

Geremia, R., Rossi, P., Mocini, D., Pezzotti, R., and Conti, M. (1984). Characterization of a calmodulin-dependent high affinity cyclic AMP and cyclic GMP phosphodiesterase from male mouse germ cells. *Biochem. J.* **217,** 693–700.

Gillespie, P. G. (1990). Phosphodiesterases in visual transduction by rods and cones. *In* "Cyclic Nucleotide Phosphodiesterases: Structure, Regulation and Drug Action" (J. Beavo and M. D. Houslay , eds.), Vol. 2, pp. 163–184. Wiley, Chichester.

Godfraind, T., Mennig, D., Morel, N., and Wibo, M. (1989). Effect of endothelin-1 on calcium channel gating by agonists in vascular smooth muscle. *J. Cardiovasc. Pharmacol.* **13,** Suppl. 5, 5112–5117.

Grant, P. G., Mannarino, A. F., and Colman, R. W. (1990). Purification and characterization of a cyclic GMP-stimulated cyclic nucleotide phosphodiesterase from the cytosol of human platelets. *Thromb. Res.* **59,** 105–119.

Grewal, J., Karuppiah, N., and Mutus, B. (1989). A comparative kinetic study of bovine calmodulin-dependent cyclic nucleotide phosphodiesterase isozymes utilizing cAMP, cGMP and their 2′-O-anthraniloyl-,2′-O-(N-methylanthraniloyl)-derivatives as substrates. *Biochem. Int.* **19,** 1287–1295.

Hagiwara, M., Endo, T., and Hidaka, H. (1984). Effects of vinpocetine on cyclic nucleotide metabolism in vascular smooth muscle. *Biochem. Pharmacol.* **33,** 453–457.

Hansen, R. S., and Beavo, J. A. (1986). Differential recognition of calmodulin-enzyme complexes by a conformation-specific anti-calmodulin monoclonal antibody. *J. Biol. Chem.* **261,** 14636–14645.

Harrison, S. A., Reifsnyder, D. H., Gallis, B., Cadd, G. G., and Beavo, J. A. (1986). Isolation and characterization of bovine cardiac muscle cGMP-inhibited phosphodiesterase: A receptor for new cardiotonic drugs. *Mol. Pharmacol.* **29,** 506–514.

Harrison, S. A., Beier, N., Martins, T. J., and Beavo, J. A. (1988). Isolation and comparison of bovine heart cGMP-inhibited and cGMP-stimulated phosphodiesterases. *In* "Methods in Enzymology" (J. Corbin and R. Johnson, eds.), Vol. 159, pp. 685–702. Academic Press, San Diego, CA.

Hartzell, H. C., and Fischmeister, R. (1986). Opposite effects of cyclic GMP and cyclic AMP on Ca2+ current in single heart cells. *Nature* (*London*) **323,** 273–275.

Hashimoto, Y., Sharma, R. K., and Soderling, T. R. (1989). Regulation of Ca2+/calmodulin-dependent cyclic nucleotide phosphodiesterase by the autophosphorylated form of Ca2+/calmodulin-dependent protein kinase II. *J. Biol. Chem.* **264,** 10884–10887.

Hurwitz, R. L., Hirsch, K. M., Clark, D. J., Holcombe, V. N., and Hurwitz, M. Y. (1990). Induction of a calcium/calmodulin-dependent phosphodiesterase during phytohemagglutinin-stimulated lymphocyte mitogenesis. *J. Biol. Chem.* **265,** 8901–8907.

Jarrett, H. W., and Madhavan, R. (1991). Calmodulin-binding proteins also have a calmodulin-like binding site within their structure. The flip-flop model. *J. Biol. Chem.* **266,** 362–371.

Kauffman, R. F., Crowe, V. G., Utterback, B. G., and Robertson, D. W. (1987). LY 195115: A potent, selective inhibitor of cyclic nucleotide phosphodiesterase located in the sarcoplasmic reticulum. *Mol. Pharmacol.* **30,** 609–616.

Kemp, B. E., Pearson, R. B., Guerriero, V. J., Bagchi, I. C., and Means, A. R. (1987). The calmodulin binding domain of chicken smooth muscle myosin light chain kinase contains a pseudosubstrate sequence. *J. Biol. Chem.* **262,** 2542–2548.

Kincaid, R. L., Stith, C. I. E., and Vaughan, M. (1985). Proteolytic activation of calmodulin-dependent cyclic nucleotide phosphodiesterase. *J. Biol. Chem.* **260,** 9009–9015.

Kincaid, R. L., Balaban, C. D., and Billingsley, M. L. (1987). Differential localization of calmodulin-dependent enzymes in rat brain: Evidence for selective expression of cyclic nucleotide phosphodiesterase in specific neurons. *Proc. Natl. Acad. Sci. U.S.A.* **84,** 1118–1122.

Klee, C. B. (1980). Calmodulin: The coupling factor of the two second messengers Ca^{2+} and cAMP. *In* "Protein Phosphorylation and Bioregulation" (T. G. Podesta and E. J. Gordon , eds.), pp. 61–69. Karger, Basel.

Krinks, M. H., Haiech, J., Rhoads, A., and Klee, C. B. (1984). Reversible and irreversible activation of cyclic nucleotide phosphodiesterase:separation of the regulatory and catalytic domains by limited proteolysis. *Adv. Cyclic Nucleotide Res.* **16,** 31–47.

Langlands, J. M., Rodger, I. W., and Diamond, J. (1989). The effect of M&B 22948 on methacholine- and histamine-induced contraction and inositol 1,4,5-trisphosphate levels in guinea-pig tracheal tissue. *Br. J. Pharmacol.* **98,** 336–338.

Lee, M. A., West, R. J., and Moss, J. (1988). Atrial natriuretic factor reduces cyclic adenosine monophosphate content of human fibroblasts by enhancing phosphodiesterase activity. *J. Clin. Invest.* **82,** 388–393.

Lindgren, S., Rascón, A., Andersson, K. E., Manganiello, V., and Degerman, E. (1991). Selective inhibition of cGMP-inhibited and cGMP-noninhibited cyclic nucleotide phosphodiesterases and relaxation of rat aorta. *Biochem. Pharmacol.* **42,** 545–552.

Lorenz, K. L., and Wells, J. N. (1983). Potentiation of the effects of sodium nitroprusside and of isoproterenol by selective phosphodiesterase inhibitors. *Mol. Pharmacol.* **23,** 424–430.

Loten, E. G., and Snyed, J. G. T. (1970). An effect of insulin on adipose tissue phosphodiesterase. *Biochem. J.* **120,** 187–193.

Lugnier, C., and Schini, V. B. (1990). Characterization of cyclic nucleotide phosphodiesterases from cultured bovine aortic endothelial cells. *Biochem. Pharmacol.* **39,** 75–84.

Lugnier, C., Schoeffter, P., Le, B. A., Strouthou, E., and Stoclet, J. C. (1986). Selective inhibition of cyclic nucleotide phosphodiesterases of human, bovine and rat aorta. *Biochem. Pharmacol.* **35,** 1743–1751.

MacFarland, R. T., Zelus, B. D., and Beavo, J. A. (1991). High concentrations of a cGMP-stimulated phosphodiesterase mediate ANP-induced decreases in cAMP and steroidogenesis in adrenal glomerulosa cells. *J. Biol. Chem.* **266,** 136–142.

Macphee, C. H., Harrison, S. H., and Beavo, J. A. (1986). Immunological identification of the major platelet low-Km cAMP phosphodiesterase: Probable target for anti-thrombotic agents. *Proc. Natl. Acad. Sci. U.S.A.* **83,** 6660–6663.

Macphee, C. H., Reifsnyder, D. H., Moore, T. A., Lerea, K. M., and Beavo, J. A. (1988). Phosphorylation results in activation of a cAMP phosphodiesterase in human platelets. *J. Biol. Chem.* **263,** 10353–10358.

Malta, E. (1989). Biphasic relaxant curves to glyceryl trinitrate in rat aortic rings. Evidence for two mechanisms of action. *Naunyn-Schmiedeberg's Arch. Pharmacol.* **339,** 236–243.

Manganiello, V. C., Tanaka, T., and Murashima, S. (1990a). Cyclic GMP-stimulated cyclic nucleotide phosphodiesterases. *In* "Cyclic Nucleotide Phosphodiesterases: Structure, Regulation and Drug Action" (J. Beavo and M. D. Houslay, eds.), Vol. 2, pp. 61–86. Wiley, Chichester.

Manganiello, V. C., Smith, C. J., Degerman, E., and Belfrage, P. (1990b). Cyclic GMP-inhibited cyclic nucleotide phosphodiesterases. *In* "Cyclic Nucleotide Phosphodiesterases: Structure, Regulation and Drug Action" (J. Beavo and M. D. Houslay , eds.), Vol. 2, pp. 87–116. Wiley, Chichester.

Marcus, R., and Grant, B. F. (1983). Parathyroid hormone induces a calmodulin-dependent alteration in phosphodiesterase activity of rat kidney in vivo. *Endocrinology (Baltimore)* **112,** 1065–1069.

Martins, T. J. (1984). Cyclic GMP-stimulated cyclic nucleotide phosphdoiesterase: Purification from bovine tissues and characterization. Ph.D., Thesis, University of Washington, Seattle.

Martins, T. J., Mumby, M. C., and Beavo, J. A. (1982). Purification and characterization of a cyclic GMP-stimulated cyclic nucleotide phosphodiesterase from bovine tissues. *J. Biol. Chem.* **257,** 1973–1979.

Maurice, D. H., and Haslam, R. J. (1990). Molecular basis of the synergistic inhibition of platelet function by nitrovasodilators and activators of adenylate cyclase: Inhibition of cyclic AMP breakdown by cyclic GMP. *Mol. Pharmacol.* **37,** 671–681.

McAllister-Lucas, L. M.,Kadlecek, A. T.,Sonnenburg, W. K.,Seger, D.,Francis, S. H.,Thomas, M. K.,Corbin, J. D., & Beavo, J. A. (1993). Isolation of a cDNA clone for the bovine lung cGMP-binding cGMP-specific phosphodiesterase. *J. Biol. Chem.*, **268,** 22863–22873.

Meacci, E., Moos, M., Taira, M., Smith, C. J., Vasta, V., Degerman, E., Movsesian, M. A., Belfrage, P., and Manganiello, V. C. (1991). Cloning of cDNA for a human cardiac cGMP-inhibited low Km cAMP phosphodiesterase (CGI PDE). *FASEB J.* **5,** A454.

Mery, P. F., Lohmann, S. M., Walter, U., and Fischmeister, R. (1991). Ca2+ current is regulated by cyclic GMP-dependent protein kinase in mammalian cardiac myocytes. *Proc. Natl. Acad. Sci. U.S.A.* **88,** 1197–1201.

Miot, F., Dumont, J. E., and Erneux, C. (1983). The involvement of a calmodulin-dependent phosphodiesterase in the negative control of carbamylcholine on cyclic AMP levels in dog thyroid slices. *FEBS Lett* **151,** 273–276.

Miot, F., Van, H. P. J., and Erneux, C. (1985). Specificity of cGMP binding to a purified cGMP-stimulated phosphodiesterase from bovine adrenal tissue. *Eur. J. Biochem.* **149,** 59–65.

Mumby, M. C., Martins, T. J., Chang, M. L., and Beavo, J. A. (1982). Identification of cGMP-stimulated cyclic nucleotide phosphodiesterase in lung Tissue with monoclonal antibodies. *J. Biol. Chem.* **257,** 13283–13290.

Murashima, S., Tanaka, T., Hockman, S., and Manganiello, V. (1990). Characterization of particulate cyclic nucleotide phosphodiesterases from bovine brain: Purification of a distinct cGMP-stimulated isoenzyme. *Biochemistry* **29,** 5285–5292.

Nawy, S., and Jahr, C. E. (1990). Suppression by glutamate of cGMP-activated conductance in retinal bipolar cells. *Nature (London)* **346,** 269–271.

Nawy, S., and Jahr, C. E. (1991). cGMP-gated conductance in retinal bipolar cells is suppressed by the photoreceptor transmitter. *Neuron* **7,** 677–683.

Novack, J. P., Charbonneau, H., Bentley, J. K., Walsh, K. A., and Beavo, J. A. (1991). Sequence comparison of the 63-, 61-, and 59-kDa calmodulin-dependent cyclic nucleotide phosphodiesterases. *Biochemistry* **30,** 7940–7947.

Pittler, S. J., and Baehr, W. (1991). The molecular genetics of retinal photoreceptor proteins involved in cGMP metabolism. *Prog. Clin. Biol. Res.* **362,** 33–66.

Purvis, K., and Rui, H. (1988). High-affinity, calmodulin-dependent isoforms of cyclic nucleotide phosphodiesterase in rat testis. *In* "Methods in Enzymology" (J. Corbin and R. Johnson, eds.), Vol. 159, pp. 675–685. Academic Press, San Diego, CA.

Pyne, N. J., Cooper, M. E., and Houslay, M. D. (1986). Identification and characterization of both the cytosolic and particulate forms of cyclic GMP-stimulated cyclic AMP phosphodiesterase from rat liver. *Biochem. J.* **234,** 325–334.

Pyne, N. J., Cooper, M. E., and Houslay, M. D. (1987a). The insulin- and glucagon-stimulated 'dense-vesicle' high-affinity cyclic AMP phosphodiesterase from rat liver. Purification, characterization and inhibitor sensitivity. *Biochem. J.* **242,** 33–42.

Pyne, N. J., Anderson, N., Lavan, B. E., Milligan, G., Nimmo, H. G., and Houslay, M. D. (1987b). Specific antibodies and the selective inhibitor ICI 118233 demonstrate that the hormonally stimulated 'dense-vesicle' and peripheral-plasma-membrane cyclic AMP phosphodiesterases display distinct tissue distributions in the rat. *Biochem. J.* **248,** 897–901.

Pyne, N. J., Cushley, W., Nimmo, H. G., and Houslay, M. D. (1989). Insulin stimulates the tyrosyl phosphorylation and activation of the 52 kDa peripheral plasma-membrane cyclic AMP phosphodiesterase in intact hepatocytes. *Biochem. J.* **261,** 897–904.

Radomski, M. W., Palmer, R. M., and Moncada, S. (1990a). Characterization of the L-arginine:nitric oxide pathway in human platelets. *Br. J. Pharmacol.* **101,** 325–328.

Radomski, M. W., Palmer, R. M., and Moncada, S. (1990b). An L-arginine/nitric oxide pathway present in human platelets regulates aggregation. *Proc. Natl. Acad. Sci. U.S.A.* **87,** 5193–5197.

Rapoport, R. M., Ashraf, M., and Murad, F. (1989). Effects of melittin on endothelium-dependent relaxation and cyclic GMP levels in rat aorta. *Circ. Res.* **64,** 463–473.

Repaske, D. R., Swinnen, J. V., Jin, S. L., Judson, J., Wyk, V., and Conti, M. (1992). A polymerase chain reaction strategy to identify and clone cyclic nucleotide phosphodiesterases: Molecular cloning of the 63-kDa calmodulin-dependent phosphodiesterase. *J. Biol. Chem.* **267,** 18683–18688.

Rimele, T. J., Sturm, R. J., Adams, L. M., Henry, D. E., Heaslip, R. J., Weichman, B. M., and Grimes, D. (1988). Interaction of neutrophils with vascular smooth muscle: Identification of a neutrophil-derived relaxing factor. *J. Pharmacol. Exp. Ther.* **245,** 102–111.

Rossi, P., Giorgi, M., Geremia, R., and Kincaid, R. L. (1988). Testis-specific calmodulin-dependent phosphodiesterase. A distinct high affinity cAMP isoenzyme immunologically related to brain calmodulin-dependent cGMP phosphodiesterase. *J. Biol. Chem.* **263,** 15521–15527.

Salvemini, D., Radziszewski, W., Korbut, R., and Vane, J. (1990). The use of oxyhaemoglobin to explore the events underlying inhibition of platelet aggregation induced by NO or NO-donors. *Br. J. Pharmacol.* **101,** 991–995.

Schoeffter, P., Lugnier, C., Demesy, W. F., and Stoclet, J. C. (1987). Role of cyclic AMP- and cyclic GMP-phosphodiesterases in the control of cyclic nucleotide levels and smooth muscle tone in rat isolated aorta. A study with selective inhibitors. *Biochem. Pharmacol.* **36,** 3965–3972.

Sharma, R. K. (1991). Phosphorylation and characterization of bovine heart calmodulin-dependent phosphodiesterase. *Biochemistry* **30,** 5963–5968.

Sharma, R. K., and Wang, J. H. (1985). Differential regulation of bovine brain calmodulin-dependent cyclic nucleotide phosphodiesterase isoenzymes by cyclic AMP-dependent protein kinase and calmodulin-dependent phosphatase. *Proc. Natl. Acad. Sci. U.S.A.* **82,** 2603–2607.

Sharma, R. K., and Wang, J. H. (1986a). Calmodulin and Ca2+-dependent phosphorylation and dephosphorylation of the 63-kDa subunit-containing bovine brain calmodulin-stimulated cyclic nucleotide phosphodiesterase isozyme. *J. Biol. Chem.* **261,** 1322–1328.

Sharma, R. K., and Wang, J. H. (1986b). Purification and characterization of bovine lung calmodulin-dependent cyclic nucleotide phosphodiesterase. An enzyme containing calmodulin as a subunit. *J. Biol. Chem.* **261,** 14160–14166.

Sharma, R. K., and Wang, J. H. (1986c). Regulation of cAMP concentration by calmodulin-dependent cyclic nucleotide phosphodiesterase. *Biochem. Cell Biol.* **64,** 1072–1080.

Sharma, R. K., Adachi, A. M., Adachi, K., and Wang, J. H. (1984). Demonstration of bovine brain calmodulin-dependent cyclic nucleotide phosphodiesterase isozymes by monoclonal antibodies. *J. Biol. Chem.* **259,** 9248–9254.

Shenolikar, S., Thompson, W. J., and Strada, S. J. (1985). Characterization of a Ca2+-calmodulin-stimulated cyclic GMP phosphodiesterase from bovine brain. *Biochemistry* **24,** 672–678.

Shoepp, D. D., and Johnson, B. G. (1988) Excitatory amino acid agonist-antagonist interactions at 2-amino-4-phosphonobutyric acid-sensitive quisqualate receptors coupled to phosphoinositide hydrolysis in slices of rat hippocampus. *J. Neurochem.* **50,** 1605–1613.

Simmons, M. A., and Hartzell, H. C. (1988). Role of phosphodiesterase in regulation of calcium current in isolated cardiac myocytes. *Mol. Pharmacol.* **33,** 664–671; erratum: **34**(4), 604 (1988).

Smith, C. J., Vasta, V., Degerman, E., Belfrage, P., and Manganiello, V. C. (1991). Hormone-sensitive cyclic GMP-inhibited cyclic AMP phosphodiesterase in rat adipocytes. Regulation of insulin- and cAMP-dependent activation by phosphorylation. *J. Biol. Chem.* **266,** 13385–13390.

Snyder, S. H., and Bredt, D. S. (1991). Nitric oxide as a neuronal messenger. *Trends Pharmacol. Sci.* **12,** 125–128.

Sonnenburg, W. K., Mullaney, P. J., and Beavo, J. A. (1991). Molecular cloning of a cyclic GMP-stimulated cyclic nucleotide phosphodiesterase cDNA. Identification and distribution of isozyme variants. *J. Biol. Chem.* **266,** 17655–17661.

Sonnenburg, W. K.,Seger, D., & Beavo, J. A. (1993). Molecular cloning of a cDNA encoding the "61-kDa" calmodulin-stimulated cyclic nucleotide phosphodiesterase: Tissue-specific expression of structurally related isoforms. *J. Biol. Chem.* **268**(1), 645–652.

Souness, J. E., Brazdil, R., Diocee, B. K., and Jordan, R. (1989). Role of selective cyclic GMP phosphodiesterase inhibition in the myorelaxant actions of M&B 22,948, MY-5445, vinpocetine and 1-methyl-3-isobutyl-8-(methylamino)xanthine. *Br. J. Pharmacol.* **98,** 725–734.

Souness, J. E., Diocee, B. K., Martin, W., and Moodie, S. A. (1990). Pig aortic endothelial-cell cyclic nucleotide phosphodiesterases. Use of phosphodiesterase inhibitors to evaluate their roles in regulating cyclic nucleotide levels in intact cells. *Biochem. J.* **266,** 127–132.

Stroop, S. D., and Beavo, J. A. (1991). Structure and function studies of the cGMP-stimulated phosphodiesterase. *J. Biol Chem.* **266,** 23802–23809.

Stroop, S. D., Charbonneau, H., and Beavo, J. A. (1989). Direct photolabeling of the cGMP-stimulated cyclic nucleotide phosphodiesterase. *J. Biol. Chem.* **264,** 13718–13725.

Sugden, M. C., and Ashcroft, S. J. (1981). Cyclic nucleotide phosphodiesterase of rat pancreatic islets. Effects of Ca2+, calmodulin and trifluoperazine. *Biochem. J.* **197,** 459–464.

Sumners, C., and Myers, L. M. (1991). Angiotensin II decreases cGMP levels in neuronal cultures from rat brain. *Am. J. Physiol.* **260,** C79–C87.

Taira, M., Meacci, E., and Manganiello, V. C. (1991). Two distinct cDNAs of cGMP-inhibited low Km cAMP phosphodiesterase (cGI-PDE) from rat adipose tissue and human heart. *Pharmacologist* **33,** 190.

Thomas, M. K., Francis, S. H., and Corbin, J. D. (1990). Substrate- and kinase-directed regulation of phosphorylation of a cGMP-binding phosphodiesterase by cGMP. *J. Biol. Chem.* **265,** 14971–14978.

Torphy, T. J., Zhou, H. L., Burman, M., and Huang, L. B. (1991). Role of cyclic nucleotide phosphodiesterase isozymes in intact canine trachealis. *Mol. Pharmacol.* **39,** 376–384.

Trong, H. L., Beier, N., Sonnenburg, W. K., Stroop, S. D., Walsh, K. A., Beavo, J. A., and Charbonneau, H. (1990). Amino acid sequence of the cyclic GMP stimulated cyclic nucleotide phosphodiesterase from bovine heart. *Biochemistry* **29,** 10280–10288.

Uzunov, P., Gnegy, M. E., Revuelta, A., and Costa, E. (1976). Regulation of the high Km cyclic nucleotide phosphodiesterase of adrenal medulla by the endogenous calcium-dependent protein activator. *Biochem. Biophys. Res. Commun.* **70,** 132–138.

Vandermeers, A., Vandermeers, P. M. C., Rathe, J., and Christophe, J. (1983). Purification and kinetic properties of two soluble forms of calmodulin-dependent cyclic nucleotide phosphodiesterase from rat pancreas. *Biochem. J.* **211,** 341–347.

Wang, J. H., Sharma, R. K., and Mooibroek, M. J. (1990). Calmodulin-stimulated cyclic nucleotide phosphodiesterases. *In* "Cyclic Nucleotide Phosphodiesterases: Structure, Regulation and Drug Action" (J. Beavo and M. D. Houslay , eds.), Vol. 2, pp. 19–60. Wiley, Chichester.

Whalin, M. E., Strada, S. J., and Thompson, W. J. (1988). Purification and partial characterization of membrane-associated type II (cGMP-activatable) cyclic nucleotide phosphodiesterase from rabbit brain. *Biochim. Biophys. Acta* **972,** 79–94.

Whalin, M. E., Scammell, J. G., Strada, S. J., and Thompson, W. J. (1991). Phosphodiesterase II, the cGMP-activatable cyclic nucleotide phosphodiesterase, regulates cyclic AMP metabolism in PC12 cells. *Mol. Pharmacol.* **39,** 711–717.

Wilkins, M. R., Settle, S. L., and Needleman, P. (1990a). Augmentation of the natriuretic activity of exogenous and endogenous atriopeptin in rats by inhibition of guanosine 3′,5′-cyclic monophosphate degradation. *J. Clin. Invest.* **85,** 1274–1279.

Wilkins, M. R., Settle, S. L., Stockmann, P. T., and Needleman, P. (1990b). Maximizing the natriuretic effect of endogenous atriopeptin in a rat model of heart failure. *Proc. Natl. Acad. Sci. U.S.A.* **87,** 6465–6469.

Yamamoto, T., Manganiello, V. C., and Vaughan, M. (1983a). Purification and characterization of cyclic GMP-stimulated cyclic nucleotide phosphodiesterase from calf liver. *J. Biol. Chem.* **258,** 12526–12533.

Yamamoto, T., Yamamoto, S., Osborne, J. J., Manganiello, V. C., Vaughan, M., and Hidaka, H. (1983b). Complex effects of inhibitors on cyclic GMP-stimulated cyclic nucleotide phosphodiesterase. *J. Biol. Chem.* **258,** 14173–14177.

Progress in Understanding the Mechanism and Function of Cyclic GMP-Dependent Protein Kinase

Sharron H. Francis and Jackie D. Corbin

Department of Molecular Physiology and Biophysics
Vanderbilt University School of Medicine
Nashville, Tennessee 37232

I. Introduction

The discovery of cGMP in 1963 (Ashman *et al.*, 1963) and the subsequent identification of a cGMP-dependent protein kinase (cGMP kinase) in arthropods (Kuo and Greengard, 1970) and mammalian tissue (Hofmann and Sold, 1972) have led to many studies addressing the physiological processes that are modulated by cGMP-regulated phosphorylation events. The cGMP kinase is present in numerous mammalian tissues (Walter, 1981, 1989; Walter *et al.*, 1981; Lohmann and Walter, 1984), and, because of experience with the cAMP cascade, it was believed for many years to be the main mediator of cGMP effects. The enzyme, a cytosolic serine/threonine kinase, has been purified to homogeneity (Kuo and Greengard, 1970; Gill *et al.*, 1976; Lincoln *et al.*, 1976; Wolfe *et al.*, 1989b) by several groups. However, despite considerable effort, the importance of the cGMP system as a regulator of specific metabolic processes has only recently been clearly documented. In contrast to the cAMP system, where the cAMP kinase appears to be the predominant, if not the exclusive intracellular receptor for cAMP, it is now clear that numerous receptors for cGMP are present in mammalian cells (Lincoln *et al.*, 1976); those described to date include the cGMP kinases (Hofmann and Sold, 1972; Lincoln *et al.*,

Advances in Pharmacology, Volume 26

1977, 1988; Wolfe *et al.*, 1989b), cation channel proteins (Fesenko *et al.*, 1985; Koch and Kaupp, 1985; Nakamura and Gold, 1987; Kaupp *et al.*, 1989), and a family of cGMP-binding phosphodiesterases (Beavo *et al.*, 1971; Miki *et al.*, 1973; Hamet and Coquil, 1978; Francis *et al.*, 1980; Beavo, 1988). The presence of these numerous and varied receptors complicate interpretation of effects attributed to elevations of intracellular cGMP and require a rigorous analysis of the receptor protein that is responsible for eliciting the specific response(s) (Corbin *et al.*, 1990). Even so, the cGMP kinase may be considered to be a prototype for these receptor subtypes, which exhibit many homologies and analogies.

Fortunately, despite the absence of a clearly delineated biological role, many of the biochemical features of the cGMP kinase have been well studied (Glass and Krebs, 1979, 1982; Lincoln and Corbin, 1983; Lohmann and Walter, 1984; Beebe and Corbin, 1986; Edelman *et al.*, 1987; Scott, 1991; Landgraf *et al.*, 1991). This information regarding the functional parameters of the enzyme can now be used in determining whether the kinase serves as the intracellular receptor for changes in cGMP levels. Before addressing this important problem, an examination of the potential roles of cGMP itself is in order. A critical role for cGMP in visual transduction (Miki *et al.*, 1973; Hurley, 1977; Stryer, 1986; Kaupp and Koch, 1990) is well established. The implication of cGMP as an important regulator of smooth muscle tension in response to atrial natriuretic peptide, nitrovasodilators, and endothelium-derived agents has also heightened the interest in the cGMP field (Rapoport and Murad, 1983; Lincoln, 1983; Waldman *et al.*, 1984; Lincoln and Johnson, 1984; Fiscus *et al.*, 1985; Ignarro and Kadowitz, 1985; Murad, 1986; Furchgott, 1987; Tremblay *et al.*, 1988). More recent developments in the field suggest that cGMP may also be a critical second messenger for regulation of processes as diverse as long-term potentiation in hippocampal neurons (Schuman and Madison, 1991; O'Dell *et al.*, 1991), olfaction (Nakamura and Gold, 1987), insulin secretion (Laychock *et al.*, 1991; Schmidt *et al.*, 1992), amylase secretion (Rogers *et al.*, 1988), chloride secretion in the gut (Forte *et al.*, 1992), smooth muscle proliferation (Garg and Hassid, 1989), and mesangial cell mitogenesis and proliferation (Appel, 1990). Elevation of intracellular cGMP has also been associated with a shift of mRNAs from the ribonucleoprotein fraction into a translationally active membrane-associated polysomal pool (Stockert *et al.*, 1992). These observations have prompted renewed interest in the specific catalytic and regulatory features of the proteins involved in the synthesis and degradation of cGMP, and in those proteins that act as receptors for cGMP action.

II. Tissue Distribution of Cyclic GMP Kinase

The tissue distribution of cGMP kinase is rather restricted compared to the ubiquitous distribution of cAMP kinase. Significant concentrations of the enzyme are found in adult and fetal lung, cerebellum, platelets, smooth muscle and smooth muscle-like tissues (Kuo, 1975; Ives *et al.*, 1980; Boyles *et al.*, 1984; Joyce *et al.*, 1984, 1986), and intestinal epithelial cells (DeJonge, 1981). Immunocytochemical studies provide evidence for localization of cGMP kinase to the Purkinje fibers in the cerebellum (Bandle and Guidotti, 1979; Lohmann *et al.*, 1981; DeCamilli *et al.*, 1984), where the enzyme is distributed throughout the cytosol of dendrites, axons, perikarya, and nerve terminals. However, in many organs, the detectable cGMP kinase may be largely attributable to the smooth muscle of the vasculature traversing that tissue. Immunocytochemical studies show vascular smooth muscle to be particularly rich in cGMP kinase activity of which 75% is distributed throughout the cytoplasm and 25% is associated with the membrane fractions (Ives *et al.*, 1980). Direct enzymatic analysis in porcine vascular smooth muscle based on both kinase activity and cGMP-binding activity estimates the intracellular cGMP kinase concentration to be 0.13 μM compared to 0.17 μM for the cAMP kinase; this assumes random distribution of these enzymes (Francis *et al.*, 1988). Tracheal smooth muscle and human and bovine aortic smooth muscle also contain relatively large amounts of cGMP kinase activity (Wolfe *et al.*, 1989b; Sekhar *et al.*, 1992). Unusually high levels ($\sim 1\ \mu M$) of cGMP kinase have been estimated in platelets (Waldmann *et al.*, 1986; Walter, 1989), with the majority of this kinase being associated with the particulate fraction (perhaps the cytoskeleton).

cGMP kinase is also present in cells closely related to smooth muscle. It is found in the cytosol of pericytes (Joyce *et al.*, 1984), cells that are closely related to smooth muscle and that encircle capillaries and postcapillary venules in the microvasculature. The enzyme is also found in the contractile mesangial cells of the kidney glomeruli (Joyce *et al.*, 1986), as well as in the vascular smooth muscle, and may prove to be important in mediating the natriuretic and diuretic effects of atrial natriuretic peptide in the kidney (Nonoguchi *et al.*, 1987; Tremblay *et al.*, 1988).

However, cGMP kinase activity is extremely low in liver, fibroblasts, adipose tissue, macrophages, cardiac and skeletal muscle, endothelial cells, neutrophils, and secretory tissues (Walter, 1988, 1989; Walter *et al.*, 1981). However, the presence of low levels of the enzyme in a particular cell does not necessarily exclude the possibility of important functions in that cell. Low levels of cGMP kinase have been demonstrated in adherent neutrophils through the use of Western blot analysis and immunofluorescence microscopy. In adherent neutrophils, cGMP kinase is localized to the euchromatin of the nucleus and the microtubule organizing center, as well as being diffusely distributed throughout the cytoplasm (Pryzwansky *et al.*, 1990). In response to chemotactic stimuli, the distribution of the enzyme within subcellular compartments exhibits a time-dependent shift (Wyatt *et al.*, 1991; Cornwell *et al.*, 1991). This suggests that the relocaton allows cGMP kinase to phosphorylate a target substrate(s) specifically in that microenvironment despite low levels of the kinase in that tissue (Wyatt *et al.*, 1991; Cornwell *et al.*, 1991). This suggests that the absolute amount of this enzyme in a tissue may be a poor indicator of its importance in regulating cellular processes.

III. Isozymes

Several forms of cGMP kinase have been reported in mammalian tissues and have been designated types I and II. Type I is a principally cytosolic form of the enzyme (Gill *et al.*, 1976; Lincoln *et al.*, 1977, 1988; Lincoln, 1983; Wolfe *et al.*, 1989b), and type II is a membrane-bound form, which was described originally in intestinal epithelial cells (DeJonge, 1981). Biochemical features of the type I cGMP kinase have been studied more extensively than those of type II. Two isoenzymes (types Iα and Iβ) of the cytoplasmic cGMP kinase have been purified and characterized in mammalian systems (Lincoln *et al.*, 1988; Wolfe *et al.*, 1989b). Either of these could also be present to some extent in particulate fractions. Separation of types Iα and Iβ is achieved by DEAE-Sephacel chromatography; the Iβ elutes at a significantly higher salt concentration (0.25 *M* compared with 0.15 *M* for type Iα) (Wolfe *et al.*, 1989b), consistent with greater surface electronegativity for this isoform. Both types Iα and Iβ are homodimers with two identical subunits of approximately 76 kDa each (Lincoln *et al.*, 1977; Lincoln and Corbin, 1983; Wolfe *et al.*, 1989b); type Iα is 670 amino acids in length (Takio *et al.*, 1984b) and is blocked at the amino terminus by an acetyl group. Type Iβ has 684 amino acids and also appears to have a blocked amino terminus (Francis *et al.*, 1988–1989; Wolfe *et al.*, 1989b; Sandberg *et al.*, 1989; Wernet *et al.*, 1989). These isoenzymes are identical in sequence from Ser-89 (type Iα) and Ser-104 (type Iβ) through the carboxyl terminus [Phe-670 (Iα) and Phe-684 (Iβ)], but differ in their amino termini where only 36% of the first 103 amino acids in type Iβ are identical to those in type Iα (Fig. 1). It is possible that these two distinct sequences in the isoforms provide for different functions or

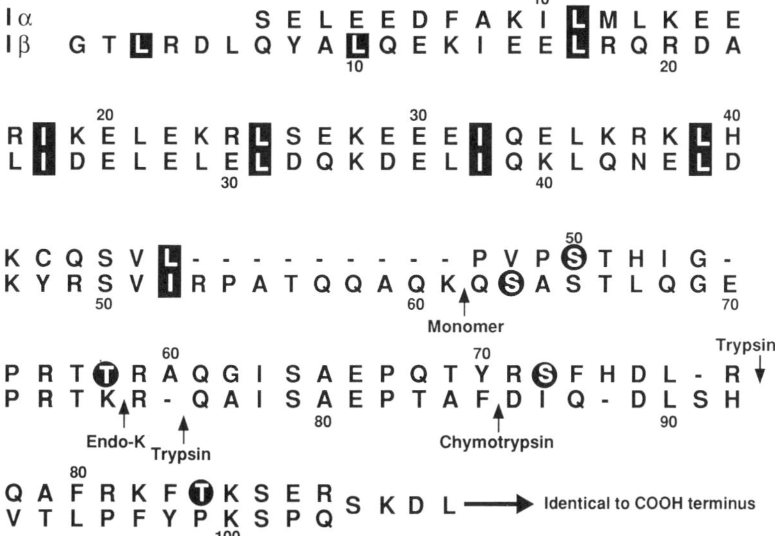

Fig. 1 Alignment of the amino acid sequences of amino terminal segments of bovine type Iα (1–90) and type Iβ (1–105) cGMP kinases. Leucine and isoleucine residues, which may assume a "leucine zipper" motif involved in the homo-dimerization of the respective monomers, are indicated by the darkened squares. The major phosphorylation sites (serines 50 and 72, and threonines 58 and 84) in type Iα cGMP kinase and serine 63 in type Iβ are denoted by the darkened circles. Positions in the sequence where partial proteolytic cleavage of the cGMP kinases by various endoproteases occurs are indicated by the arrows. (Endo-K, endoproteinase Lys-C; monomer denotes cleavage site in the type Iβ by an unknown endogenous protease during enzyme purification.

localizations within the cell. The amino acid sequence of the type Iα, originally determined by Takio *et al.* (1984b), has recently been confirmed by the nucleotide sequence of a partial cDNA derived from a bovine smooth muscle cDNA library for the type Iα isoform (Wernet *et al.*, 1989). Partial amino acid sequence has been directly determined for the bovine type Iβ (Francis *et al.*, 1988–1989; Wolfe *et al.*, 1989a) and is in agreement with that derived from a partial cDNA from bovine trachealis that encodes the sequence of the amino-terminal 293 amino acids of type Iβ, including the methionine at the initiation site (Wernet *et al.*, 1989). The primary sequence of human type Iβ (Sandberg *et al.*, 1989) differs from that of the bovine enzyme (Wernet *et al.*, 1989) at only two amino acids; Thr-280 and Ser-290 in the human type Iβ are replaced by Lys and Asn, respectively, in the bovine enzyme. Both type I isoforms have recently been transiently expressed in COS-7 cells (Ruth *et al.*, 1991). The type

II isozyme has recently been cloned using a mouse brain cDNA library, and it has been expressed in COS-7 cells (Uhler, 1993). The primary sequence of the type II cGMP kinase predicts a protein of 761 amino acids with a molecular mass of 87 kDa. The predicted type II kinase sequence has highest homology (~66%) with type I isoforms in the carboxyl-terminal catalytic domain and approximately 45% homology in the cGMP binding domain. Little homology is evident in the amino-terminal region that contains the autoinhibitory/autophosphorylation domain.

A. Type I Isoforms

Both type I isoforms bind two cGMP molecules per monomer of enzyme (MacKenzie, 1982; Corbin and Doskeland, 1983; Wolfe *et al.,* 1989b), and no differences in substrate specificities, catalytic rates, and immunoreactivities to polyclonal antibodies have been found. The relative tissue distribution of types Iα and Iβ has not been thoroughly investigated, but vascular smooth muscle from pig coronary arteries, human, bovine, and rat aorta contain roughly equal quantities of the two isoforms (Lincoln *et al.*, 1988; Wolfe *et al.*, 1989b; Sekhar *et al.*, 1992). Type Iα is the predominant isoform in bovine trachealis smooth muscle (70% Iα vs 30% Iβ), lung (>90% Iα), uterine smooth muscle, and cerebellum (Wolfe *et al.*, 1989b). The high level of cGMP kinase in bovine lung is thought to be largely derived from blood vessel smooth muscle in that tissue (Lincoln and Corbin, 1983), but greater than 90% is the type Iα isoform. This might suggest that the relative abundance of type Iα and type Iβ in vascular beds may vary considerably.

B. Type II Isoform

Type II cGMP kinase was originally identified in intestinal microvilli as an 86-kDa monomeric enzyme as estimated by SDS–PAGE (DeJonge, 1981). In this tissue, it is membrane-bound, and high salt and detergent are required for its solubilization. These characteristics are consistent with that of an intrinsic membrane protein in the microvillous. Membrane association occurs via a 15-kDa peptide that can be removed by limited proteolysis and which apparently contains an autophosphorylation site. In contrast, the type II cGMP kinase recently expressed in COS-7 cells is soluble (Uhler, 1993) despite its similarity in size and predicted isoelectric point to that of the rat intestinal enzyme (DeJonge, 1981). Differences in subcellular locations could relate to selective differences in tissue expression of anchoring proteins, such as cytoskeletal elements, for type II cGMP kinase.

Approximately 1 mol of cGMP is bound per mole of enzyme at a single class of high affinity cGMP-binding sites that do not show cooperativity. Type II, like type I, has ~100-fold greater affinity for cGMP than for cAMP and can be photoaffinity labeled by 8-N_3-[^{32}P]cIMP. Purified type II has a catalytic rate similar to that of the type I and, like the type I, undergoes slow autophosphorylation. However, endogenous autophosphorylation in brush border preparations is rapid. Rat intestinal type II cGMP kinase has an isoelectric point of pH 7.5 compared to 5.6 for the type Iα enzyme, and this difference in pI is maintained in the 81-kDa fragment released from the membrane by limited proteolysis (DeJonge, 1981). However, antibodies against the type Iα lung enzyme inhibit the catalytic activity of the intestinal type II isoform, consistent with similarities in structure between the two families of isoforms.

C. mRNA Size and Distribution

The mRNAs for type Iα and type Iβ have been studied in several species, and estimates of their sizes greatly exceed the length of the coding regions for the expressed enzymes (Sandberg *et al.*, 1989). The mRNAs for these isozymes have been identified and characterized in extracts of bovine, human, porcine, and rat tissues. In bovine trachealis, two species of mRNA of approximately 6.2 kb, representing the types Iα and Iβ, respectively, have been identified, but the mRNA level for type Iβ represents only 10% of that for the type Iα. Using Northern blot analysis, Sandberg *et al.* (1989) have identified a type Iβ cGMP kinase mRNA in human tissue estimated to be 7 kb. This is the predominant form in human uterus and placenta and has been shown to bind to two cDNA probes, the first of which is specific for the type Iβ and the second of which is capable of binding mRNA for both isoenzymes. The mRNA for cGMP kinase is very low in placenta. A 7-kb mRNA for cGMP kinase has also been identified in rat lung, cerebellum, adrenal gland, cerebrum, and kidney. A doublet of 7.5- and 6.5-kb mRNA is seen in rat heart, whereas no mRNA is detected in rat liver. In rat tissues, developmental changes in cGMP kinase mRNA occur between 5 to 30 days after birth with notable increases in cerebellum, a slight increase in cerebrum, and a decline in both the 7.5- and the 6.5-kb mRNA in heart, although the change in levels of the 6.5-kb species occurs much more quickly. cGMP kinase is detected in all of those tissues by Western blot analysis. A 6-kb mRNA for type II cGMP kinase has been observed in mouse tissues, including strong signals in brain and lung and a weak signal in testis (Uhler, 1993). In *Drosophila* two distinct genes (DG1 and DG2) that putatively encode for cGMP kinases have been identified (Kalderon and Rubin, 1989) and have been mapped to separate chromosomes (21D and 24D). Three mRNA products of varying lengths derived from transcription of the DG2 have been detected.

The type I isoenzymes have been proposed to be products of alternative splicing of mRNA (Fig. 2) (Francis *et al.*, 1988–1989), but it cannot be ruled out that they are products of separate genes. This hypothesis was originally based on the variability of amino acid sequences in the amino-terminal 100 amino acids of the enzymes (Fig. 1) (Francis *et al.*, 1988–1989; Wolfe *et al.*, 1989a; Wernet *et al.*, 1989; Sandberg *et al.*, 1989), compared to the identical sequences in the remainder of the molecules. The possibility of an alternative splicing mechanism is supported by several lines of evidence. First, at the junction of the amino-terminal segment of human type I with the sequence conserved (SKDLIKEAIL. .) between type Iβ (Ser-104) and type Iα (Ser-89), the nucleotide sequence of the type Iβ mRNA (Sandberg *et al.*, 1991) immediately preceding that coding for the Ser-104 (CAG) is the most common sequence at the 5′-end of an exon/intron junction (Shapiro and Senepathy, 1987). Second, the nucleotide sequence at the exon/intron junction of the H1 exon from the human type Iβ cGMP kinase gene has a corresponding lead sequence (CA); the third

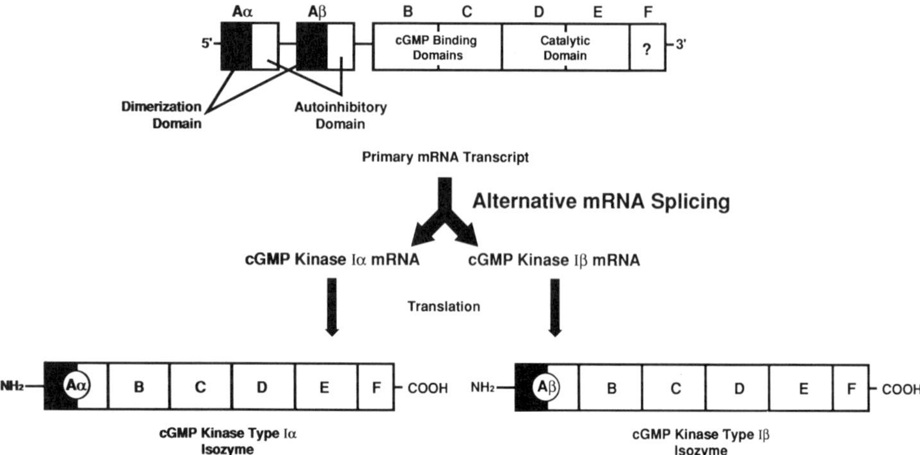

Fig. 2 Putative mechanism by which type Iα and type Iβ cGMP kinases are generated through alternative splicing of the mRNA. Aα and Aβ indicate the segments that encode the unique sequences of the amino termini of type Iα and type Iβ cGMP kinases and which, respectively, contain the dimerization domains (dark gray region) and autophosophorylation and autoinhibitory domains of both enzymes. The precise boundaries of these functional regions within the A segment are not known. The exact alignment of Aα and Aβ in the primary mRNA transcript is not known.

base in the codon would be G in order to code for the glutamine found at this position in type Iβ. Third, the corresponding sequence in the *Drosophila* cGK DG2 gene occurs at an exon/intron junction and is also consistent with the likelihood of an alternative splicing mechanism (Kalderon and Rubin, 1989). Thus, the Gln-103 in type Iβ and the Arg-88 in type Iα (Fig. 1) would be the last amino acids partially encoded (i.e., the first two bases in the codon) in the specific exons for each of the amino termini for the two isoenzymes.

D. Chromosomal Location/Exon–Intron Organization

In studies of the human genome, type I cGMP kinase(s) has been localized to chromosome 10 in the p11.2–q11.2 region (Sandberg *et al.*, 1991). These studies estimate the gene(s) for the type I cGMP kinase to be larger than 100 kb with a minimum of 8 exons. A human genomic library has been screened using restriction fragments derived from cDNA clones for type Iβ cGMP kinase, and 10 different clones (each ~15 kb in length) that contain sequences for 8 exons have been isolated. The 8 exons are derived from different regions in the gene(s), and 7 of these represent regions conserved between type Iα and type Iβ, with the eighth being specific for type Iβ. Kalderon and Rubin (1989), have reported the isolation of two genes (DG1 and DG2) from *Drosophila* that appear to be strongly related to the genes of mammalian cGMP kinases. The amino acid sequences deduced from these genes would encode chimeric proteins containing putative cGMP-binding and protein kinase domains on the same polypeptide chain. These predicted sequences have significant homology (>55%) with the bovine type I cGMP kinases except in the extreme amino terminus, which would include the dimerization/inhibitory domain. The protein products of these two cGMP kinase-like genes have not been detected, but the predicted products of the mRNAs from these genes would vary considerably in primary structure. Relatively recent gene duplication has been proposed for these two *Drosophila* genes since the intron/exon arrangement in the cGMP-binding and kinase domains is similar.

On comparison with the exon/intron organization of the *Drosophila* cGMP kinase (DG2) (Kalderon and Rubin, 1989), the human cGMP kinase gene contains at least four of the seven splice junctions (Sandberg *et al.*, 1991) found in *Drosophila*. The remaining three splice junctions have not been characterized in the human gene. However, despite significant similarities in the organization of the cGMP kinase genes between human and *Drosophila*, the human gene contains multiple exons in 2 regions that are encoded by only one exon in *Drosophila*. In contrast, comparison of the splice junction of the exons of the type I cGMP kinase to exons that have been characterized for the RIα and RIβ subunits of the cAMP kinase reveals no similarities in organization (Sandberg *et al.,* 1991).

IV. Nonmammalian Cyclic GMP Kinases

In addition to the putative cGMP kinases in *Drosophila* (Kalderon and Rubin, 1989), different forms of this enzyme have been identified in several other nonmammalian species, including nematodes (Hofer and Thalfofer, 1989), the slime mold *Dictyostelium discoideum* (Wanner and Wurster, 1990), *Paramecium* (Miglietta and Nelson, 1988), and grasshopper pupae (Vardanis, 1980). Some of these enzymes have proven to be monomeric. A monomeric cGMP kinase of 77–80 kDa has been purified from the cilia of *Paramecium* where it is present in unusually high concentration (approximately equal to that of cAMP kinase) (Miglietta and Nelson, 1988). This enzyme has a molecular weight similar to the cGMP kinase subunits found in mammalian tissues. Like the mammalian enzymes, this enzyme contains two kinetically distinct cGMP binding sites (a fast site and a slow site with k_d values of $5–10 \times 10^{-3}$ s^{-1} and 0.44×10^{-3} s^{-1}, respectively), but the binding sites do not show cooperativity (Hill coefficient = 0.8–1.1). The cGMP analog specificity of the binding is reported to be similar to that reported for type Iα, and, like the mammalian enzyme, the cGMP kinase from *Paramecium* is slowly autophosphorylated. Histone II-A mixture serves as a good substrate for this form, and both threonine and serine residues are phosphorylated. However, the enzyme differs significantly from the mammalian form since both ATP and GTP can be used as phosphate donors in the phosphotransferase catalysis, and Kemptide is phosphorylated poorly. The presence of a cGMP kinase in *Paramecium,* which contains the two cGMP-binding sites and the catalytic domain in one protein, suggests that the proposed fusion of the genes encoding the domains with these functions occurred relatively early in evolution. The monomeric cGMP kinase appears to predominate in lower species, and the functional significance of the dimerization of the mammalian cGMP kinase is not clear. In fact, a monomeric form of the bovine type Iβ enzyme (Wolfe *et al.,* 1989a) generated by partial proteolysis retains most of the salient functional features of the dimer (see below).

V. General Structure

The mammalian cGMP kinases are highly asymmetric proteins composed of two identical monomers. Each monomer in the dimeric cGMP kinase (types Iα and Iβ) contains at least five types of functional domains (Fig. 3): (*a*) a dimerization domain (A_1) located at the extreme amino terminus; (*b*) an autophosphorylation (A_2)/autoinhibitory (A_3) domain that lies just carboxyl terminal to the dimerization domain and includes the multiple autophosphorylation sites; (*c*) two cGMP-binding sites (B and C) arranged in tandem; (*d*) the catalytic domain (D and E); and (*e*) a carboxyl-terminal domain (F) of unknown function. cGMP binds to the two homologous binding sites on the kinase [a high affinity site from which cGMP dissociates slowly (slow site) and a site from which cGMP dissociates rapidly (fast site)] (Corbin and Doskeland, 1983; Wolfe *et al.*, 1989b).[1] The binding of cGMP to the kinase induces a conformational change that relieves the inhibition of the catalytic site. In contrast to the cAMP kinases (Cobb *et al.*, 1987), the type Iα isoform is partially active when only one cGMP is

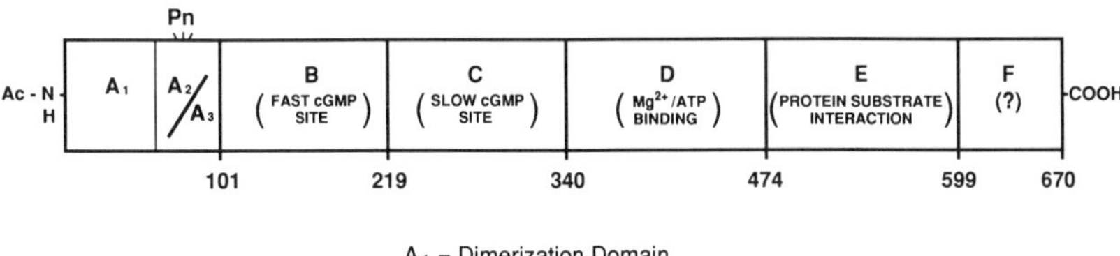

A_1 = Dimerization Domain
A_2 / A_3 = Autophosphorylation sites/ Autoinhibitory Region
Pn = Multiple Sites of Autophosphorylation

Fig. 3 Segmental depiction of established functional domains in type Iα and type Iβ cGMP kinase. Amino acid residue numbers representing the somewhat arbitrary boundaries of the various domains are taken from the description provided by Takio *et al.* (1984b) and refer to the type Iα cGMP kinase. Comparable numbers for the type Iβ cGMP kinase would be 116, 234, 355, 489, 614, and 685. The "A" segment is subdivided as A_1 and A_2/A_3 to reflect a partial separation of functions within this region. A portion of the region in A_1 is critical for dimerization. Multiple sites for autophosphorylation indicated by P_n in both type Iα and type Iβ cGMP kinase occur throughout the A_2/A_3 region and overlap with the minimal amino acid sequence for maintaining full cGMP dependency, i.e., the autoinhibitory domain. The full potency of the intact autoinhibitory domain (A_3) is likely to involve more extensive structural features throughout this region. B and C domains represent the "fast" and "slow" cGMP binding sites also known as sites A and B, as sites 2 and 1, and as the low affinity and the high affinity cGMP-binding sites, respectively. Segments D and E constitute the catalytic domain with ATP/Mg^{2+} binding primarily in segment D and interaction with the protein substrate occurring in segment E. The function of segment F is unknown.

[1] The high affinity "slow" cGMP-binding site and the low affinity "fast" cGMP-binding site have also been described as site 1 and site 2 (Corbin and Doskeland, 1983), sites I and II (MacKenzie, 1982), and sites B and A (Doskeland *et al.*, 1987).

bound per subunit (at the slow site) (Wolfe *et al.*, 1987), but expression of full catalytic activity requires that both binding sites on each subunit be filled. However, the division of the enzyme structure into discrete functional domains as depicted in Fig. 3 is an overly simplistic approach since the manner in which a segment functions is strongly influenced by the myriad of interactions with other regions in the enzyme molecule.

VI. Microheterogeneity

In addition to multiple isozymic forms, the cGMP kinase has been shown to exhibit heterogeneity due to covalent modifications of the enzyme and due to the interactions of the enzyme with cGMP. Wolfe *et al.* (1987) have shown that binding of one cGMP per type Iα monomer shifts the elution position of the kinase on DEAE-Sephacel chromatography, consistent with the induction of a conformational change in the enzyme that produces increased surface electronegativity. Since binding of cGMP is positively cooperative (Corbin and Doskeland, 1983; Doskeland *et al.*, 1987), the presence of one cGMP per subunit would not only produce partial activity but would also increase the rapidity of binding of the second cGMP, thereby facilitating full and rapid activation of the catalytic activity. Based on measured intracellular cGMP concentrations in smooth muscle, it is probable that the type Iα enzyme exists in a partially active state under most physiological conditions (Francis *et al.*, 1988). However, the extent of phosphorylation of specific cGMP kinase substrates would depend on a balance between the levels of kinase activity and phosphoprotein phosphatase activity, respectively.

A second type of microheterogeneity in both isoenzymes of cGMP kinase is brought about by autophosphorylation (Foster *et al.*, 1981) of multiple sites (Aitken *et al.*, 1984) in the region containing the putative inhibitory domains of the enzymes (residues 40–100) (Fig. 1). Variation in the extent of autophosphorylation in the population of enzyme molecules could affect enzymatic function (Hofmann and Flockerzi, 1983; Hofmann *et al.*, 1983, 1985; Landgraf *et al.*, 1986; J. A. Smith *et al.*, 1992). The characteristics of autophosphorylation are similar to those for other catalytic functions of this enzyme (K_m for ATP = 10 μM, and for Mg^{2+} = 2–5 mM), and the rate of phosphorylation of either type Iα (Hofmann and Flockerzi, 1983; J. A. Smith *et al.*, 1992) or type Iβ (J. A. Smith *et al.*, 1992) is increased in the presence of either cGMP or cAMP. However, the rate of autophosphorylation is very slow and requires several hours at 30° for the incorporation of 1–3 phosphate groups. The type Iα cGMP kinase purified from bovine lung contains 1.4 mol phosphate per mole of enzyme monomer (Hofmann and Flockerzi, 1983), but the specific location(s) of this preexisting phosphate has not been determined. This endogenous phosphate is acid stable and base labile, suggesting that a serine or threonine residue(s) is modified. The catalytic domain of cGMP kinase contains a threonine (Thr-516) in the position homologous to that of threonine-phosphate (Thr-197) in the cAMP kinase catalytic subunit. It is possible that a major portion of the endogenous phosphate in cGMP kinase could be located at this position. As suggested for cAMP kinase (Knighton *et al.*, 1991b), this phosphorylated threonine may be important in the maintenance of structural features of the protein and may also be important for high affinity interactions between the catalytic domain and the regulatory domain (Levin and Zoller, 1990).

In addition to the phosphate in the catalytic domain, four major sites of autophosphorylation in the amino terminus of type Iα have been identified (Fig. 1); these include serines 50 and 72, and threonines 58 and 84 (Aitken *et al.*, 1984). Two other sites (serines 1 and 64) are modified to a very limited extent. Threonine-58 of type Iα is phosphorylated most rapidly in the presence of either cGMP or cAMP. In the presence of cAMP, autophosphorylation modifies serines 50 and 72, and threonine 84, in addition to threonine 58, and concomitantly increases the affinity of the enzyme for cAMP 10-fold, as well as increasing the basal activity (-cGMP) of the enzyme (Foster *et al.*, 1981; Hofmann and Flockerzi, 1983); there is no change in the K_a for cGMP. The K_d for cAMP of the slow binding site is decreased (1.9 to 0.17 μM) (Landgraf *et al.*, 1986), but there is no measurable effect on the fast site. Autophosphorylation abolishes the inhibitory effects of Mg^{2+}/ATP on the affinity of cGMP binding to the slow site (Hofmann and Flockerzi, 1983; Hofmann *et al.*, 1985).

Type Iβ is autophosphorylated at multiple serines in the amino-terminal region corresponding to the autophosphorylation domain of type Iα, and Ser-63 is the primary site phosphorylated in the presence of cGMP or cAMP (S. Francis, unpublished results). Autophosphorylation of type Iβ in the presence of cGMP or cAMP elevates the basal activity of the enzyme and increases the affinity for both cGMP and cAMP (J. A. Smith *et al.*, 1992).

The phosphorylation sites on either isozyme are apparently not phosphorylated by a 10-fold excess of the purified catalytic subunit of the cAMP kinase (Hofmann and Gensheimer, 1983; J. A. Smith *et al.*, 1992), which suggests that these sequences are not recognized as substrates by the catalytic subunit. Furthermore, the observation that the monomeric cGMP kinase undergoes autophosphorylation (Wolfe *et al.*, 1989a) and the kinetics of the autophosphorylation reaction for the native enzyme support an intramolecular process (J. A. Smith *et al.*, 1992). In the latter instance, it is not possible to determine whether autophosphorylation could occur between monomers within a given dimeric structure. The demonstrated increase in the affinity of both isoenzymes for cAMP upon autophosphorylation would facilitate cAMP cross-activation of the cGMP kinase in response to cAMP-elevating agents in certain cells (see below). Thus, autophosphorylation might be a mechanism whereby the cGMP kinase is "sensitized" to activation by cAMP in the cell.

Thus, it is clear that, in addition to the different isoenzymes, the cGMP kinase within a cell may exist in several forms, including the nucleotide-free enzyme, enzyme partially saturated with bound cyclic nucleotide, enzyme with saturated cyclic nucleotide binding sites, and phospho-/dephospho-forms.

VII. Domain Structures and Functions

cGMP kinase can be divided into discrete domains that provide for specific functions of the enzyme, and the amino acid sequences of these regions are homologous to related functional domains in other kinases. In the original analysis of the primary structure (Takio *et al.*, 1984b), the sequence was divided into six segments corresponding to functional features in the enzyme (Fig. 3). However, a particular function may not be completely confined to a specific region but may include or be heavily influenced by other domains. The catalytic domain that catalyzes the transfer of the γ-phosphate of ATP to a serine or threonine of a protein (or peptide) is located in the carboxyl-terminal portion of the protein. The amino acid sequence of this catalytic domain is highly homologous (43% identity) to that of the catalytic subunit of cAMP kinase (Takio *et al.*, 1984b) and exhibits homology with the catalytic domains of all mammalian protein kinases that have been sequenced thus far, including tyrosine protein kinases (Edelman *et al.*, 1987; Hanks *et al.*, 1988; Taylor, 1989; Taylor *et al.*, 1990).

The cGMP-binding domains are also highly homologous to each other (Takio *et al.*, 1984a,b) and to the cAMP-binding domains of the regulatory subunit of cAMP kinase, the cAMP-binding domain of the catabolite gene activator protein (CAP) from *Escherichia coli* (Weber *et al.*, 1982), the olfactory cyclic nucleotide-gated channel (Ludwig *et al.*, 1990), and the cGMP-gated cation channel from photoreceptors (Kaupp *et al.*, 1989). Evolutionary relationships among the various cyclic nucleotide binding domains have recently been analyzed (Shabb and Corbin, 1992). The dimerization and inhibitory domains of the cGMP kinases are less conserved and show weak homology with these domains of cAMP kinase.

Analysis of the structure of cGMP kinase by far-ultraviolet circular dichroism shows the enzyme to contain both α-helical and β-pleated sheet components (Landgraf *et al.*, 1990). cGMP binding causes an increase in the amount of β-pleated sheet and a reduction in the amount of random coil. The crystal structure for the catalytic subunit of cAMP kinase has recently been determined (Knighton *et al.*, 1991a,b). This structure will be valuable in analyzing the biochemical similarities and differences in the catalytic functions of these two closely related kinases.

A. Dimerization Domain

Despite major differences in the primary structures of types Iα and Iβ isoenzymes in the first ~100 amino-terminal amino acids, this region contains critical functional components, including the dimerization domain, the autoinhibitory domain, and the autophosphorylation sites for both isozymes (Fig. 3). The exact sequence required for dimerization is not known. Chymotryptic cleavage of type Iα generates a 65-kDa monomeric kinase and a 16-kDa fragment that is linked by a disulfide bridge at Cys-42 (Monken and Gill, 1980). A disulfide bond in the dimerization fragment might suggest its involvement in dimer formation, but the dimeric type Iβ cGMP kinase lacks cysteine in the amino-terminal segment (Fig. 1). A monomeric form of type Iβ that is missing the amino-terminal 61 amino acids (Fig. 1) further defines the limits of the dimerization domain (Wolfe *et al.*, 1989a).

Studies of this region in type Iα using ^{1}H NMR spectroscopy and circular dichroism suggest that a leucine zipper motif formed by 6 repeating heptads involving leucines and isoleucines extends from Leu-11 through Leu-46 of type Iα (Atkinson *et al.*, 1991); in type Iβ this alignment is longer and may involve 7–8 repeating heptads extending from Leu-3 through Ile-52 (Fig. 1). The synthetic peptide corresponding to amino acids 1–39 of this region in type Iα has a strong α-helical content (75–80%) that is stable under a wide range of conditions (Atkinson *et al.*, 1991). A modeled structure for this "dimerization" peptide suggests that the peptides (and by extrapolation, the monomers of cGMP kinase) interact along a hydrophobic face formed by the repeating pattern of leucine/isoleucine residues. Based on modeling of this peptide, monomers in the dimeric cGMP kinase are suggested to align in a parallel (head-to-head) arrangement. This is further supported by the fact that the alignment of the Cys-42 from each monomer is sufficiently close for formation of a disulfide bond (Atkinson *et al.*, 1991). Thus, structural and functional features in the amino-terminal region may be retained for the isozymes despite minimal primary sequence homology. Likewise, the predicted amino acid sequence for type II cGMP kinase (Uhler, 1993) includes a leucine zipper motif of at least 8 heptad repeats in this region. By analogy with type I cGMP kinase this would predict a dimeric structure. The tertiary structure of the soluble type II enzyme has not been determined, but in intestinal epithelial cells, type II is reportedly monomeric. Therefore, the tertiary structure of type II GMP kinase awaits resolution.

The amino acid sequence deduced from the *Drosophila* DG2-T1 gene (Kalderon and Rubin, 1989) contains a similar leucine/isoleucine zipper motif (5 repeating heptads) and predicts that the expressed enzyme would be dimeric. The predicted sequence for this region in the DG1 gene product contains only 3 repeating heptads of leucine residues and 3 other repeating positions are composed of valines. This motif is not present in the regulatory subunits of the cAMP kinase (Takio *et al.*, 1984a; Titani *et al.*, 1984), suggesting that the cGMP kinases have developed a different strategy to achieve dimerization. No heterodimers of type Iα and type Iβ subunits have been identified despite the presence of a putative leucine/isoleucine zipper in each.

Although the extreme amino terminus may account for a major portion of the elements that stabilize dimerization, there are likely to be additional points of contact between the monomers. DEAE chromatography of aged type Iβ cGMP kinase containing both intact enzyme and enzyme missing the 64 amino-terminal amino acids does not separate the proteolyzed fragment from the intact enzyme, as would be predicted by the electronegativities of the respective fragments (S. H. Francis, unpublished data). Similar behavior has been observed for proteolyzed fragments of type IIβ regulatory subunits of cAMP kinase on gel filtration chromatography in the presence of 50 m*M* potassium phosphate (J. D. Corbin, unpublished results). Thus, stabilization of a dimeric structure may involve the amino terminus as well as sites distant from that region.

B. Autoinhibitory Domain

The region of the cGMP kinase just carboxyl terminal to the dimerization domain contains the autophosphorylation sites of the enzyme, A_2, and the "autoinhibitory" domain (A_3) (Fig. 3). This region is poorly understood with regard to the specific residues that account for the inhibition of catalytic activity, the length of sequence required to effectively block catalysis, and the specificity determinants required for autophosphorylation at the various sites.

Recent observations suggest that autoinhibition of the catalytic sites of protein kinases to produce and maintain the latent form of the enzyme involves at least two interactions. These include (*a*) the interaction of the catalytic site with a region within the inhibitory domain that contains a short primary sequence having a substrate-like motif and (*b*) interaction with regions exhibiting a higher order of structure. Studies with the bovine type II regulatory subunit (R_{II}) of the cAMP kinase led Corbin *et al.* (1978) to propose that inhibition of the catalytic activities of the cAMP kinase and cGMP kinase is produced in part by a substrate-like sequence (RRXSX) in the inhibitory domain of the regulatory component of either protein. The substrate-like sequences in the regulatory component could directly compete with substrates for the catalytic site and thereby inhibit catalytic activity. In the case of bovine type II regulatory subunit (R_{II}) of cAMP kinase, the serine at the site of interaction (RRVSV) can be phosphorylated, but in the type I isoform, the phosphorylatable serine is replaced in the consensus sequence by an alanine (RRGAI), thus producing a "pseudo-substrate" sequence (Robinson-Steiner and Corbin, 1986; Titani *et al.*, 1984). The inhibitory capacity of a given sequence might be strengthened by the veracity with which the preferred substrate site sequence is duplicated since the affinity for the initial interaction should be increased (Corbin *et al.*, 1978; Hardie, 1988; Soderling, 1991; Kemp and Pearson, 1990, 1991). This concept of kinase regulation has been extended to encompass most serine/threonine kinase regulatory mechanisms (Edelman *et al.*, 1987; House and Kemp, 1987; Kemp and Pearson, 1990).

Studies of bovine R_{II} (Flockhart *et al.*, 1980) suggest that a structure(s) other than the substrate-like sequence is also critically involved in autoinhibition and may include the cyclic nucleotide binding sites. Interaction of the catalytic center with the consensus substrate sequence may be an important step, perhaps even the initial contact, in autoinhibition by fostering a contact between catalytic and regulatory domains, but full inhibition may be effected by an additional interaction between other regions of the proteins. Heat-denaturation studies with R_{II} show a close correlation between the loss of R_{II} binding of cAMP and the progressive decrease in its inhibitory potency for the catalytic subunit. However, the rate and extent of phosphorylation of the R_{II} are unaffected; i.e., the ability of the catalytic subunit to recognize and phosphorylate the substrate-like region in the regulatory subunit is unaltered. These studies emphasize that components of the secondary or tertiary structure of the R_{II} must account for a major portion of the inhibitory capacity exhibited toward the catalytic subunit. It is also possible that the inhibitory domain competes with the ATP- as well as with the substrate-binding site of the catalytic domain. Competition for the ATP-binding site is a part of the mechanism of the inhibitory domain of calcium/calmodulin-dependent protein kinase (Colbran *et al.*, 1989; M. K. Smith *et al.*, 1992).

1. Autophosphorylation Sites

If the mechanism of autoinhibition for cGMP kinase is similar to that for cAMP kinase, then a substrate-like sequence should exist in the autoinhibitory domain. The substrate sequence preferred by cGMP kinase has been studied extensively and is similar to that for the cAMP kinase (i.e., two basic amino acids linked amino-terminally to the phosphorylatable serine by a hydrophobic residue) (Lincoln and Corbin, 1977; Glass and Krebs, 1979, 1982; Glass and Smith, 1983; Roskoski *et al.*, 1987; Glass, 1990; Kennelly and Krebs, 1991; Colbran *et al.*, 1992) and recently reviewed in detail (Glass, 1990). However, substrate-like sequences in the putative autoinhibitory domains for the type I cGMP kinases have been difficult to identify on this basis. In contrast to type II cAMP kinase, which autophosphorylates at one specific serine of a typical substrate consensus sequence of its autoinhibitory domain, the catalytic site of cGMP kinase autophosphorylates multiple sites of its autoinhibitory region (Fig. 1), suggesting a less rigid type of interaction. In type Iα, the multiple phosphorylation sites have been determinated (Ser-50, Thr-58, Ser-72, Thr-84) (Aitken *et al.*, 1984), and none of these sites contain the "optimal" substrate consensus sequence (see Fig. 8). One of the minor sites of phosphorylation (Thr-84) is preceded amino terminally by a sequence containing two basic amino acids (Arg–Lys); Ser-72 has an arginine immediately adjacent, and Ser-50 has no basic amino acids nearby (within 8 residues on its amino-terminal side and 6 amino acids on its carboxyl-terminal side). Clearly, identification of phosphorylation site sequences that are involved in autoinhibition through recognition of typical consensus sequences will prove difficult in some instances. A synthetic peptide whose sequence encompasses the Thr-58 is a very poor substrate for cGMP kinase (Glass and Smith, 1983), suggesting that the interaction of the cGMP kinase catalytic site with this region in the protein is not dictated by optimum substrate recognition. Lastly, synthetic peptides that mimic pseudo-substrate sequences from cAMP kinase and putative pseudo-substrate sequences for cGMP kinases are weak inhibitors of catalysis (Glass and Smith, 1983; Kemp *et al.*, 1987, 1989; Francis *et al.*, 1992).

2. Cyclic GMP-Dependent Monomeric Protein Kinases

Using cGMP kinase isolated from silkworm pupae, Inoue *et al.* (1976) could proteolytically cleave the enzyme with trypsin to separate the catalytic domain from the cGMP-binding domain. The 34-kDa catalytic fragment was constitutively active, thus indicating that the catalytic and regulatory features of the cGMP kinase are separable. Subsequent studies by Heil *et al.* (1987) have demonstrated that limited trypsin treatment of the bovine type Iα cleaves the enzyme carboxyl terminal to Arg-77 (Fig. 1) and generates a 65-kDa monomeric fragment that is fully active in the absence of exogenous cGMP. This observation suggests that the autoinhibitory domain is localized to the first 77 amino acids in the amino-terminal sequence. It was thought at the time that the cGMP kinase, like cAMP kinase, had an inhibitory domain on one subunit that interacted with the catalytic domain of the other subunit in the pair to block catalysis. However, recent work with the type Iβ isoform suggests otherwise. During purification of the bovine aorta type Iβ isoform, a cGMP-dependent monomeric species of type Iβ is produced by cleavage between Lys-61 and Glu-62 by an endogenous protease (Fig. 1) (Wolfe *et al.*, 1989a). Subsequently, a cGMP-dependent monomer produced by aging of type Iβ has been sequenced and shown to start at Ser-65. The cGMP-dependencies of these monomeric enzymes support a revised structural model for cGMP kinase (Fig. 4) wherein the catalytic site is inhibited by the autoinhibitory domain of the same polypeptide chain. This model is also revised from previous models to reflect the predicted symmetrical alignment of the monomers along the leucine zipper of the dimerization domain (Atkinson *et al.*, 1991).

The specific segment of the sequence that provides for autoinhibition of the catalytic site is not known for either type Iα or type Iβ (Heil *et al.*,

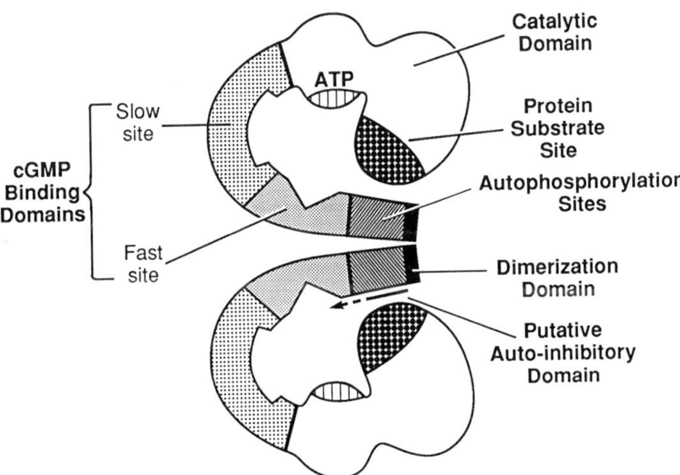

Fig. 4 Model of cGMP kinase dimer. Proposed new model for structure of dimeric cGMP kinase demonstrating that autoinhibition of the catalytic site is due to interaction with the autoinhibitory domain on the same polypeptide chain. The parallel arrangement of the monomers reflects the putative leucine/isoleucine zipper, which may provide a significant portion of the dimerization energy.

1987; Wolfe *et al.*, 1989a; Francis *et al.*, 1992). Limited endoproteinase K cleavage of type Iβ between Lys-74 and Arg-75 (Fig. 1) produces an approximately 65-kDa fragment that again is fully dependent on cGMP for activation (10- to 12-fold stimulation) (Francis *et al.*, 1992). However, limited trypsin treatment cleaves type Iβ between Arg-75 and Glu-76 to produce a fragment that is partially dependent on cGMP (1.7- to 3.3-fold stimulation). Partial digestion of type Iβ with chymotrypsin cleaves the type Iβ sequence carboxyl terminal to Phe-85 (Fig. 1) and generates a monomer that is still partially cGMP-dependent (1.4-fold stimulation). Thus, the main components of the autoinhibitory domain must be located carboxyl terminal to Lys-74 with significant contributions coming from the region carboxyl terminal to Phe-85. The absence of two juxtaposed basic residues in the remaining sequence constituting the putative autoinhibitory domain (the sequence preceding the cGMP-binding domains) (Fig. 1) suggests that this feature may not be so crucial for inhibition of the cGMP kinase, as is apparently the case for cAMP kinase. There is little amino acid sequence homology between the types Iα and Iβ throughout this region except for the sequence ISAEP (Iα 62–66 and Iβ 78–82), which could prove to contain important elements of the autoinhibitory domain, although in type Iα the Ser-63 is a very minor site of autophosphorylation. Endoproteinase K treatment of autophosphorylated Iβ also removes a major autophosphorylation site (Ser-63), suggesting that, although potentially important, these do not comprise critical elements for inhibition of the catalytic site.

The studies described above, among others, suggest that the inhibition of the catalytic activity of cGMP kinase, like the cAMP kinase, may depend in large part on a conserved component of secondary structure rather than on particular substrate-like sequence of amino acids in this region. Although the kinase activities of these various fragments of cGMP kinase are cGMP-dependent, the affinity with which the catalytic site and the inhibitory domain interact could be diminished. Such a decrease in the affinity of interaction would be difficult to quantitate since cGMP kinase inhibitory and catalytic elements do not physically dissociate upon activation, as occurs with cAMP kinase. In a similar vein, studies with smooth muscle myosin light-chain kinase suggest that the entire pseudosubstrate sequence is not necessary for inhibition of the enzyme, but its presence greatly improves the potency of the interaction (Pearson *et al.*, 1991). The evolutionary pressure to conserve rigidly a high affinity substrate-like site in the cGMP kinase autoinhibitory domain may have been less intense than in the cAMP kinase regulatory subunit. The physical proximity of the catalytic and autoinhibitory domains within the same polypeptide chain would immeasurably enhance the likelihood of association.

3. Interaction of the Autoinhibitory/Autophosphorylation Domain with Other Domains in cGMP Kinases

On binding of cGMP, the inhibition of type Iα or type Iβ by their distinctly different autoinhibitory domains is relieved, but there is ample evidence for complex interactions between this region and other domains of the kinase molecule. This assertion is supported by several observations. First, the cyclic-nucleotide binding domains of both of the isoenzymes are identical in amino acid sequence (Sandberg *et al.*, 1989; Wernet *et al.*, 1989) and would be presumed to possess the same secondary structure. However, as shown in Table I, the affinity of type Iβ for cGMP is somewhat lower than that of type Iα (K_a = 250 nM for Iβ vs K_a = 110 nM for Iα at 30°) (Corbin *et al.*, 1986; Wolfe *et al.*, 1989b; Sekhar *et al.*, 1992). Second, and more notably, however, is the weak activation of type Iβ by most cGMP analogs derivatized at the 8-position of the guanine ring, whereas these same compounds bind more tightly to the type Iα isoform than does cGMP itself. For instance, 8-(2-aminophenylthio)-cGMP (Table I) is an excellent activator of the type Iα isoform (K_a = 7 nM), but for the type Iβ the K_a is 195-fold higher for this compound, despite the fact that its binding sites are identical in sequence to those of type Iα. A number of other cGMP analogs show similar patterns of selectivity between type Iα and type Iβ, but to a lesser degree (Wolfe *et al.*, 1989b; Sekhar *et al.*, 1992). These results suggest that the amino termini of the cGMP kinases either (*a*) induce indirect conformational changes in the binding sites that

Table I

Potencies of cGMP Analogs as Activators of cGMP Kinases

	K_a (nM)	
Cyclic nucleotide	Type Iα	Type Iβ
cGMP	110	250
8-I-cGMP	9	122
8-Br-cGMP	26	210
1,N^2-PET-cGMP	26	20
8-Br-1,N^2-PET-cGMP	13	9
8-(4-OH-Ph-*S*)-1,N^2-PET-cGMP	17	23
8-(4-OH-Ph-*S*)-cGMP	50	440
8-(2,4-Di-OH-Ph-S)-cGMP	5	360
8-(2-NH_2-Ph-*S*)-cGMP	7	1370
8-Br-1,N^2-β-NET-cGMP	54	3000

alter the ability to bind nucleotides or (*b*) directly contact the binding sites to cause steric or physical constraints that affect the ability to bind cGMP or cGMP analogs (Francis *et al.*, 1988–1989).

C. Cyclic GMP-Binding Domains

1. Characteristics of cGMP Binding

Two tandem cyclic nucleotide-binding domains comprising approximately 120 amino acids each are located amino terminal to the catalytic domain (Fig. 3) (Takio *et al.*, 1984b). Occupation of both binding sites is required for full activation, but partial activation of catalysis occurs when

$$\underset{\text{(inactive)}}{E_2} + 2\ \text{cGMP} \leftrightarrow \underset{\text{(50\% active)}}{E_2\text{-cGMP}_2} + 2\ \text{cGMP} \leftrightarrow \underset{\text{(fully active)}}{E_2\text{-cGMP}_4}$$

cGMP binds to only one site (slow site) (Corbin and Doskeland, 1983; Wolfe *et al.*, 1987). In type Iα the two cyclic nucleotide-binding sites extend from amino acid 102–219 (fast site) and 220–340 (slow site), and the amino acid sequences of the two sites are similar (28% homology) to each other. Duplication of the ancestral gene encoding the cyclic nucleotide-binding sites (Takio *et al.*, 1984b) appears to have preceded the subsequent divergence of the cGMP kinase and the cAMP kinase (Fig. 5) (Shabb and Corbin, 1992). The fast sites in cGMP kinase and cAMP kinase are generally more homologous to each other than to the slow sites in each of the respective proteins, and, conversely, the same relatedness is present in the slow site binding domains of the two proteins. Structural homology translates into functional homologies since cGMP kinase binds and is activated by cAMP, albeit with a 50- to 200-fold lower affinity. Likewise, the cAMP kinase can be activated by cGMP at a concentration 80-fold greater than that required for activation by cAMP (Corbin and Lincoln, 1978; Miller *et al.*, 1981; Ogreid *et al.*, 1985).

2. Structural Features of cGMP-Binding Sites

The structures of the cyclic nucleotide-binding sites have been predicted using computer modeling based on the known three-dimensional structure of the homologous bacterial protein, CAP (Fig. 6) (Weber *et al.*, 1982, 1989). The overall structures of the cyclic nucleotide-binding sites in CAP, cGMP kinases, and cAMP kinase are similar despite differences in cyclic nucleotide specificities and binding affinities (Shabb *et al.*, 1990, 1991; Shabb and Corbin, 1992). The general features of the binding sites include three α-helices and an eight-stranded, anti-parallel β-barrel that in combi-

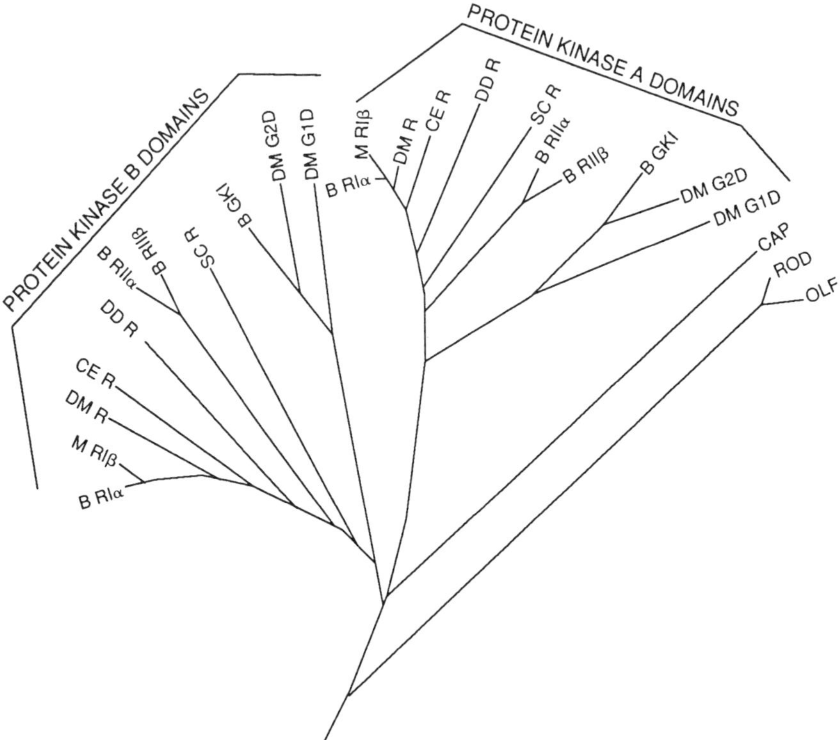

Fig. 5 Phylogenetic tree of cyclic nucleotide-binding domains related to those found in cGMP-dependent protein kinase. A distance matrix tree was constructed using programs designed by Feng and Doolittle (1990) and run on a VAX 3400 microcomputer. Analyses were done with the amino acid sequences and alignments previously described (Weber *et al.*, 1989; Shabb and Corbin, 1992), except the carboxyl termini of all sequences were arbitrarily truncated to match the shortest cyclic nucleotide-binding domain (yeast R subunit B domain). Percentage standard deviation of branch lengths was 6.6. The branching order was similar to a tree of the same sequences generated by a parsimony algorithm (Shabb and Corbin, 1992) except the yeast B domain branched after the B domains of the type II R subunits. cAMP kinase R subunits: B RIα, bovine RIα; MRIβ, murine RIβ; DM R, *Drosophila melanogaster* R; CE R, *Caenorhabditis elegans* R; DD R, *D. discoideum* R; RIIα, bovine RIIα; B RIIβ, bovine RIIβ; SC R, *Saccharomyces cerevisiae* R. cGMP kinases: B GKI, bovine cGMP kinase I; DM G1d and DM G2D, *D. melanogaster* G1D and G2D. Ion channels: ROD is bovine rod photoreceptor cGMP-gated channel and OLF is bovine olfactory epithelium cyclic nucleotide-gated channel.

nation with the αC-helix forms a pocket in which the cyclic nucleotide binds. All of these binding sites have retained six invariant amino acids that apparently provide key elements for the binding of cyclic nucleotides. These include 3 glycine residues that are thought to play structural roles in the sites, an alanine of unknown function, an arginine that interacts with an exocyclic phosphate oxygen, and a glutamic acid that forms a hydrogen bond with the ribose 2′-OH (Figs. 6 and 7). Since these residues are present in all of the CAP-related cyclic nucleotide-binding sites, the specificity determinants that dictate the preference of the cGMP kinase for cGMP over cAMP must reside at other positions in the sites. Both binding sites in cGMP kinase bind cGMP with strong specificity and with high affinity. Even though cyclic nucleotides exist in both *syn* and *anti* conformations in solution, cGMP analog studies suggest that the cGMP kinase binds the nucleotides in the *syn* conformation.

A modeled structure of the "cGMP-binding pocket" of the cGMP-binding sites is shown in Fig. 7. In both cGMP- and cAMP-binding sites of the respective kinases, the cyclic phosphate moiety of the nucleotide is a critical feature providing for high affinity binding to the enzymes since

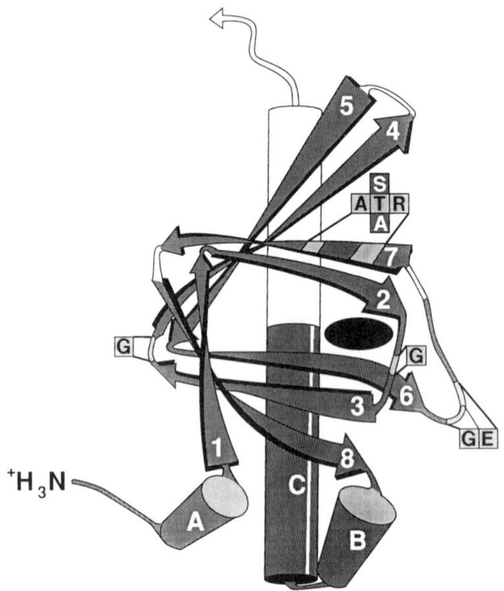

Fig. 6 Cartoon depiction of structural features common to CAP-related cyclic nucleotide-binding proteins such as cGMP kinase. The cGMP is indicated by the darkened elipse at the center right of the structure. The α-helices are denoted as A, B, and C, and β-strands (1–8) are indicated by the arrows. The threonine providing for cGMP selectivity by the two cGMP binding sites in cGMP kinase resides in β-strand 7 and is flanked by an alanine and an arginine, which are invariant in all CAP-related cyclic nucleotide-binding proteins. The threonine is an alanine in cAMP kinases and a serine residue in CAP.

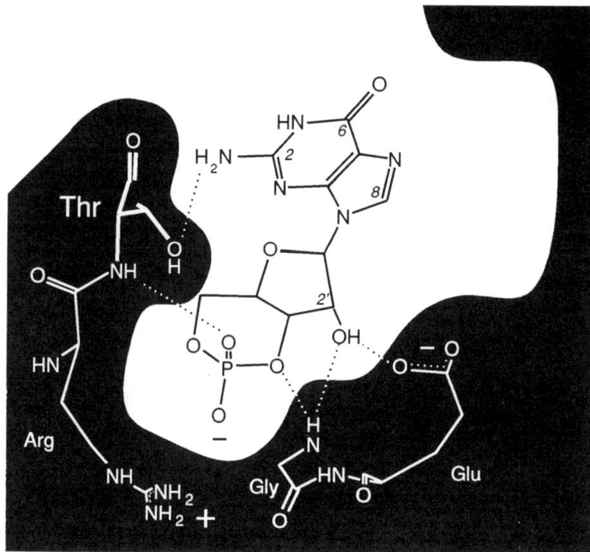

Fig. 7 Depiction of important contacts between cGMP and amino acids forming the respective cGMP binding sites of cGMP kinase. The hydrogen bond that is formed between the Thr-OH and the amino group at C-2 of the guanine ring and that provides for the specificity of cGMP binding compared to cAMP is shown. Other interactions between the cGMP and amino acids within the site are likely to be common to all the binding sites in CAP-related proteins. The cGMP is depicted as being bound in the *syn* conformation in the kinase site, although cAMP is bound to CAP in the *anti* confomer.

5′-GMP or 5′-AMP do not bind to the sites. The negative charge in the cyclized phosphate is also critical since substitution of a nitrogen for either oxygen abolishes binding. Furthermore, substitution of a sulfur atom for the axial exocyclic oxygen in cGMP (Sp)-cGMPS reduces the affinity ~100-fold, but this compound still activates cGMP kinase fully. However, on incorporation of a sulfur at the equatorial exocyclic oxygen, (Rp)-cGMPS, the ability of the analog to activate cGMP kinase is lost, although binding of the analog to cGMP kinase is equipotent with the (Sp)-cGMPS (Butt *et al.*, 1990). This suggests not only that the cyclic phosphate moiety provides for a major portion of the binding energy between cGMP and cGMP kinase, but also that the nature of this bond is critical to produce the conformational change required for enzyme activation.

Another important contact point for interaction in the sites is the 2′-OH of the cyclic nucleotide (Fig. 7). Replacement of the OH with a H or esterification of the oxygen with an aliphatic group greatly diminishes the affinity of interaction with either cGMP kinase or cAMP kinase. Thus, it is likely that the amino acids that interact with the cyclized phosphate and those that form a bond with the 2′-OH have been rigidly conserved in the cGMP-binding sites and in the cAMP-binding sites throughout evolution. The amino group at the C-2 position of the guanine ring is a particularly important contact point between the guanine nucleotide ring and both cGMP-binding sites. cIMP, which lacks this amino group, binds to the cGMP kinase with a 200-fold lower affinity than does cGMP. The interaction of cGMP with the binding sites through the amino at C-2, in combination with the ribose and cyclic phosphate bonds, is likely to provide for most of the binding energy.

3. Determinants of cGMP-Binding Specificity

Using comparisons of the known amino acid sequences in numerous cAMP- and cGMP-binding sites, Shabb *et al.* (1991) have identified an invariant threonine in the sequence of cGMP-binding sites of several related proteins. In type Iα cGMP kinase these are Thr-177 and Thr-301 in the fast and slow sites, respectively. This invariant threonine is present in the sequences of cGMP-binding sites in mammalian kinases, the putative *Drosophila* cGMP kinases (Kalderon and Rubin, 1989), the photoreceptor cGMP-gated cation channel (Kaupp *et al.*, 1989), and the cyclic nucleotide-gated olfactory channel (Ludwig *et al.*, 1990). In the cAMP-binding sites of the regulatory subunit of cAMP kinase, this invariant threonine is replaced by an alanine (Takio *et al.*, 1984a; Titani *et al.*, 1984), a mutation requiring one base change.

Computer modeling of the binding sites, based on the known X-ray crystallographic structure of the related CAP from *E. coli*, places the threonines in the respective cGMP-binding sites in close proximity to the C-2 amino group of the guanine ring of bound cGMP (Figs. 6 and 7). The positioning of the threonine hydroxyl group suggests that hydrogen bonding between this group and the C-2 amino group of the guanine ring could provide a specificity determinant for binding guanine nucleotides as opposed to adenine nucleotides in this site (Weber *et al.*, 1989). The importance of this hydrogen bond is supported experimentally by the results of site-directed mutagenesis of the corresponding alanine residues to threonines in the fast and slow sites of the bovine type I regulatory subunit of the cAMP kinase (Shabb *et al.*, 1990, 1991). These changes produce a 200-fold increase in the affinity of the regulatory subunit for cGMP with little change in the affinity for cAMP. The energy change for binding cGMP is consistent with the calculated free energy that would be derived from the formation of a hydrogen bond in the interaction with each binding site (Shabb *et al.*, 1991). These results strongly suggest that

Thr-177 and Thr-301 located at homologous positions in the binding sites of cGMP kinase provide a major component of the cGMP vs cAMP selectivity. Furthermore, when electron-donating substituents are appended to the guanine moiety of the cGMP molecule to create analogs with increased hydrogen bonding capacity at the C-2 amino group (Table I), the binding affinity of these compounds for cGMP kinase is relatively high.

The cyclic nucleotide-gated cation channels in photoreceptor and olfactory sensory neurons are activated by both cAMP and cGMP, but both channels have 30- to 40-fold higher affinities for cGMP than for cAMP (Kaupp, 1991). The amino acid sequences of the cyclic nucleotide binding portions of both channels are homologous to the sequences of the cyclic nucleotide-binding sites in cGMP- and cAMP-dependent protein kinases (Kaupp *et al.*, 1989), and computer modeling of the photoreceptor channel supports a similar secondary structure (Kumar and Weber, 1992). The cyclic nucleotide-binding domains of both channels have threonine residues in the same positions in their amino acid sequences (Thr-560 in the rod and Thr-537 in the olfactory proteins, respectively) as the invariant threonines proposed to be critical in determining cGMP vs cAMP specificity in cGMP kinase (Thr-177 and Thr-301 in bovine type Iα cGMP kinase) (Weber *et al.*, 1989; Shabb *et al.*, 1991). Substitution of alanine residues for the respective threonines in the channel proteins results in a large reduction in their affinities for cGMP with little change in the affinities for cAMP (Altenhofen *et al.*, 1992). Thus, in this family of proteins, a threonine residue in this position in the binding sites provides a major portion of the discrimination in selectivity between cGMP and cAMP.

4. Two Kinetically Distinct cGMP-Binding Sites

The two cGMP-binding sites in both type Iα and type Iβ exhibit distinctly different kinetic parameters (Lincoln and Corbin, 1983; Corbin and Doskeland, 1983; Doskeland *et al.*, 1987; Wolfe *et al.*, 1989b; Ruth *et al.*, 1991). Extensive studies with the purified type Iα cGMP kinase demonstrate different analog specificities for each binding site that could relate to varying degrees of hydrophobicity of the sites. The slow site contains multiple charged groups (Glu-270, Asp-271, Arg-281, Asp-287, Asp-337, Lys-341), whereas the side chain of Phe-214 projects into the fast site, giving it a more hydrophobic character (Takio *et al.*, 1984b; Weber *et al.*, 1989). The slow site, or the "high affinity site," is so named since cGMP dissociates from this site very slowly ($t_{1/2}$ = 0.04 min^{-1} at 30° in type Iα and 0.057 min^{-1} in type Iβ). Likewise, cGMP dissociates relatively rapidly from the fast site, or "low affinity site" in both isoforms ($t_{1/2}$ = 3–5 s at 0°). Binding affinity of type Iα cGMP kinase for cGMP is ~100 fold greater at 0° compared to 30° (Francis *et al.*, 1988), and the two binding sites exhibit positive cooperativity (McCune and Gill, 1979), as reflected by a relatively high Hill constant (1.5–1.8) for activation of the enzyme. Although the slow site can bind cGMP independently of the fast site (Corbin and Doskeland, 1983; Corbin *et al.*, 1986; Doskeland *et al.*, 1987) cGMP analog binding to one of the sites is stimulated on addition of an analog selective for the other site. Furthermore, activation of the enzyme by cGMP or by its analogs is best correlated with the relative affinities of both binding sites, rather than with the affinity for either the slow or the fast site. When cGMP is bound to the fast site, cGMP dissociates 10 times more slowly from the slow site (Corbin and Doskeland, 1983; Corbin *et al.*, 1986), which could in large part explain the observed positive cooperativity. By the principle of reciprocity, cGMP binding to the slow site should enhance cGMP binding at the fast site, although this has not been experimentally demonstrated.

Whether the positive cooperativity in cGMP binding is due to interaction of the binding sites of two separate cGMP monomers or between the two sites within a given monomer is not known, although the homologous cAMP kinase is known to exhibit intrasubunit positive cooperativity. Dissociation kinetics for [^{3}H]cGMP from the monomer of the type Iβ show a rapid loss of nucleotide from the protein (Hill coefficient = 1.0), although 2 mol of [^{3}H]cGMP are bound per mole of monomer (Wolfe *et al.*, 1989b). Thus, removal of the amino-terminal 62 amino acids in type Iβ alters the positive cooperativity between the two binding sites and converts the high affinity slow site to a lower affinity site. However, these changes in kinetic parameters could be due either to structural changes induced by monomerization of the enzyme or to removal of the direct influence of the amino terminus on the binding sites (Francis *et al.*, 1988–1989; Landgraf and Hofmann, 1989; Wolfe *et al.*, 1989b; Ruth *et al.*, 1991). Since the monomeric Iβ is almost completely inactive in the absence of cGMP, it seems unlikely that the inhibitory domain is responsible for the positive cooperativity. Although the monomeric cGMP kinase from *Paramecium* contains a kinetically defined "slow cGMP-binding site" and a "fast cGMP-binding site," there is no evidence for cooperativity between the binding sites (Miglietta and Nelson, 1988). Thus, it is tempting to suggest that the cooperativity results from dimerization of the enzyme (Landgraf and Hofmann, 1989; Wolfe *et al.*, 1989a).

5. Interaction of cGMP-Binding Sites with the Autoinhibitory Domain and the Catalytic Domain

Although the cGMP-binding domains appear to contain the elements that directly provide for interaction with the cyclic nucleotide, the kinetic features of binding indicate subtle influences of the catalytic domain and the inhibitory domain on cGMP binding. Both the Mg^{2+}/ATP site and the peptide-binding site in the catalytic domain influence cGMP binding. In the presence of Mg^{2+}/ATP, the affinity of the slow site for cGMP is decreased in type Iα (Corbin *et al.*, 1986; Landgraf and Hofmann, 1989). The dissociation rate for cGMP from the slow site is increased approximately 10-fold, and the association rate with either site is lowered by 50%. These effects occur at concentrations of ATP near the K_m for ATP binding (10 μM) to the catalytic site, suggesting that when Mg^{2+}/ATP is bound to the catalytic site, the association/dissociation of cGMP with/from the enzyme is significantly altered. Since there is no evidence for autophosphorylation of the cGMP kinase under the conditions of the experiment, the authors propose that a conformational effect on the binding sites is induced by the presence of Mg^{2+}/ATP at the catalytic site. Under normal physiological concentrations of Mg^{2+}/ATP, the enzyme would be saturated with these ligands so that this lower affinity for cGMP binding may reflect the physiological status. Likewise, high concentrations of substrate peptide increase the binding affinity of the fast site ~5-fold (Landgraf and Hofmann, 1989). These changes clearly reflect communication, either direct or indirect, between the catalytic center and the cyclic nucleotide-binding sites. However, the effects of peptide substrate are not present either in the autophosphorylated type Iα or in its constitutively active fragment that is missing the amino-terminal 77 amino acids. Thus, the characteristics of the binding sites (kinetics of binding and dissociation, spatial features of the sites and interaction with the catalytic centers) are significantly altered by structural features of the amino termini. The complexity of the interactions of the binding sites with other elements within the enzyme structure cannot be overemphasized.

6. Analog Specificities

In addition to the differing kinetic characteristics, the cyclic nucleotide analog specificities of the two binding sites of type Iα cGMP kinase also differ (Corbin and Doskeland, 1983; Corbin *et al.*, 1986; Doskeland *et al.*, 1987; Landgraf *et al.*, 1988). This has not been thoroughly investigated for the type Iβ isoenzyme. In general, modifications of the guanine ring in the pyrimidine portion (N-1 and C-6 positions) provide increased selectivity for the fast site, whereas analogs with modifications in the imidazole portion (C-7 and C-8 positions) interact more selectively with the slow site. The selectivity is relative since most cGMP analogs interact with both binding sites, albeit with significantly different affinities. Simultaneous addition of two cGMP analogs, each selective for one or the other of the two binding sites, produces a synergistic activation of the cGMP kinase catalytic activity since binding to both sites is required for expression of full catalytic activity.

The differences in the interaction of cyclic nucleotide analogs with the slow and fast sites in type Iα can be ascribed to differences in their amino acid sequences. However, the different analog specificities of type Iα and type Iβ (Table I) are difficult to explain (Wolfe *et al.*, 1989b; Sekhar *et al.*, 1992). The addition of a large substitution at the 1,2 position of the guanine ring provides for a high affinity interaction with the binding sites in both isoforms (Wolfe *et al.*, 1989b), which is largely attributable to the stabilization of a hydrogen bond with the C-2 amino group in the binding site. The poor binding of the C-8 derivatives to type Iβ compared to type Iα (Wolfe *et al.*, 1989b) can be overcome by the introduction of a phenyletheno group at the 1,2 position on the purine ring (Sekhar *et al.*, 1992). For example, 8-bromo-1,N^2-PET-cGMP binds equally well to both type Iα and type Iβ (K_a = 13 and 9 nM, respectively) compared to the disparate activation constants of the parent compound, 8-bromo-cGMP, for the respective isozymes (K_a = 26 and 210 nM, respectively) (Table I). That is, addition of the phenyletheno group to 8-bromo-cGMP improves binding to type Iα 2-fold and to type Iβ, 23-fold. 8-(4-OH-Ph-*S*)-1,N^2-PET-cGMP is also bound equally well by either enzyme (K_a = 17 and 23 nM, for types Iα and Iβ, respectively) compared to the parent compound 8-(4-OH-Ph-S-)cGMP, (K_a = 50 and 440 nM, respectively). Since analogs with large substitutions on either end of the purine ring bind to the sites with high affinities, it is clear that there are few rigidly fixed steric constraints in the regions of the binding sites that abut the C-1/C-2 of the pyrimidine ring or the C-8 of the imidazole ring on the purine.

D. Catalytic Domain

1. Subdomains for Mg^{2+}/ATP and Protein Substrate Binding

The catalytic domain of the cGMP kinase is located in the carboxyl-terminal segment of the enzyme, and the boundaries are approximated to extend from Ala-340 through Ser-599 (Takio *et al.*, 1984b). The amino acid sequence of this region is strongly homologous to that of the catalytic subunit of the cAMP kinase. The crystal structure of the catalytic subunit of the cAMP kinase has recently been determined (Knighton *et al.*, 1991a,b), and, using the X-ray coordinates derived from the structure, the catalytic domain of cGMP kinase has been modeled (Hofmann *et al.*, 1992). The structure of the catalytic subunit of cAMP kinase is characterized by the presence of two lobes (unequal in size) separated by a deep cleft. Studies based on protein modification and sequence analyses have shown the catalytic domain to be arbitrarily subdivided into an ATP-binding site that is more amino terminal in the sequence and the catalytic center where the peptide is bound. The crystal structure confirms this location and indicates that Mg^{2+}/ATP is associated with the smaller of the two lobes, which is rich in anti-parallel β-sheet structure. The more carboxyl-terminal catalytic center is associated with the larger lobe, which contains a predominance of α-helical structure and provides multiple contact points to bind and position the substrate peptide in the site.

Covalent modification of the ATP-binding site of cGMP kinase by the ATP analog [^{14}C]FSBA labels a specific lysine (Lys-389) (Hashimoto *et al.*, 1982) that is homologous to an invariant lysine in the ATP-binding site of all kinases studied thus far (Hanks *et al.*, 1988). Comparison of the amino acids surrounding the lysine with the same region in cAMP kinase (Takio *et al.*, 1984a) indicates a striking homology, with identities in 17 residues and 12 other residues where only 1 base change would have been required. Beginning at Gly-366 there is a consensus sequence for ATP binding (GVGGFG) that is commonly present in nucleotide-binding proteins (Rossmann *et al.*, 1974). Val-373 is also a highly conserved residue that is thought to provide important structural constraints on the site. Based on crystal studies of the ternary complex of the catalytic subunit with Mg^{2+}/ATP and the cAMP kinase inhibitor peptide (5–24) (Knighton *et al.*, 1991a,b), the Mg^{2+}/ATP is localized to the base of the cleft between the two lobes, with the adenine portion located near the base of the cleft; the γ-phosphate is closely positioned with the conserved Lys-72 (Lys-389 in type Iα cGMP kinase), Asp-184 (Asp-501 in cGMP kinase), and Glu-91 (Glu-408 in cGMP kinase), and extends outward toward the edge. Asp-501 may chelate Mg^{2+}. The structure of the ATP-binding region differs significantly from the characteristic Rossmann fold for nucleotide binding in many other proteins. Regions associated with substrate peptide binding are likely to be less well conserved between the two kinases.

2. Substrate Specificity

The catalytic function of the cGMP kinase is similar in many respects to that of cAMP kinase although distinct differences are clear. Both enzymes transfer the γ-phosphate of ATP to a serine or threonine residue in a variety of proteins and synthetic peptides, and in the catalytic subunit, Asp-166 (Asp-482 in cGMP kinse) is proposed to function as the catalytic base for phosphate transfer. There is no evidence for a phospho-enzyme intermediate (Ho *et al.*, 1988). The phosphorylated residue is usually in a typical consensus sequence (RRXSX) containing two tandem amino-terminal basic residues, one spacing residue followed by the serine or threonine residue and a hydrophobic residue at the carboxyl terminus (Lincoln and Corbin, 1983; Glass, 1990; Kennelly and Krebs, 1991). The amino acids juxtaposed on either side of the serine profoundly influence the ability of the respective kinases to phosphorylate that site. Although the presence of the two amino-terminal basic residues in the consensus sequence seems to be a general feature of cGMP kinase substrates (Fig. 8), they are not absolutely mandatory, as evidenced by the sequences surrounding the autophosphorylation sites of the enzyme described above. Whether the phosphorylation of these atypical sequences is unique to the autophosphorylation process or whether other proteins can also be phosphorylated at unpredictable sites is not known.

The efficiency of cGMP kinase as a catalyst is less than that of the cAMP kinase for most substrates studied, but there are notable exceptions (Glass and Krebs, 1979; Geahlen and Krebs, 1980; Aitken *et al.*, 1981; Aswad and Greengard, 1981a,b; Hashimoto *et al.*, 1981; Thomas *et al.*, 1990). Substrates that appear to exhibit selective preferences for the cGMP kinase (Fig. 8) include the α-subunit of skeletal muscle phosphorylase b kinase (Yeaman *et al.*, 1977), which is phosphorylated at approximately a threefold higher rate than with the cAMP kinase. The purified type I regulatory subunit of bovine skeletal muscle cAMP kinase is phosphorylated at Ser-99 by the cGMP kinase, but not by the catalytic subunit of the cAMP kinase (Geahlen and Krebs, 1980; Hashimoto *et al.*, 1981), and is one of the very few substrates that has an apparently absolute specificity for the cGMP kinase. However, the physiological importance of the slow phosphorylation of phosphorylase b kinase or regulatory subunit of type I cAMP kinase is doubtful. Calf thymus histone H1 contains a site in the carboxyl-terminal portion that is also very selective toward cGMP kinase, although the sequence of the site is unknown. In mammalian cerebellum, cGMP kinase phosphorylates two threonines in a 23-kDa protein (also known as G-substrate) of unknown function (Aitken *et al.*, 1981; Schlichter *et al.*, 1978; Aswad and Greengard, 1981a,b). The K_m of cGMP kinase for this protein (0.2 μM) is ~30 times lower than the K_m of cAMP kinase for this protein (Aswad and Greengard, 1981b). The presence of this protein in cerebellum along with the localization of cGMP kinase in cerebellar Purkinje fibers increases the possibility that this is a physiologically relevant substrate.

Compared to cAMP kinase, cGMP kinase is generally in low concentration in most mammalian tissues. This has traditionally brought into question whether there is sufficient catalytic activity in these tissues to account for significant physiological effects. However, recent work by Wyatt *et*

```
cGMP-BPDE          R K I S A S E F D R P L R

cGMP KINASE
     (46-59)     L P V P S T H I G P R T T R
     (54-67)     G P R T T R A Q G I S A E P
     (68-81)     Q T Y R S F H D L R Q A F R
     (80-93)     F R K F T K S E R S K D L I

cAMP KINASE
          RI     R G A I S A E V Y T E E D A

G-SUBSTRATE      R R K D T P A L H I P P F I
                 R R K D T P A L H T S P F Q

H2B  (29-39)       R K R S R K E S Y S V

PHOS K           F R R L S I S T E S E P D G

TYR H            G R R Q S L I E D A R K

HMG 14           K R K V S S A E G A A K

HSL              P M R R S V S E A A L T Q P
                 S M R R S V S E A A L A Q P
```

Fig. 8 Substrates selectively phosphorylated by cGMP kinase. Substrates that are preferentially phosphorylated by cGMP kinase compared to cAMP kinase include cGMP-BPDE (cGMP-binding cGMP-specific phosphodiesterase), cGMP kinase Iα (autophosphorylation sites in bovine heart cGMP-dependent protein kinase), bovine skeletal muscle type I regulatory subunit of cAMP kinase, G-substrate (rabbit cerebellar G-substrate), H2B (calf thymus histone H2B), Phos k (rabbit skeletal muscle phosphorylase kinase, α subunit), Tyr H (rat pheochromacytoma tyrosine hydroxylase), HMG 14 (calf thumus chromosomal high mobility group protein 14), and HSL (bovine adipose tissue and rat adipose tissue hormone-sensitive lipase). Taken from Thomas *et al.* (1990).

al. (1991) using adherent neutrophils demonstrates a cGMP-induced colocalization of the cGMP kinase with vimentin, a cytoskeletal protein of unknown function that is phosphorylated by cGMP kinase. Likewise, these same workers have provided evidence supporting colocalization of cGMP kinase with another substrate, phospholamban, located in the sarcoplasmic reticulum (Cornwell *et al.*, 1991). Thus, it is possible that the overall concentration of cGMP kinase in a tissue does not reflect its true importance in regulating physiological processes in that cell type. Concentration of cGMP kinase within a specific microenvironment in the cell may provide the required access to targeted protein substrates.

In histone H2B, two serine residues (Ser-32 and Ser-36) are phosphorylated by both cGMP kinase and cAMP kinase (Fig. 8) (Hashimoto *et al.*, 1976; Glass and Krebs, 1979), but the rate of catalysis of phosphorylation by cGMP kinase is greater at Ser-32, which has the sequence -DGKKRKRSRKE. From studies of a synthetic peptide derived from this sequence, the cGMP kinase has a lower K_m and a higher V_{max} (Glass and

Krebs, 1982; Glass, 1990) compared to the phosphorylation by cAMP kinase. Using synthetic peptide analogs, Glass and co-workers have determined that Arg-33 immediately adjacent to Ser-32 makes a major contribution to the favorable characteristics of this peptide as a cGMP kinase substrate compared to cAMP kinase. The Lys-34 has a deleterious effect on the phosphorylation of this peptide by either kinase, and, as in other substrates, substitution of threonine for the Ser-32 position produces a poorer substrate for cGMP kinase. A heptapeptide (RKRSRAE), modeled after the H2B peptide, is available commercially and is widely used for cGMP kinase assays.

Studies of a purified cGMP-binding phosphodiesterase (BPDE) have demonstrated phosphorylation of a single serine on this protein by cGMP kinase or cAMP kinase (Thomas *et al.*, 1990). The phosphorylation of this protein is unique since the modification either by the cGMP kinase or by the catalytic subunit of the cAMP kinase occurs only when cGMP is bound to the BPDE, thus exposing the phosphorylation site. The phosphorylation of the BPDE by cGMP kinase is estimated to proceed at a rate 10 times that by cAMP kinase. Thus, it seems likely that BPDE is specifically phosphorylated by cGMP kinase in intact tissues since the site is revealed for modification only in the presence of elevated cGMP, and, once exposed, this site is a better substrate for the activated cGMP kinase than for the cAMP kinase.

The sequence of the phosphorylated tryptic peptide from the BPDE has been determined (RKISASEFDRPLR) (Thomas *et al.*, 1990), and only the more amino-terminal serine in this sequence contains phosphate (Fig. 8). The full-length synthetic peptide (BPDEtide) retains a high selectivity for cGMP kinase (Colbran *et al.*, 1992) as exhibited in the intact protein, showing a $K_m = 68\ \mu M$ and a $V_{\max} = 11\ \mu$mol/min/mg compared to the values for cAMP kinase ($K_m = 320\ \mu M$ and $V_{\max} = 3.2\ \mu$mol/min/mg). A truncated peptide derived from the BPDEtide sequence (RKISASEF) shows the same pattern of selectivity between the kinases, but removal of the carboxyl-terminal phenylalanine abolishes the selectivity. This suggests that the phenylalanine is a negative determinant for cAMP kinase. Similar positioning of phenylalanine relative to serine in other phosphorylation sites may also act as a negative determinant for phosphorylation by cAMP kinase. Thus, the presence of a single phenylalanine (or perhaps other aromatic amino acids) in a phosphorylation site sequence may provide for preferential phosphorylation by cGMP kinase. Thus, despite strong homologies in the structures of the catalytic centers of the cGMP kinase and cAMP kinase, and the fact that they exhibit overlapping substrate specificities, the sites are likely to contain structural features that may account for preferred substrate specificities in the cell.

The potency and selectivity of the heat-stable protein kinase inhibitor (PKI) or the peptide [PKI-tide (5-24)] (Cheng *et al.*, 1986; Walsh *et al.*, 1990) derived therefrom for inhibition of cAMP kinase (K_i = 2 nM) as opposed to cGMP kinase (as well as other serine/threonine kinases) are difficult to explain since PKI-tide contains a typical pseudosubstrate consensus sequence -RRNSAI- (Scott *et al.*, 1985; Walsh *et al.*, 1990). Using crystallographic analysis, multiple points of interaction between the catalytic site of cAMP kinase and the PKI-tide (6-22) (Knighton *et al.*, 1991a,b) have been established. Several amino acids making specific contact with residues in PKI-tide (Phe-10, Arg-15, Arg-18, Arg-19, and Ile-22) are absent in the homologous region of the cGMP kinase. In particular, Glass *et al.* (1992) have noted the absence in cGMP kinase of 2 of the 3 residues making contact with the Phe-10 in PKI-tide that resides in the α-helical portion of the structure and accounts for much of the high affinity binding of PKI to cAMP kinase. These authors also suggest that electrostatic interaction between the cGMP kinase and one of the arginines in the substrate peptides as well as in the PKI-tide may be lessened, compared to cAMP kinase, by the absence of residues homologous to Asp-329 and Glu-331 of cAMP kinase. A more complete understanding of elements that contribute to the variations in the specificities of the catalytic sites of these two closely related kinases must await determination of the crystal structure of cGMP kinase or more extensive computer modeling of the catalytic center of this enzyme based on the coordinates determined for the cAMP kinase structure. The peptide [Ser-21] PKI-tide (14-22)amide has recently been shown to be an excellent substrate for both cGMP kinase and cAMP kinase, but cAMP kinase has a 10-fold higher affinity for the peptide than does cGMP kinase (Hofmann *et al.*, 1992).

E. Carboxyl-Terminal Domain

At the carboxyl-terminal extreme of cGMP kinase is a region of unknown function. It is 70 amino acids in length and is ~40% identical to the same region in cAMP kinase. There is little or no homology with similar regions in other kinases. Takio and colleagues (1984b) suggested that this region may constitute another folding domain in the protein. This region may still prove to be important in enzymatic function.

VII. Physiological Function

Nitrovasodilators generate nitric oxide (NO), which activates the soluble form of guanylate cyclase and increases intracellular cGMP (Rapoport and Murad, 1983; Holzmann, 1982; Ignarro *et al.*, 1984; Murad,

1986). Recently, NO has been determined to be a naturally occurring substance that is produced in numerous tissues throughout the body. The endothelium-derived relaxation factor (EDRF) has been identified as NO and is released from endothelial cells under a variety of conditions (Ignarro *et al.*, 1987; Palmer *et al.*, 1987). In adherent neutrophils an increase in NO synthesis in response to stimulation by the chemotactic peptide *N*-formyl-methionyl-leucyl-phenylalanine (fMLP) is postulated to occur; the increase in intracellular calcium in response to fMLP may stimulate the activity of a Ca^{2+}/calmodulin-sensitive nitric oxide synthase. This sequence of events could explain the transient rise in cGMP that occurs immediately after exposure of the cells to fMLP (Wyatt *et al.*, 1991). In neural tissue, long-term potentiation has also been associated with NO generation in the postsynaptic neuron. The NO thus produced is proposed to diffuse retrogradely to the presynaptic neuron to increase cGMP synthesis by activating the soluble form of guanylate cyclase. It is possible that other physiological agents in addition to NO will be identified that activate the soluble guanylate cyclase thereby increasing cGMP production and eliciting a panoply of physiological responses.

The membrane-bound form of guanylate cyclase is also a target for physiological regulation of cGMP levels. Atrial natriuretic peptide (ANP) activates the membrane-associated (or receptor-linked) guanylate cyclase (Waldman *et al.*, 1984), and the associated elevations in cGMP correlate with the effects of the hormone on phosphorylatin of vascular smooth muscle proteins (Sarevic *et al.*, 1989) as well as its natriuretic and diuretic effects in kidney. A recently discovered 15 amino acid gastrointestinal peptide, guanylin (Currie *et al.*, 1992), which has structural similarity to the heat-stable enterotoxin from *E. coli,* activates the membrane-bound form of guanylate cyclase in T_{84} human intestinal cells (Forte *et al.*, 1992). As research interests in this area increase, it is likely that a broader family of agents that use cGMP as their second messenger will be identified. cGMP kinase is likely to mediate many of these effects.

With increasing frequency, cGMP is considered a potential intracellular second messenger in the regulation of smooth muscle tone, inhibition of platelet aggregation, and neuronal long-term potentiation. These processes will be mentioned briefly in this chapter as they may relate to cGMP kinase function.

A. Regulation of Smooth Muscle Tone

Elevation of cGMP in vascular smooth muscle by agents such as ANP, EDRF, and exogenous nitrovasodilators is known to cause relaxation of tension in smooth muscles (Schultz *et al.*, 1977; Katsuki *et al.*, 1977; Lincoln, 1983, 1989; Hardman, 1984; Lincoln and Johnson, 1984; Waldman *et al.*, 1984; Fiscus *et al.*, 1985; Murad, 1986). The cyclic nucleotide receptor(s) that mediates these effects is now being intensively investigated (Furukawa and Nakamura, 1987; Francis *et al.*, 1988; Furukawa *et al.*, 1988; Lincoln *et al.*, 1990; Yoshida *et al.*, 1991). Cyclic nucleotide analogs that potently activate the purified cGMP kinase *in vitro* cause relaxation of vascular and tracheal smooth muscle with a similar pattern of potency. These studies provide strong evidence that activation of cGMP kinase mediates the initiation of relaxation in response to increased levels of cGMP (Francis *et al.*, 1988). The potencies of the cyclic nucleotides in relaxing smooth muscle correlate well with the established K_a values of the analogs for activation of the type Iα isoform of cGMP kinase (Fig. 9) (Sekhar *et al.*, 1992). Type Iβ isoform may also be involved in activating

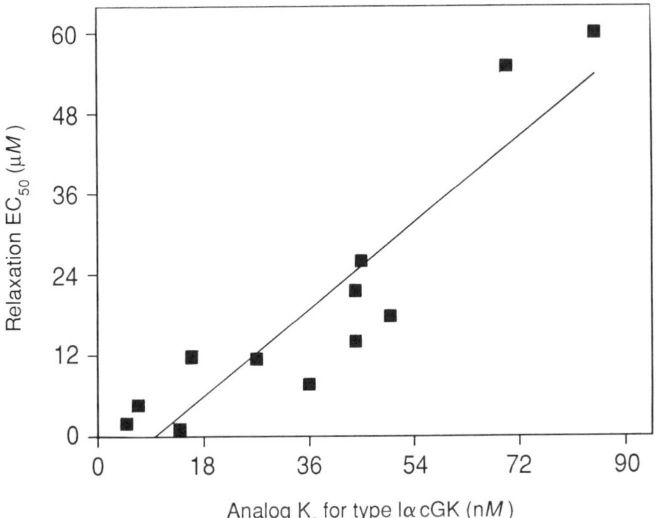

Fig. 9 Correlation between the potencies of cGMP analogs to activate purified cGMP kinase and the potencies with which they elicit smooth muscle relaxation. The potencies of various phenylthio- derivatives of cGMP in relaxing pig coronary artery strips (EC_{50}) are correlated with the K_a of these analogs in activating cGMP kinase type Iα. The curve was generated by a linear regression analysis program. Specific analogs used in the figure, including the K_a and EC_{50}, respectively are: 8-(2,4-dihydroxyphenylthio)-cGMP (5 n*M*, 2 μM), 8-(2-aminophenylthio)-cGMP (7 n*M*, 4.7 μM), 8-(4-hydroxyphenylthio)-cGMP (14 n*M*, 1.05 μM), 8-(4-aminophenylthio)-cGMP (16 n*M*, 11.8 μM), 8-(β-napthylthio)-cGMP (27 n*M*, 11.5 μM), 8-(3-bromophenylthio)-cGMP (36 n*M*, 7.7 μM), 8-(4-methoxyphenylthio)-cGMP (44 n*M*, 14 μM), 8-(2,6-dichlorophenylthio)-cGMP (44 n*M*, 21.6 μM), 8-(2-bromophenylthio)-cGMP (45 n*M*, 26 μM), 8-(4-chlorophenylthio)-cGMP (50 n*M*, 17.8 μM), 8-(4-fluorophenylthio)-cGMP (70 n*M*, 55 μM), and 8-benzylthio-cGMP (85 n*M*, 60 μM).

the relaxation process, but a similar comparison cannot be made since analogs highly specific for the type Iβ are not yet available. Although smooth muscle contains a number of proteins that appear to be preferentially phosphorylated by cGMP kinase (Ives *et al.*, 1980; Parks *et al.*, 1987), the specific protein(s) targeted for action by the kinase is not known, but cyclic nucleotide analogs of known specificites provide a powerful tool in defining the proteins that mediate the intracellular effects of cGMP and cAMP.

B. Inhibition of Platelet Aggregation

Platelet aggregation is inhibited by agents such as EDRF and sodium nitroprusside that elevate intracellular cGMP and by cAMP-elevating agents such as prostaglandin E_1 and prostacyclin (PGI_2) (Haslam, 1987). Relatively high levels of cGMP kinase and cAMP kinase have been demonstrated in human platelets, although a clear role for these kinases in mediating the effects of the cyclic nucleotides has not been established. Distinct patterns of phosphorylation of proteins in intact platelets are observed following activation of the respective kinase systems (Waldmann *et al.*, 1987). Activation of the cAMP kinase system results in the phosphorylation of numerous bands, whereas activation of the cGMP kinase either in intact platelets or in extracts of platelets primarily phosphorylates a 46/50-kDa membrane-associated protein [vasodilator-stimulated phosphoprotein (VASP)]. VASP appears to be the only protein in platelets that is phosphorylated in response to elevation of either cAMP or cGMP (Waldmann *et al.*, 1987; Halbrugge and Walter, 1989; Waldmann and Walter, 1989; Halbrugge *et al.*, 1990), and the time course and extent of its phosphorylation have been shown to correlate with increases in intracellular cGMP and with the inhibition of platelet activation (Nolte *et al.*, 1991; Eigenthaler *et al.*, 1992). However, the function of VASP is unknown, and the mechanism by which cyclic nucleotides block platelet activation is not known.

C. Regulation of Intracellular Calcium Levels

Many studies of cGMP and smooth muscle relaxation have focused on proteins associated with Ca^{2+} homeostasis since elevation of endogenous cGMP or cyclic nucleotide analogs, such as 8-bromo-cGMP, have long been known to produce marked reductions in intracellular Ca^{2+}, or to block Ca^{2+} transients (Johnson and Lincoln, 1985; Kobayashi *et al.*, 1985; Collins et al., 1986; Rashatwar *et al.*, 1987; Feibel *et al.*, 1988; Lincoln, 1989). Similarly, in human platelets either nitrovasodilators or 8-pCPT-cGMP (a potent activator of cGMP kinase) inhibits Ca^{2+} mobilization from intracellular compartments in response to ADP or thrombin as well as diminishing the influx of Ca^{2+} from the medium (Geiger *et al.*, 1992). However, Ca^{2+} influx mediated by the ADP receptor cation channel is not affected. Lowering of Ca^{2+} levels could be achieved by (*a*) more efficient sequestration of cytosolic Ca^{2+} within the cell, (*b*) extrusion of Ca^{2+} into the surrounding medium, (*c*) reduction in Ca^{2+} mobilization in response to agents such as IP_3, or (*d*) a combination of these processes.

Rapoport *et al.* first proposed that ANP, through a cGMP-mediated process, reduces phosphatidylinositol hydrolysis (Rapoport, 1986). cGMP-elevating agents, such as nitroglycerine and sodium nitroprusside or 8-bromo-cGMP, prevent inositol triphosphate accumulation in response to norepinephrine treatment. Using a broken cell preparation, Hirata *et al.* (1990) have recently presented evidence that cGMP in the presence of ATP lowers inositol phosphates in response to either GTPγS alone or GTPγS in combination with a vasopressin analog. By comparison, a 100-fold greater concentration of cAMP is required to elicit the same effect. Thus, pretreatment of cell extracts with cGMP and ATP completely blocks G-protein activation (as measured by an AVP-induced increase in GTPase activity) and subsequent coupling between the G-protein and phospholipase C. The binding of vasopressin to its receptor is unaffected. cGMP kinase is likely to be the catalyst in this process because (*a*) the process is ATP-dependent, (*b*) the EC_{50} for cGMP is 0.01 μM, which is well within the range of cGMP intracellular concentrations and is also a concentration sufficient for cGMP kinase activation; and (*c*) the effect is not blocked by the cAMP protein kinase inhibitor even though cAMP at 100-fold higher concentration is also effective. However, the role of cGMP kinase in this process has not been proven.

Recent work has provided evidence for direct effects of cGMP kinase on Ca^{2+}/ATPase activities from both the plasma membrane and the sarcoplasmic reticulum. Purified plasma membrane Ca^{2+} pump/ATPase from pig aorta that contains two proteins (240 and 135 kDa)(Furukawa and Nakamura, 1987; Furukawa *et al.*, 1988; Yoshida *et al.*, 1991) is activated on phosphorylation of the 240-kDa species by cGMP kinase; two isoforms of the Ca^{2+}-pump ATPase (135 and 145 kDa) are not phosphorylated by cGMP kinase (Yoshida *et al.*, 1992), but the increase in the activity of the Ca^{2+} pump in reconstituted vesicles correlates with the phosphorylation of the associated 240-kDa protein. These results strongly argue for an indirect role for cGMP kinase in regulating the Ca^{2+} pump via phosphorylation of an associated protein. However, the most common isoform of the plasma membrane Ca^{2+}-pump ATPase in smooth muscle (PCMA1b) contains a consensus sequence for phosphorylation by cyclic nucleotide-dependent protein kinases, which may still prove to be important in the regulation of Ca^{2+}-pump activity (James *et al.*, 1989; De Jaegere *et al.*, 1990). Phosphorylation of a single serine in this consensus sequence in the erythrocyte plasma membrane Ca^{2+}-pump (James *et al.*, 1989) increases the V_{max} and lowers the K_m of the pump for Ca^{2+}.

Recent work by Cornwell *et al.* (1991) has led these workers to conclude that cGMP kinase phosphorylates phospholamban in aortic smooth muscle and leads to increased activity of the Ca^{2+}/ATPase localized in the sarcoplasmic reticulum. Phospholamban purified from the sarcoplasmic reticulum of cardiac muscle and smooth muscle can be phosphorylated at comparable rates by either cAMP kinase or cGMP kinase (Raeymaekers *et al.*, 1988). The phosphorylation of a specific serine in phospholamban is correlated with an increase in the affinity of the Ca^{2+} pump for Ca^{2+} and thereby, depending on the state of its phosphorylation, can increase Ca^{2+} uptake in isolated vesicles of sarcoplasmic reticulum. However, the role of phospholamban in the expression of cGMP effects in smooth muscle remains controversial. Although the correlation between increases in cytoplasmic cGMP levels and the lowering of cytoplasmic Ca^{2+} could be explained in part by increased phosphorylation of phospholamban, the reduction in tension in rings of rabbit aorta by treatment with sodium nitroprusside is not correlated with increased phosphorylation of phospholamban (Huggins *et al.*, 1989). However, the work of Cornwell *et al.* (1991), using a broken cell preparation derived from rabbit aortic smooth muscle cells, demonstrates that phospholamban is phosphorylated in response to cGMP in the extract, but not when cAMP is added. These workers also provide immunocytochemical evidence supporting colocalization of the cGMP kinase with the phospholamban in the sarcoplasmic reticulum. However, in CHO cells, which apparently lack phospholamban, activation of cGMP kinase expressed following transfection with the type Iα cDNA lowers thrombin-induced Ca^{2+} transients (Ruth *et al.*, 1991). Therefore, whether phospholamban is a major target for modification by cGMP kinase *in vivo* and for eliciting the effects of cGMP in smooth muscle cells is still unclear.

cGMP and cGMP analogs have also been shown to inhibit L-type Ca^{2+} channel current in cardiac cells. In patch-clamp studies of mammalian ventricular cells, Mery et al. (1991), have demonstrated that exposure of the membrane to constitutively activated cGMP kinase also causes an inhibition of the Ca^{2+} current. The inhibitory effects of cGMP and cGMP kinase appear to occur only after the channel activity has been increased by elevation of cAMP. Using Western blot analysis, cGMP kinase has been detected in rat heart and in isolated pure rat cardiomyocytes (although in relatively low levels), (Lohmann *et al.*, 1991) further supporting the potential role of this system in regulating the Ca^{2+} current in heart. Physiologically, these effects may relate to regulation of heart function by ANP and acetylcholine.

A unique role for cGMP kinase in regulating Ca^{2+} levels in vascular smooth muscle cells is strongly supported by a classical study by Cornwell and Lincoln (1989). Cells that have been repeatedly passaged have large reductions in cGMP kinase levels (Cornwell and Lincoln, 1989; Lincoln *et al.*, 1990) and have a diminished ability to decrease intracellular Ca^{2+} in response to ANP or 8-bromo-cGMP. However, the ability of these cGMP-kinase depleted cells to lower intracellular Ca^{2+} in response to elevation of either cGMP or cAMP is restored on reintroduction of purified cGMP kinase to the cells. Since cAMP kinase levels are not lowered significantly on passage of these cells, these studies strongly argue for a unique and pivotal role of cGMP kinase in modulating Ca^{2+} metabolism.

D. Other Possible Functions

Whether cGMP-dependent phosphorylations are important in regulating a wide range of cellular processes remains to be determined. Elevation of cGMP in vascular smooth muscle has been implicated in the modulation of the Na^+, K^+, Cl^- cotransport system (O'Donnell and Owen, 1986), but the involvement of cGMP kinase has not been established. Several studies suggest that cGMP kinase may be important in regulating components of the cytoskeleton in response to various stimuli. The major cGMP kinase substrate in smooth muscle is a 120-kDa protein (G_1) that binds actin (Ives *et al.*, 1980). It has been suggested that this protein may be important in anchoring other proteins to the plasma membrane and/or the cytoskeletal matrix in smooth muscle cells (Baltensperger *et al.*, 1990). Phosphorylation of G_1 might then regulate changes in the cytoskeletal structure and cellular morphology, although there is no definitive evidence to support this. Further evidence supporting involvement with cytoskeletal components is derived from studies in adherent neutrophils where in the unstimulated state cGMP kinase is primarily localized to the cytoplasm (Pryzwansky *et al.*, 1990), as well as being associated to some extent with the microtubule organizing center and with the euchromatin of the nucleus. On activation of the neutrophils with the chemotactic peptide fMLP, cGMP kinase is translocated to cytoskeletal structures and the nucleus coincident with changes in the cellular morphology (Pryzwansky *et al.*, 1990; Wyatt *et al.*, 1991). Stimulation by fMLP causes a slight increase in cGMP levels and cGMP kinase can be shown to phosphorylate vimentin, an intermediate filament protein that is colocalized with the cGMP kinase in the stimulated neutrophil. Another system in which cGMP may also modulate cytoskeltal function is in the release of storage granules from cells. Cholecystokinin-8-stimulated amylase secretion from rat pancreatic acini is inhibited by cGMP analogs and by elevation of intracellular cGMP by sodium nitroprusside and phosphodiesterase inhibitors (Rogers *et al.*, 1988). It is possible that cGMP is important in modulating granule exocytosis (Laychock *et al.*, 1991; Schmidt *et al.*, 1992) in a variety of tissues, but the involvement of cGMP kinase in these processes remains a question.

The rapidly growing field involving NO production and release from a variety of cell types (Laychock *et al.*, 1991; O'Dell *et al.*, 1991; Schuman and Madison, 1991; Schmidt *et al.*, 1992; Ozaki *et al.*, 1992) raises the question whether all of the nontoxic effects of NO are mediated through activation of guanylate cyclase to produce increased levels of cGMP. If so, a reexamination of the tissue distribution of cGMP kinase and careful evaluation of its role in a broad range of physiological processes will be warranted.

IX. Cross-Activation

The possibility of the "cross-activation" of cGMP kinase by cAMP (Fig. 10) was first suggested by Foster *et al.* (1981), when in studies of the autophosphorylation of cGMP kinase *in vitro*, an increase in the affinity of the enzyme for cAMP was noted. Subsequently, Landgraf *et al.* (1986) have demonstrated that autophosphorylation of type Iα cGMP kinase

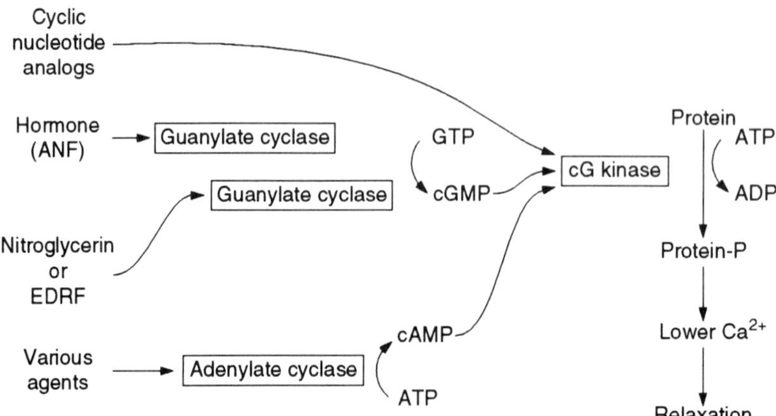

Fig. 10 Putative mechanism for cGMP and cAMP action to elicit relaxation in smooth muscle. In this scheme cAMP effects are attributable to its ability to bind to cGMP kinase and thereby "cross-activate" this enzyme in smooth muscle. ANF, atrial natriuretic factor; EDRF, endothelium-derived relaxation factor.

produces even greater changes in the affinity for cAMP (6- to 10-fold), and recent studies of the type Iβ cGMP kinase show that autophosphorylation of this isoenzyme also lowers the K_a for cAMP (J. A. Smith *et al.*, 1992). However, the physiological importance of this affinity change is not understood.

Elevation of cAMP, as well as cGMP, in smooth muscle (Hardman, 1984) has long been known to cause relaxation, and the respective cAMP and cGMP kinases have been presumed to serve as the intracellular mediators. Though it seems that cGMP kinase is the mediator of cGMP for relaxation (Francis *et al.*, 1988),there is now evidence that cAMP could also interact with the cGMP kinase to cause relaxation, an effect referred to as cross-activation. For instance, the poor potencies of cAMP kinase-specific analogs in producing relaxation in pig coronary arteries and guinea pig trachealis suggest that cAMP kinase is not involved in this process (Francis *et al.*, 1988), whereas similar cGMP kinase-specific analogs are highly potent. Other workers have also noted a weak correlation between the potencies with which cAMP analogs activate cAMP kinase and the potencies with which they induce smooth muscle relaxation; similarly, some agents that elevate cAMP are only weakly correlated with smooth muscle relaxation (Schultz *et al.*, 1977; Lincoln, 1983; Vegesna and Diamond, 1986; Hei *et al.*, 1991). As mentioned above, passaged smooth muscle cells that retain cAMP kinase activity but are deficient in cGMP kinase do not lower Ca^{2+} transients in response to elevation of cAMP (Lincoln *et al.*, 1990). However, following reintroduction of cGMP kinase to these cells, Ca^{2+} transients induced by either arginine vasopressin or depolarizing concentrations of K^+ are lowered in response to increases in either cGMP or cAMP. In studies of crude extracts of pig coronary arteries (Jiang *et al.*, 1992), elevation of cGMP by either atrial natriuretic factor or sodium nitroprusside increases the activity ratio of the cGMP kinase 2.3- and 2.6-fold, respectively, but cAMP kinase is not activated (Table II). However, when cAMP levels are increased by incubating the tissues with isoproterenol, both cAMP kinase and cGMP kinase are activated (2.3- and 1.6-fold, respectively.) These studies provide the first direct evidence that physiological concentrations of either cGMP or cAMP can bring about smooth muscle relaxation by activation of cGMP kinase. Thus, in this particular instance, one second messenger, i.e., cAMP, specific for activation of one pathway, apparently targets the receptor for another second messenger system.

It is too early to state whether cross-activation is a mechanism that is extremely rare or commonplace in biological systems. If it is commonplace, then it can be imagined that nature has evolved cGMP kinase to have a relatively high affinity for cGMP and low affinity for cAMP because

Table II

Effects of Agents That Elevate cGMP or cAMP on the Activity Ratios of cGMP Kinase and cAMP Kinase

Treatment	Activity ratio	
	cGMP Kinase −cGMP/+cGMP	cAMP Kinase −cAMP/+cAMP
Control	0.19	0.37
Sodium nitroprusside (10 μM)	0.49	0.35
Control	0.13	0.26
Atrial natriuretic factor (10 nM)	0.30	0.27
Control	0.12	0.22
Isoproterenol (100 μM)	0.18	0.50
Control	0.13	0.20
Forskolin (1 μM)	0.25	0.61
Control	0.09	0.32
Forskolin (10 μM)	0.44	0.93

of the presence of a high cAMP/cGMP concentration ratio in certain tissues such as smooth muscle. In other words, the relative affinities are appropriate to allow regulation of cGMP kinase by either cGMP or cAMP.

The converse mechanism, i.e., the activation of the cAMP kinase by cGMP, may function in some tissues as well. Enterotoxins that induce secretory diarrhea largely through increased Cl^- secretion cause increases in cyclic nucleotide content of intestinal epithelial cells (Field *et al.*, 1978). The heat-labile enterotoxins, such as cholera toxin, activate adenylate cyclase to increase cAMP levels, and the heat-stable enterotoxins activate guanylate cyclase to increase cGMP levels. These particular changes in cyclic nucleotide levels, i.e., increases in either cAMP or cGMP, appear to be very specific for the particular enterotoxin. Studies by Forte *et al.* (1992) in cultured T84 intestinal cells have provided strong evidence that the heat-stable enterotoxin stimulation of Cl^- secretion elicited by increases in intracellular cGMP is mediated by activation of cAMP kinase. Cyclic nucleotide analogs known to be specific activators of cAMP kinase also increase Cl^- secretion in these cells, but analogs exhibiting a similar affinity and specificity for cGMP kinases are ineffective. Although the cAMP kinase has been demonstrated to be present in these cells, there is no detectable cGMP kinase. These studies suggest that in T84 cells, elevation of either cAMP or cGMP may serve to increase the activity of the cAMP kinase, which could then phosphorylate the Cl^- channel itself or cellular proteins important in the regulation of the Cl^- channel.

Thus, despite the ability of cGMP kinase and cAMP kinase to catalyze specific phosphorylations in response to elevation of the respective nucleotides, in certain tissues the concentrations of the nucleotides can achieve levels sufficient to activate both kinases significantly.

X. Concluding Remarks

The cGMP kinases exhibit unique structural and kinetic features that could allow for distinct physiological functions. This also applies to the different isoforms of this kinase, although more evidence is needed to establish different cellular compartmentalization or functional properties. The best established role for cGMP kinase is smooth muscle relaxation in response to cGMP elevation by agents such as ANP, NO, or various drugs, although the precise pathway for this effect needs to be delineated. Similarly, inhibition of platelet activation by activators of cGMP kinase suggests that it may play an important role in this system. In some tissues cGMP kinase could serve as mediator for either cGMP or cAMP. However, cGMP kinase is only one of several cGMP-binding proteins in cells, and caution must be exercised in interpreting physiological responses induced by increases in cGMP levels. Many of the biochemical features of the cGMP kinase are known, and these can be used to more confidently ascribe physiological roles for the kinase, as opposed to other cGMP-binding proteins.

References

Aitken, A., Bilham, T., Cohen, P., Aswad, D., and Greengard, P. (1981). A specific substrate from rabbit cerebellum for guanosine-3′ : 5′ monophosphate-dependent protein kinase. Amino acid sequences at the two phosphorylation sites. *J. Biol. Chem.* **256,** 3501–3506.

Aitken, A., Hemmings, B., and Hofmann, F. (1984). Identification of the residues on cyclic GMP-dependent protein kinase that are autophosphorylated in the presence of cyclic AMP and cyclic GMP. *Biochim. Biophys. Acta* **790,** 219–225.

Altenhofen, W., Ludwig, J., Eismann, E., Kraus, W., Bonigk, W., and Kaupp, U. B. (1991). Control of ligand specificity in cyclic nucleotide-gated channels from rod photoreceptors and olfactory epithelium. *Proc. Natl. Acad. Sci. U.S.A.* **88,** 9868–9872.

Appel, R. G. (1990). Mechanism of atrial natriuretic factor-induced inhibition of rat mesangial cell mitogenesis. *Am. J. Physiol.* **259,** E312–E318.

Ashman, D. F., Lipton, R., Milicow, M. M., and Price, T. D. (1963). Isolation of cAMP and cGMP from rat urine. *Biochem. Biophys. Res. Commun.* **11,** 330–334.

Aswad, D. W., and Greengard, P. (1981a). A specific substrate from rabbit cerebellum for guanosine 3′ : 5′-monophosphate-dependent protein kinase: Purification and characterization. *J. Biol. Chem.* **256,** 3487–3493.

Aswad, D. W., and Greengard, P. (1981b). A specific substrate from rabbit cerebellum for guanosine 3′ : 5′-monophosphate-dependent protein kinase. Kinetic studies on its phosphorylation by guanosine 3′:5′-monophosphate-dependent and adenosine 3′:5′-monophosphate-dependent protein kinases. *J. Biol. Chem.* **256,** 3494–3500.

Atkinson, R. A., Saudek, V., Huggins, J. P., and Pelton, J. T. (1991). ^{1}H NMR and circular dichroism studies of the N-terminal domain of cyclic GMP dependent protein kinase: A leucine/isoleucine zipper. *Biochemistry* **30,** 9387–9395.

Baltensperger, K., Chiesi, M., and Carafoli, E. (1990). Substrates of cGMP kinase in vascular smooth muscle and their role in the relaxation process. *Biochemistry* **29,** 9753–9760.

Bandle, E., and Guidotti, A. (1979). Ontogenic studies of cGMP-dependent protein kinase in rat cerebellum. *J. Neurochem.* **32,** 1343–1347.

Beavo, J. A. (1988). Multiple isozymes of cyclic nucleotide phosphodiesterase. *Adv. Second Messenger Phosphoprotein Res.* **22,** 1–38.

Beavo, J. A., Hardman, J. G., and Sutherland, E. W. (1971). Stimulation of adenosine 3′,5′-monophosphate hydrolysis by guanosine 3′,5′-monophosphate. *J. Biol. Chem.* **246,** 3841–3846.

Beebe, S. J., and Corbin, J. D. (1986). Cyclic nucleotide-dependent protein kinases. *Enzymes* **17,** 43–111.

Boyles, J., Joyce, N., DeCamilli, P., Walter, U., and Mentone, S. (1984). Immunocytochemical localization of high levels of cGMP-dependent protein kinase in vascular and somatic smooth muscle cells, myofibroblasts and myoepithelial cells. *Adv. Cyclic Nucleotide Protein Phosphorylation Res.* **17,** A:65.

Butt, E., van Bemmelen, M., Fischer, L., Walter, U., and Jastorff, B. (1990). Inhibition of cGMP-dependent protein kinase by (Rp) guanosine 3′,5′-monophosphorothioates. *FEBS Lett.* **263,** 47–50.

Cheng, H. C., Kemp, B. E., Pearson, R. B., Smith, A. J., Misconi, L., Van Patten, S. M., and Walsh, D. A. (1986). A potent synthetic peptide inhibitor of the cAMP-dependent protein kinase. *J. Biol. Chem.* **261,** 989–992.

Cobb, C. E., Beth, A. H., and Corbin, J. D. (1987). Purification and characterization of an inactive form of cyclic AMP-dependent protein kinase containing bound cyclic AMP. *J. Biol. Chem.* **262,** 16566–16574.

Colbran, J. L., Francis, S. H., Leach, A. B., Thomas, M. K., Jiang, H., McAllister, L. M., and Corbin, J. D. (1992). A phenylalanine in peptide substrates provides for selectivity between cGMP- and cAMP-dependent protein kinases. *J. Biol. Chem.* **267,** 9589–9594.

Colbran, R. J., Smith, M. K., Schworer, C. M., Fong, Y.-L., and Soderling, T. R. (1989). Regulatory domain of calcium/calmodulin-dependent protein kinase II: Mechanism of inhibition and regulation by phosphorylation. *J. Biol. Chem.* **264,** 4800–4804.

Collins, P., Griffith, T. M., Henderson, R. H., and Lewis, M. J. (1986). Endothelium-derived relaxing factor alters Ca^{2+} fluxes in rabbit aorta: A cyclic guanosine monophosphate effect. *J. Physiol.* (*London*) **381,** 427–437.

Corbin, J. D., and Doskeland, S. O. (1983). Studies of two different intrachain cGMP-binding sites of cGMP-dependent protein kinase. *J. Biol. Chem.* **258,** 11391–11397.

Corbin, J. D., and Lincoln, T. M. (1978). Comparison of cAMP- and cGMP-dependent protein kinases. *Adv. Cyclic Nucleotide Res.* **9,** 159–170.

Corbin, J. D., Sugden, P. H., West, L., Flockhart, D. A., Lincoln, T. M., and McCarthy, D. (1978). Studies on the properties and mode of action of the purified regulatory subunit of bovine heart adenosine 3′,5′-monophosphate-dependent protein kinase. *J. Biol. Chem.* **253,** 3997–4003.

Corbin, J. D., Ogreid, D., Miller, J. P., Suva, R. H., Jastorff, B., and Doskeland, S. O. (1986). Studies of cGMP analog specificity and function of the two intrasubunit binding sites of cGMP-dependent protein kinase, *J. Biol. Chem.* **261,** 1208–1214.

Corbin, J. D., Thomas, M. K., Wolfe, L., Shabb, J. B., Woodford, T. A., and Francis, S. H. (1990). New insights into cGMP action. *Adv. Second Mess. and Phosphoprot. Res.* **24,** 411–418.

Cornwell, T. L., and Lincoln, T. M. (1989). Regulation of intracellular Ca^{2+} levels in cultured vascular smooth muscle cells. Reduction of Ca^{2+} by atriopeptin and 8-bromo-cyclic GMP is mediated by cyclic GMP-dependent protein kinase. *J. Biol. Chem.* **264,** 1146–1155.

Cornwell, T. L., Pryzwansky, K. B., Wyatt, T. A., and Lincoln, T. M. (1991). Regulation of sarcoplasmic reticulum protein phosphorylation by localized cGMP-dependent protein kinase in vascular smooth muscle cells. *Mol. Pharmacol.* **40,** 923–931.

Currie, M. G., Fok, K. F., Kato, J., Moore, R. J., Hamra, F. K., Duffin, K. L., and Smith, C. E. (1992). Guanylin: An endogenous activator of intestinal guanylate cyclase. *Proc. Natl. Acad. Sci. U.S.A.* **89,** 947–952.

DeCamilli, P., Miller, P. E., Levit, O., and Walter, U. (1984). Anatony of cerebellar Purkinje cells in the rat determined by a specific immunohistochemical marker. *Neuroscience* **11,** 761–817.

De Jaegere, S., Wuytack, F., Eggermont, J. A., Verboomen, H., and Casteels, R. (1990). Molecular cloning and sequencing of the plasma-membrane Ca^{2+} pump of pig smooth muscle. *Biochem. J.* **271,** 655–660.

DeJonge, H. R. (1981). Cyclic GMP-dependent protein kinase in intestinal brushborders. *Adv. Cyclic Nucleotide Res.* **14,** 315–333.

Doskeland, S. O., Vintermyr, O. K., Corbin, J. D., and Ogreid, D. (1987). Studies on the interactions between the cyclic nucleotide-binding sites of cGMP-dependent protein kinase. *J. Biol. Chem.* **262,** 3534–3540.

Edelman, A. M., Blumenthal, D. K., and Krebs, E. G. (1987). Protein serine/threonine kinases. *Annu. Rev. Biochem.* **56,** 567–613.

Eigenthaler, M., Nolte, C., Halbrugge, M., Walter, U. (1992). Concentration and regulation of cyclic nucleotides, cyclic nucleotide-dependent protein kinases and one of their major substrates in human platelets. Estimating the rate of cAMP regulated and cGMP-regulated protein phosphorylation in intact cells. *Eur. J. Biochem.* **205,** 471–481.

Felbel, J., Trockur, B., Ecker, T., Landgraf, W., and Hofmann, F. (1988). Regulation of cytosolic calcium by cAMP and cGMP in freshly isolated smooth muscle cells from bovine trachea. *J. Biol. Chem.* **263,** 16764–16771.

Feng, D.-F., and Doolittle, R. F. (1990). Progressive alignment and phylogenetic tree construction of protein sequences. *In* "Methods in Enzymology" (R. Doolittle, ed.), Vol. 183, pp. 375–387. Academic Press, San Diego, CA.

Fesenko, E. E., Kolesnikov, S. S., and Lyubarsky, A. L. (1985). Induction by cyclic GMP of cationic conductance in plasma membrane of retinal rod outer segment. *Nature (London)* **313,** 310–313.

Field, M., Graf, L. H., Laird, W. J., and Smith, P. L. (1978). Heat stable enterotoxin of *E. coli: In vitro* effects on guanylate cyclase activity, cyclic GMP concentration and ion transport in small intestine. *Proc. Natl. Acad. Sci. U.S.A.* **75,** 2800–2804.

Fiscus, R. R., Rapoport, R. M., and Murad, F. (1985). Atriopeptin II elevates cyclic GMP, activates cyclic GMP-dependent protein kinase and causes relaxation in rat thorax aorta. *Biochim. Biophys. Acta* **846,** 179–184.

Flockhart, D. A., Watterson, D. M., and Corbin, J. D. (1980). Studies on functional domains of the regulatory suynbunit of bovine heart adenosine 3′ : 5′-monophosphate dependent protein kinase. *J. Biol. Chem.* **255,** 4435–4440.

Forte, L. R., Thorne, P. K., Eber, S. L., Krause, W. J., Freeman, R. H., Francis, S. H., and Corbin, J. D. (1992). Stimulation of intestinal Cl^- transport by heat-stable enterotoxin: Activation of cAMP-dependent protein kinase by cGMP. *Am. J. Physiol.* **263,** C607–C615.

Foster, J. L., Guttman, J., and Rosen, O. M. (1981). Autophosphorylation of cGMP-dependent protein kinase. *J. Biol. Chem.* **256,** 5029–5036.

Francis, S. H., Lincoln, T. M., and Corbin, J. D. (1980). Characterization of a novel cGMP binding protein from rat lung. *J. Biol. Chem.* **255,** 620–626.

Francis, S. H., Noblett, B. D., Todd, B. W., Wells, J. N., and Corbin, J. D. (1988). Relaxation of vascular and tracheal smooth muscle by cyclic nucleotide analogs that preferentially activate purified cGMP-dependent protein kinase. *Mol. Pharmacol.* **34,** 506–517.

Francis, S. H., Woodford, T. A., Wolfe, L., and Corbin, J. D. (1988–1989). Types Iα and Iβ isozymes of cGMP-dependent protein kinase: Alternative mRNA splicing may produce different inhibitory domains. *Second Messengers Phosphoproteins* **12,** 301–310.

Francis, S. H., Wilson, C. P., and Corbin, J. D. (1992). On the question of the requirement for basic amino acids in the pseudosubstrate autoinhibitory domain of cGMP-dependent protein kinase. *FASEB J.* **6,** A315.

Furchgott, R. F. (1987). Studies on relaxation of rabbit aorta by sodium nitrite: The basis for the proposal that the acid-activatable inhibitory factor from bovine retractor penis is inorganic nitrite and the EDRF is NO. *In* "Mechanism of Vasodilation" (P. M. Vanhoutte, ed.). Raven Press, New York.

Furukawa, K., and Nakamura, H. (1987). cGMP regulation of the plasma membrane (Ca^{2+}/Mg^{2+})ATPase in vascular smooth muscle. *J. Biochem.* (*Tokyo*) **101,** 287–290.

Furukawa, K., Tawada, Y., and Shigekawa, M. (1988). Regulation of the plasma membrane Ca^{2+}-pump by cyclic nucleotides in cultured vascular smooth muscle cells. *J. Biol. Chem.* **263,** 8058–8065.

Garg, U. C., and Hassid, A. (1989). Nitric oxide-generating vasodilators and 8-bromo-cyclic guanosine monophosphate inhibit mitogenesis and proliferation of cultured rat vascular smooth muscle cells. *J. Clin. Invest.* **83,** 1774–1777.

Geahlen, R. L., and Krebs, E. G. (1980). Regulatory subunit of the type I cAMP-dependent protein kinase as an inhibitor and substrate of the cGMP-dependent protein kinase. *J. Biol. Chem.* **255,** 1164–1169.

Geiger, J., Nolte C., Butt, E., Sage, S. O., and Walter, U. (1992). Role of cGMP and cGMP-dependent protein kinase in nitrovasodilator inhibition of agonist-evoked calcium elevation in human platelets. *Proc. Natl. Acad. Sci. U.S.A.* **89,** 1031–1035.

Gill, G. N., Holdy, K. E., Walton, G. M., and Kanstein, C. B. (1976). Purification and characterization of cGMP-dependent protein kinase. *Proc. Natl. Acad. Sci. U.S.A.* **73,** 3918–3922.

Glass, D. B. (1990). Substrate specificity of the cyclic GMP-dependent protein kinase. *In* "Peptides and Protein Phosphorylation" (B. E. Kemp, ed.), pp. 210–238. CRC Press, Boca Raton, FL.

Glass, D. B., and Krebs, E. G. (1979). Comparison of the substrate specificity of adenosine 3′ : 5′-monophosphate- and guanosine 3′ : 5′-monophosphate-dependent protein kinases. *J. Biol. Chem.* **254,** 9728–9738.

Glass, D. B., and Krebs, E. G. (1982). Phosphorylation by guanosine 3′ : 5′-monophosphate-dependent protein kinase of synthetic peptide analogs of a site phosphorylated in histone H2B. *J. Biol. Chem.* **257,** 1196–1200.

Glass, D. B., and Smith, S. B. (1983). Phosphorylation by cyclic GMP-dependent protein kinase of a synthetic peptide corresponding to the autophosphorylation site in the enzyme. *J. Biol. Chem.* **258,** 14797–14803.

Glass, D. B., Feller, M. J., Levin, L. R., and Walsh, D. A. (1992). Structural basis for the low affinities of yeast cAMP-dependent and mammalian cGMP-dependent protein kinases for protein kinase inhibitor peptides. *Biochemistry* **31,** 1728–1734.

Halbrugge, M., and Walter, U. (1989). Purification of a vasodilator-regulated phosphoprotein from human platelets. *Eur. J. Biochem.* **185,** 41–50.

Halbrugge, M., Friedrich, C., Eigenthaler, M., Schanzenbacher, P., and Walter, U. (1990). Stoichiometric and reversible phosphorylation of a 46-kDa protein in human platelets in response to cGMP- and cAMP-elevating vasodilators. *J. Biol. Chem.* **265,** 3088–3093.

Hamet, P., and Coquil, J. F. (1978). Cyclic GMP binding and cyclic GMP phosphodiesterase in rat platelets. *J. Cyclic Nucleotide Res.* **4,** 281–290.

Hanks, S. K., Quinn, A. M., and Hunter, T. (1988). The protein kinase family: Conserved features and deduced phylogeny of the catalytic domains. *Science* **241,** 42–52.

Hardie, G. (1988). Pseudosubstrates turn off protein kinases. *Nature* (*London*) **335,** 592–593.

Hardman, J. G. (1984). Cyclic nucleotides and regulation of vascular smooth muscle. *J. Cardiovasc. Pharmacol.* **6,** 5639–5645.

Hashimoto, E., Takeda, M., Nishizuka, Y., Hamana, K., and Iwai, K. (1976). Studies on the sites in histones phosphorylated by adenosine 3′:5′-monophosphate-dependent and guanosine 3′:5′ monophosphate-dependent protein kinases. *J. Biol. Chem.* **251,** 6287–6293.

Hashimoto, E., Takio, K., and Krebs, E. G. (1981). Studies on the site in the regulatory subunit of type I cAMP-dependent protein kinase phosphorylated by cGMP-dependent protein kinase. *J. Biol. Chem.* **256,** 5604–5607.

Hashimoto, E., Takio, K., and Krebs, E. G. (1982). Amino acid sequence at the ATP-binding site of cGMP-dependent protein kinase. *J. Biol. Chem.* **257,** 727–733.

Haslam, R. J. (1987). Signal transduction in platelet activation. *In* "Thrombosis and Haemostasis" (M. Verstraete, J. Vermyleu, R. Lijnen, and J. Arnout, eds.), pp. 147–174. University Press, Leuven.

Hei, Y. J., MacDonnell, K. L., McNeill, J. H., and Diamond, J. (1991). Lack of *correlation* between activation of cyclic AMP-dependent protein kinase and inhibition of contraction of rat vas deferens by cyclic AMP analogs. *Mol. Pharmacol.* **39,** 233–238.

Heil, W. G., Landgraf, W., and Hofmann, F. (1987). A catalytically active fragment of cGMP-dependent protein kinase: Occupation of its cGMP-binding sites does not affect its phosphotransferase activity. *Eur. J. Biochem.* **168,** 117–121.

Hirata, M., Kohse, K. P., Chang, C. H., Ikebe, T., and Murad, F. (1990). Mechanism of cyclic GMP inhibition of inositol phosphate formation in rat aorta segments and cultured bovine aortic smooth muscle cells. *J. Biol. Chem.* **265,** 1268–1273.

Ho, M.-F., Bramson, H. N., Hansen, D. E., Knowles, J. R., and Kaiser, E. T. (1988). Stereochemical course of the phospho group transfer catalyzed by cAMP-dependent protein kinase. *J. Am. Chem. Soc.* **110,** 2680–2681.

Hofer, H. W., and Thalhofer, H. P. (1989). Purification and properties of cyclic 3′ : 5′-GMP-dependent protein kinase from the nematode *Ascaris suum. Arch. Biochem. Biophys.* **273,** 535–542.

Hofmann, F., and Flockerzi, V. (1983). Characterization of phosphorylated and native cGMP-dependent protein kinase. *Eur. J. Biochem.* **130,** 599–603.

Hofmann, F., and Gensheimer, H. P. (1983). Cyclic AMP-dependent protein kinase does not phosphorylate cyclic GMP-dependent protein kinase *in vitro*. FEBS Lett. **151,** 71–75.

Hofmann, F., and Sold, G. (1972). A protein kinase activity from rat cerebellum stimulated by guanosine-3′ : 5′-monophosphate. *Biochem. Biophys. Res. Commun.* **49,** 1100–1107.

Hofmann, F., Gensheimer, H. P., and Gobel, C. (1983). Autophosphorylation of cGMP-dependent protein kinase is stimulated only by occupancy of one of the two cGMP binding sites. *FEBS Lett.* **164,** 350–354.

Hofmann, F., Gensheimer, H. P., and Gobel, C. (1985). cGMP-dependent protein kinase: Autophosphorylation changes the characteristics of binding site 1. *Eur. J. Biochem.* **147,** 361–365.

Hofmann, F., Dostmann, W., Keilbach, A., Landgraf, W., and Ruth, P. (1992). Structure and physiological role of cGMP-dependent protein kinase. *Biochim. Biophys. Acta* **1135,** 51–60.

Holzmann, S. (1982). Endothelium-induced relaxation by acetylcholine is associated with larger rises in cyclic GMP in coronary arterial strips. *J. Cyclic Nucleotide Res.* **8,** 409–419.

House, C., and Kemp, B. E. (1987). Protein kinase C contains a pseudosubstrate prototype in its regulatory domain. *Science* **238,** 1726–1728.

Huggins, J. P., Cook, E. A., Piggott, J. R., Mattinsley, T. J., and England, P. J. (1989). Phospholamban is a good substrate for cyclic GMP-dependent protein kinase *in vitro,* but not in intact cardiac or smooth muscle. *Biochem. J.* **260,** 829–835.

Hurley, J. B. (1977). Molecular properties of the cGMP cascade of vertebrate photoreceptors. *Annu. Rev. Physiol.* **49,** 793–812.

Ignarro, L. J., and Kadowitz, P. J. (1985). The pharmacological and physiological role of cGMP in vascular smooth muscle relaxation. *Annu. Rev. Pharmacol. Toxicol.* **25,** 171–191.

Ignarro, L. J., Burke, T. M., Wood, K. S., Wolin, M. S., and Kadowitz, P. J. (1984). Association between cGMP accumulation and acetylcholine-elicited relaxation of bovine pulmonary artery. *J. Pharmacol. Exp. Ther.* **228,** 682–690.

Ignarro, L. J., Buga, G. M., Woods, K. S., Byrns, R. E., and Chaudhuri, G. (1987). EDRF produced and released from artery and vein is NO. *Proc. Natl. Acad. Sci. U.S.A.* **84,** 9265–9269.

Inoue, M., Kishimoto, A., Takai, Y., and Nishizuka, Y. (1976). Guanosine 3′:5′-monophosphate-dependent protein kinase from silkworm, properties of a catalytic fragment obtained by limited proteolysis. *J. Biol. Chem.* **251,** 4476–4478.

Ives, H. E., Casnellie, J. E., Greengard, P., and Jamieson, J. D. (1980). Subcellular localization of cyclic GMP-dependent protein kinase and its substrates in vascular smooth muscle. *J. Biol. Chem.* **255,** 3777–3785.

James, P. H., Pruschy, M., Vorherr, T. E., Penniston, J. T., and Carafoli, E. (1989). Primary structure of the cAMP-dependent phosphorylation site of the plasma membrane calcium pump. *Biochemistry* **28,** 4253–4258.

Jiang, H., Colbran, J. L., Francis, S. H., and Corbin, J. D. (1992). Direct evidence for cross-activation of cGMP-dependent protein kinase by cAMP in pig coronary arteries. *J. Biol. Chem.* **267,** 1015–1019.

Johnson, R. M., and Lincoln, T. M. (1985). Effects of nitroprusside, glyceryl trinitrate and 8-bromo-cGMP on phosphorylase *a* formation and myosin light chain phosphorylation in rat aorta. *Mol. Pharmacol.* **27,** 333–342.

Joyce, N. C., DeCamilli, P., and Boyles, J. (1984). Pericytes, like vascular smooth muscle cells, are immunocytochemically positive for cyclic GMP-dependent protein kinase. *Microvasc. Res.* **28,** 206–219.

Joyce, N. C., DeCamilli, P., Lohmann, S. M., and Walter, U. (1986). cGMP-dependent protein kinase is present in high concentrations in contractile cells of the kidney vasculature. *J. Cyclic Nucleotide Protein Phosphorylation Res.* **11,** 191–198.

Kalderon, D., and Rubin, G. M. (1989). cGMP-Dependent protein kinase genes in *Drosophila. J. Biol. Chem.* **264,** 10738–10748.

Katsuki, S., Arnold, W. P., and Murad, F. (1977). Effects of sodium nitroprusside, nitroglycerin and sodium azide on levels of cyclic nucleotides and mechanical activity of various tissues. *J. Cyclic Nucleotide Res.* **3,** 239–247.

Kaupp, U. B. (1991). The cyclic nucleotide-gated channels of vertebrate photoreceptors and olfactory epithelium. *Trends Neurosci.* **14,** 150–157.

Kaupp, U. B., and Koch, K.-W. (1990). Role of cGMP and Ca^{2+} in vertebrate photoreceptor excitation and adaptation. *Annu. Rev. Physiol.* **54,** 153–175.

Kaupp, U. B., Niidome, T., Tanabe, T., Terada, S., Bonigk, W., Stuhmer, W., Cook, N. J., Kangawa, K., Matsuo, H., Hirose, T., Miyata, T., and Numa, S. (1989). Primary structure and functional expression from complementary DNA of the rod photoreceptor cyclic GMP-gated channel. *Nature (London)* **342,** 762–766.

Kemp, B. E., and Pearson, R. B. (1990). Protein kinase recognition sequence motifs. *Trends Biochem. Sci.* **15,** 342–346.

Kemp, B. E., and Pearson, R. B. (1991). Intrasteric regulation of protein kinases and phosphatases. *Biochim. Biophys. Acta* **1094,** 67–76.

Kemp, B. E., Pearson, R. B., Guerriero, V., Jr., Bagchi, I. C., and Means, A. R. (1987). The calmodulin binding domain of chicken smooth muscle myosin light chain kinase contains a pseudosubstrate sequence. *J. Biol. Chem.* **262,** 2542–2548.

Kemp, B. E., Pearson, R. B., House, C., Robinson, P. J., and Means, A. R. (1989). Regulation of protein kinases by pseudosubstrate prototopes. *Cell. Signal.* **1,** 303–311.

Kennelly, P. J., and Krebs, E. G. (1991). Consensus sequences as substrate specificity determinants for protein kinases and protein phosphatases. *J. Biol. Chem.* **266,** 15555–15558.

Knighton, D. R., Zheng, J., Ten Eyck, L. F., Ashford, V. A., Xuong, N.-H., Taylor, S. S., and Sowadski, J. M. (1991a). Crystal structure of the catalytic subunit of cyclic adenosine monophosphate-dependent protein kinase. *Science* **253,** 407–414.

Knighton, D. R., Zheng, J., Ten Eyck, L. F., Xuong, N.-H., Taylor, S. S., and Sowadski, J. M. (1991b). Structure of a peptide inhibitor bound to the catalytic subunit of cyclic adenosine monophosphate-dependent protein kinase. *Science* **253,** 414–420.

Kobayashi, S., Kanaide, H., and Nakamura, M. (1985). Cytosolic free Ca^{2+} transients in cultured vascular smooth muscle cells: Microfluorometric measurements. *Science* **229,** 553–556.

Koch, K.-W., and Kaupp, U. B. (1985). Cyclic GMP directly regulates a cation conductance in membranes of bovine rods by a cooperative mechanism. *J. Biol. Chem.* **260,** 6788–6800.

Kumar, V. D. and Weber, I. T. (1992). Molecular model of the cyclic GMP-binding domain of the cyclic GMP-gated ion chanel. *Biochemistry* **31,** 4643–4649.

Kuo, J. F. (1975). Changes in relative levels of guanosine 3′:5′-monophosphate-dependent and adenosine 3′:5′-monophosphate-dependent protein kinases in lung, heart, and brain of developing guinea pigs. *Proc. Natl. Acad. Sci. U.S.A.* **72,** 2256–2259.

Kuo, J. F., and Greengard, P. (1970). Isolation and partial purification of a protein kinase activated by guanosine 3′,5′-monophosphate. *J. Biol. Chem.* **245,** 2493–2498.

Landgraf, W., and Hofmann, F. (1989). The amino terminus regulates binding to and activation of cGMP-dependent protein kinase. *Eur. J. Biochem.* **181,** 643–650.

Landgraf, W., Hullin, R., Gobel, C., and Hofmann, F. (1986). Phosphorylation of cGMP-dependent protein kinase increases the affinity for cyclic AMP. *Eur. J. Biochem.* **154,** 113–117.

Landgraf, W., Rack, M., Heil, W. G., and Hofmann, F. (1988). 8-(2-Carboxymethylthio)-cGMP, a site-1-selective compound for cGMP-dependent protein kinase. *Eur. J. Biochem.* **172,** 439–444.

Landgraf, W., Hofmann, F., Pelton, J. T., and Huggins, J. P. (1990). Effects of cyclic GMP on the secondary structure of cyclic GMP dependent protein kinase and analysis of the enzyme's amino-terminal domain by far-ultraviolet circular dichroism. *Biochemistry* **29,** 9921–9928.

Landgraf, W., Regulla, S., Meyer, H. E., and Hofmann, F. (1991). Oxidation of cysteines activates cGMP-dependent protein kinase. *J. Biol. Chem.* **266,** 16305–16311.

Laychock, S. G., Modica, M. E., and Cavanaugh, C. E. (1991). L-Arginine stimulates cyclic guanosine 3′,5′-monophosphate formation in rat islets of Langerhans and RINm5F insulinoma cells: Evidence for l-arginine:nitric oxide synthase. *Endocrinology (Baltimore)* **129,** 3043–3052.

Levin, L. R., and Zoller, M. J. (1990). Association of catalytic and regulatory subunits of cyclic AMP-dependent protein kinase requires a negatively charged side group at a conserved threonine. *Mol. Cell. Biol.* **10,** 1066–1075.

Lincoln, T. M. (1983). Effects of nitroprusside and 8-bromo-cGMP on the contractile activity of the rat aorta. *J. Pharmacol. Exp. Ther.* **224,** 100–107.

Lincoln, T. M. (1989). Cyclic GMP and mechanisms of vasodilation. *Pharmacol. Ther.* **41,** 479–502.

Lincoln, T. M., and Corbin, J. D. (1977). Adenosine 3′:5′-cyclic monophosphate- and guanosine 3′:5′-cyclic monophosphate-dependent protein kinases: Possible homologous proteins. *Proc. Natl. Acad. Sci. U.S.A.* **74,** 3239–3243.

Lincoln, T. M., and Corbin J. D. (1983). Characterization and biological role of the cGMP-dependent protein kinase. *Adv. Cyclic Nucleotide Res.* **15,** 139–192.

Lincoln, T. M., and Johnson, R. M. (1984). Possible role of cyclic GMP-dependent protein kinase in vascular smooth muscle function. *Adv. Cyclic Nucleotide Protein Phosphorylation Res.* **17,** 285–296.

Lincoln, T. M., Hall, C. L., Park, C. R., and Corbin, J. D. (1976). Guanosine 3′:5′-cyclic monophosphate binding proteins in rat tissues. *Proc. Natl. Acad. Sci. U.S.A.* **73,** 2559–2563.

Lincoln, T. M., Dills, W. L., Jr., and Corbin, J. D. (1977). Purification and subunit composition of guanosine 3′:5′-monophosphate-dependent protein kinase from bovine lung. *J. Biol. Chem.* **252,** 4269–4275.

Lincoln, T. M., Thompson, M., and Cornwell, T. L. (1988). Purification and characterization of two forms of cyclic GMP-dependent protein kinase from bovine aorta. *J. Biol. Chem.* **263,** 17632–17637.

Lincoln, T. M., Cornwell, T. L., and Taylor, A. E. (1990). cGMP-dependent protein kinase mediates the reduction of Ca^{2+} by cAMP in vascular smooth muscle cells. *Am. J. Physiol.* **258,** C399–C407.

Lohmann, S. M., and Walter, U. (1984). Regulation of cellular and subcellular concentrations and distribution of cyclic nucleotide-dependent protein kinases. *Adv. Cyclic Nucleotide Protein Phosphorylation Res.* **18,** 63–117.

Lohmann, S. M., Walter, U., Miller, P. E., Greengard, P., and DeCamilli, P. (1981). Immunohistochemical localization of cyclic GMP-dependent protein kinase in mammalian brain. *Proc. Natl. Acad. U.S.A.* **78,** 653–657.

Lohmann, S. M., Fischmeister, R., and Walter, U. (1991). Signal transduction by cGMP in heart. *Basic Res. Cardiol.* **86,** 503–514.

Ludwig, J., Margalit, T., Eismann, E., Lancet, D., and Kaupp, U. B. (1990). Primary structure of cAMP-gated channel from bovine olfactory epithelium. *FEBS Lett.* **270,** 24–29.

Mackenzie, C. W., III (1982). Bovine lung cyclic GMP-dependent protein kinase exhibits two types of specific cyclic GMP-binding sites. *J. Biol. Chem.* **257,** 5589–5593.

McCune, R. W., and Gill, G. N. (1979). Positive cooperativity in guanosine 3′:5′-monophosphate binding to guanosine 3′:5′-monophosphate-dependent protein kinase. *J. Biol. Chem.* **254,** 5083–5091.

Mery, P.-F., Lohmann, S. M., Walter, U., and Fischmeister, R. (1991). Ca^{2+} current is regulated by cyclic GMP-dependent protein kinase in mammalian cardiac monocytes. *Proc. Natl. Acad. Sci. U.S.A.* **88,** 1197–1201.

Miglietta, L. A. P., and Nelson, D. L. (1988). A novel cGMP-dependent protein kinase from *Paramecium*. *J. Biol. Chem.* **263,** 16096–16105.

Miki, N., Keirns, J. J., Marcus, F. R., Freeman, J., and Bitensky, M. W. (1973). Regulation of cyclic nucleotide concentration in photoreceptors: An ATP-dependent stimulation of cyclic nucleotide phosphodiesterase by light. *Proc. Natl. Acad. Sci. U.S.A.* **70,** 3820–3824.

Miller, J. P., Uno, H., Christensen, L. J., Robins, R. K., and Meyer, R. B., Jr. (1981). Effect of modification of the 1-, 2-, and 6-positions of 9-β-D-ribofuranosyl-purine cyclic 3′,5′-phosphate on the cyclic nucleotide specificity of adenosine 3′,5′ phosphate and guanosine cyclic 3′,5′ phosphate-dependent protein kinases. *Biochem. Pharmacol.* **30,** 509–515.

Monken, C. E., and Gill, G. N. (1980). Structural analysis of cGMP-dependent protein kinase using limited proteolysis. *J. Biol. Chem.* **255,** 7067–7070.

Murad, F. (1986). Cyclic guanosine monophosphate as a mediator of vasodilation. *J. Clin. Invest.* **78,** 1–5.

Nakamura, T., and Gold, G. H. (1987). A cyclic nucleotide-gated conductance in olfactory receptor cilia. *Nature (London)* **325,** 442–444.

Nolte, C., Eigenthaler, M., Schanzenbacher, P., and Walter, U. (1991). Endothelial cell-dependent phosphorylation of a platelet protein mediated by cAMP- and cGMP-elevating factors. *J. Biol. Chem.* **266,** 14808–14812.

Nonoguchi, H., Knepper, M. A., and Manganiello, V. C. (1987). Effects of atrial natriuretic factor on cyclic guanosine monophosphate and cyclic adenosine monophosphate accumulation in microdissected nephron segments from rats. *J. Clin. Invest.* **79,** 500–507.

O'Dell, T. J., Hawkins, R. D., Kandel, E. R., and Arancio, O. (1991). Tests of the roles of two diffusible substances in long-term potentiation: Evidence for nitric oxide as a possible early retrograde messenger. *Proc. Natl. Acad. Sci. U.S.A.* **88,** 11285–11289.

O'Donnell, M. E., and Owen, N. E. (1986). Role of cyclic GMP in atrial natriuretic factor stimulation of Na^+, K^+, Cl^--cotransport in vascular smooth muscle cells. *J. Biol. Chem.* **261,** 15461–15466.

Ogreid, D., Ekanger, R., Suva, R. H., Miller, J. P., Sturm, P., Corbin, J. D., and Doskeland, S. O. (1985). Activation of protein kinase isozymes by cyclic nucleotide analogs used simply or in combination. *Eur. J. Biochem.* **150,** 219–227.

Ozaki, H., Blondfield, D. P., Hori, M., Publicover, N. G., Kato, I., and Sanders, K. M. (1992). Spontaneous release of nitric oxide inhibits electrical Ca^{2+} and mechanical transients in canine gastric smooth muscle. *J. Physiol. (London)* **445,** 231–247.

Palmer, P. M. G., Ferridge, A. G., and Moncada, S. (1987). Release of nitric oxide accounts for the biological activity of endothelium-derived relaxing factor. *Nature (London)* **327,** 524–526.

Parks, T. P., Nairn, A. C., Greengard, P., and Jamieson, J. D. (1987). The cyclic nucleotide-dependent phosphorylation of aortic smooth muscle membrane proteins. *Arch. Biochem. Biophys.* **255,** 361–371.

Pearson, R. B., Ito, M., Morrice, N. A., Smith, J. J., Condran, J. R., Wettenhall, R. E. H., Kemp, B. E., and Hartshorne, D. J. (1991). Proteolytic cleavage sites in smooth muscle myosin-light-chain kinase and their relation to structural and regulatory domains. *Eur. J. Biochem.* **200,** 723–730.

Pryzwansky, K. B., Wyatt, T. A., Nichols, H., and Lincoln, T. M. (1990). Compartmentalization of cyclic GMP-dependent protein kinase in formyl-peptide stimulated neutrophils. *Blood* **76,** 612–618.

Raeymaekers, L., Hofmann, F., and Casteels, R. (1988). Cyclic GMP-dependent protein kinase phosphorylates phospholamban in isolated sarcoplasmic reticulum from cardiac and smooth muscle. *Biochem. J.* **252,** 269–273.

Rapoport, R. M. (1986). cGMP inhibition of contraction may be mediated through inhibition of phosphatidylinositol hydrolysis in rat aorta. *Circ. Res.* **58,** 407–410.

Rapoport, R. M., and Murad, F. (1983). Agonist-induced endothelium-dependent relaxation in rat thoracic aorta may be mediated through cGMP. *Circ. Res.* **52,** 352–357.

Rapoport, R. M., Draznin, M. B., and Murad, F. (1983). Endothelium-dependent relaxation in rat aorta may be mediated through cGMP-dependent protein phosphorylation. *Nature (London)* **306,** 174–176.

Rashatwar, S. S., Cornwell, T. L., and Lincoln, T. M. (1987). Effects of 8-bromo-cGMP on Ca^{2+} levels in vascular smooth muscle cells: Possible regulation of Ca^{2+}-ATPase by cGMP-dependent protein kinase. *Proc. Natl. Acad. Sci. U.S.A.* **84,** 5685–5689.

Robinson-Steiner, A. M., and Corbin, J. D. (1986). Protein phosphorylation in the heart. *In* "The Heart and Cardiovascular System" (H. A. Fozzard, R. B. Jennings, E. Haber, A. M. Katz, and H. E. Morgan, eds.), pp. 887–910. Raven Press, New York.

Rogers, J., Hughes, R. G., and Mathews, E. K. (1988). Cyclic GMP inhibits protein kinase C-mediated secretion in rat pancreatic acini. *J. Biol. Chem.* **263,** 3713–3719.

Roskoski, R., Jr., Vulliet, P. R., and Glass, D. B. (1987). Phosphorylation of tyrosine hydroxylase by cyclic GMP-dependent protein kinase. *J. Neurochem.* **48,** 840–845.

Rossman, M. G., Moras, D., and Olsen, K. (1974). Chemical and biological evolution of a nucleotide-binding protein. *Nature (London)* **250,** 194–199.

Ruth, P., Landgraf, W., Keilbach, A., May, B., Egleme, C., and Hofmann, F. (1991). The activation of expressed cGMP-dependent protein kinase isozymes Iα and Iβ is determined by the different amino-termini. *Eur. J. Biochem.* **202,** 1339–1344.

Sandberg, M., Natarajan, V., Ronander, I., Kalderon, D., Walter, U., Lohmann, S. M., and Jahnsen, T. (1989). Molecular cloning and predicted full-length amino acid sequence of the type Iβ isozyme of cGMP-dependent protein kinase from human placenta: Tissue distribution and developmental changes in rat. *FEBS Lett.* **255,** 321–329.

Sandberg, M., Natarajan, V., Orstavik, S., Lohmann, S. M., and Jahnsen, T. (1991). The human type I cGMP-dependent protein kinase gene. *NATO ASI Ser. Biol. Signal Transd.* **H-52,** 301–308.

Sarevic, B., Brookes, V., Martin, T. J., Kemp, B. E., and Robinson, P. J. (1989). Atrial natriuretic peptide-dependent phosphorylation of smooth muscle cell particulate fraction proteins is mediated by cGMP-dependent protein kinase. *J. Biol. Chem.* **264,** 20648–20654.

Schlichter, D. J., Casnellie, J. E., and Greengard, P. (1978). An endogenous substrate for cGMP-dependent protein kinase in mammalian cerebellum. *Nature (London)* **273,** 61–62.

Schmidt, H. H., Warner, T. D., Ishii, K., Sheng, H., and Murad, F. (1992). Insulin secretion from pancreatic B cells caused by L-arginine-derived nitrogen oxides. *Science* **255,** 721–723.

Schultz, K. D., Schultz, K., and Schultz, G. (1977). Sodium nitroprusside and other smooth muscle relaxants increase cGMP levels in rat ductus deferens. *Nature (London)* **265,** 750–751.

Schuman, E. M., and Madison, D. V. (1991). The intercellular messenger nitric oxide is required for long-term potentiation. *Science* **254,** 1503–1506.

Scott, J. D. (1991). Cyclic nucleotide-dependent protein kinases. *Pharmacol. Ther.* **50,** 123–145.

Scott, J. D., Fischer, E. H., Takio, K., Demaille, J. G., and Krebs, E. G. (1985). Amino acid sequence of the heat-stable inhibitor of the cAMP-dependent protein kinase from rabbit skeletal muscle. *Proc. Natl. Acad. Sci. U.S.A.* **82,** 5732–5736.

Sekhar, K. R., Hatchett, R. J., Shabb, J. B., Wolfe, L., Francis, S. H., Wells, J. N., Jastorff, B., Butt, E., Chakinala, M. M., and Corbin, J. D. (1992). Relaxation of pig coronary arteries by new and potent cGMP analogs that selectively activate type Iα, compared with type Iβ, cGMP-dependent protein kinase. *Mol. Pharmacol.* **42,** 103–108.

Shabb, J. B., and Corbin, J. D. (1992). Cyclic nucleotide-binding domains in proteins having diverse functions. *J. Biol. Chem.* **267,** 5723–5726.

Shabb, J. B., Ng, L., and Corbin, J. D. (1990). One amino acid change produces a high affinity cGMP-binding site in cAMP-dependent protein kinase. *J. Biol. Chem.* **265,** 16031–16034.

Shabb, J. B., Buzzeo, B. D., Ng, L., and Corbin, J. D. (1991). Mutating protein kinase cAMP-binding sites into cGMP-binding sites. *J. Biol. Chem.* **266,** 24320–24326.

Shapiro, M. B., and Senepathy, P. (1987). RNA splice junction of different classes of eukaryotes: Sequence statistics and functional implications in gene expression. *Nucleic Acids Res.* **15,** 7155–7174.

Smith, J. A., Francis, S. H., and Corbin, J. D. (1992). Activation of the type Iβ isozyme of cGMP-dependent protein kinase by preincubation with MgATP and cAMP. *FASEB J.* **6,** A315.

Smith, M. K., Colbran, R. J., Brickey, D. A., and Soderling T. R. (1992). Functional determinants in the autoinhibitory domain of calcium/calmodulin-dependent protein kinase II: Role of His^{282} and multiple basic residues *J. Biol. Chem.* **267,** 1761–1768.

Soderling, T. R. (1990). Protein kinases: Regulation by autoinhibitory domains. *J. Biol. Chem.* **265,** 1823–1826.

Stockert, R. J., Paietta, E., Racevskis, J., and Morell, A. G. (1992). Posttranscriptional regulation of the asialoglycoprotein receptor by cGMP. *J. Biol. Chem.* **267,** 56–59.

Stryer, L. (1986). Cyclic GMP cascade of vision. *Annu. Rev. Neurosci.* **9,** 87–119.

Takio, K., Smith, S. B., Krebs, E. G., Walsh, K. A., and Titani, K. (1984a). Amino acid sequence of the regulatory subunit of bovine type II adenosine cyclic 3′,5′-phosphate dependent protein kinase. *Biochemistry* **23,** 4200–4206.

Takio, K., Wade, R. D., Smith, S. B., Krebs, E. G., Walsh, K. A., and Titani, K. (1984b). Guanosine cyclic 3′,5′-phosphate dependent protein kinase, a chimeric protein homologous with two separate protein families. *Biochemistry* **23,** 4207–4218.

Taylor, S. S. (1989). cAMP-dependent protein kinase: Model for an enzyme family. *J. Biol. Chem.* **264,** 8443–8446.

Taylor, S. S., Buechler, J. A., and Yonemoto, W. (1990). cAMP-dependent protein kinase: Framework for a diverse family of regulatory enzymes. *Annu. Rev. Biochem.* **59,** 971–1005.

Thomas, M. K., Francis, S. H., and Corbin, J. D. (1990). Substrate- and kinase directed regulation of phosphorylation of a cGMP-binding phosphodiesterase by cGMP. *J. Biol. Chem.* **265,** 14971–14978.

Titani, K., Sasagawa, T., Ericsson, L. H., Kumar, S., Smith, S. B., Krebs, E. G., and Walsh, K. A. (1984). Amino acid sequence of the regulatory subunit of bovine type I adenosine cyclic 3′,5′-phosphate dependent protein kinase. *Biochemistry* **23,** 4193–4199.

Tremblay, J., Gerzer, R., and Hamet, P. (1988). Cyclic GMP in cell function. *Adv. Second Messenger Phosphoprotein Res.* **22,** 319–383.

Uhler, M. (1993). Cloning and expression of a novel cyclic GMP-dependent kinase from mouse brain. *J. Biol. Chem.* **268,** 13586–13591.

Vardanis, A. (1980). A unique cyclic nucleotide-dependent protein kinase. *J. Biol. Chem.* **255,** 7238–7243.

Vegesna, R. V. K., and Diamond, J. (1986). Effects of prostaglandin E_1, isoproterenol and forskolin on cyclic AMP levels and tension in rabbit aortic rings. *Life Sci.* **39,** 301–311.

Waldman, S. A., Rapoport, R. M., and Murad, F. (1984). Atrial natriuretic factor selectively activates particulate guanylate cyclase and elevates cyclic GMP in rat tissues. *J. Biol. Chem.* **259,** 14332–14334.

Waldmann, R., and Walter, U. (1989). Cyclic nucleotide-elevating vasodilators inhibit platelet aggregation at an early step of the activation cascade. *Eur. J. Pharmacol.* **159,** 317–320.

Waldmann, R., Bauer, S., Gobel, C., Hofmann, F., Jakobs, K.-H., and Walter, U. (1986). Demonstration of cGMP-dependent protein kinase and cGMP-dependent phosphorylation in cell free extracts of platelets. *Eur. J. Biochem.* **158,** 203–210.

Waldmann, R., Nieberding, M., and Walter, U. (1987). Vasodilator-stimulated protein phosphorylation in platelets is mediated by cAMP- and cGMP-dependent protein kinases. *Eur. J. Biochem.* **167,** 441–448.

Walsh, D. A., Angelos, K. L., van Patten, S. M., Glass, D. B., and Garetto, L. P. (1990). The inhibitor protein of the cAMP-dependent protein kinase. *In* "Peptides and Protein Phosphorylation" (B. E. Kemp, ed.), pp. 43–84. CRC Press, Boca Raton, FL.

Walter, U. (1981). Distribution of cyclic GMP-dependent protein kinase in various rat tissues and cell lines determined by a sensitive and specific radioimmunoassay. *Eur. J. Biochem.* **118,** 339–346.

Walter, U. (1989). Physiological role of cGMP and cGMP-dependent protein kinase in the cardiovascular system. *Rev. Physiol. Biochem. Pharmacol.* **113,** 42–88.

Walter, U., DeCamilli, P., Lohmann, S. M., and Greengard, P. (1981). Regulation and cellular localization of cAMP- and cGMP-dependent protein kinase. *Cold Spring Harbor Conf. Cell Proliferation* **8,** 141–157.

Walter, U., Waldmann, R., and Nieberding, M. (1988). Intracellular mechanism of action of vasodilators. *Eur. Heart J. 9 Supp.* **H,** 1–6.

Wanner, R., and Wurster, B. (1990). Cyclic GMP-activated protein kinase from *Dictyostelium discoideum. Biochim. Biophys. Acta* **1053,** 179–184.

Weber, I. T., Takio, K., Titani, K., and Steitz, T. A. (1982). The cAMP-binding domains of the regulatory subunit of cAMP-dependent protein kinase and the catabolite gene activator protein are homologous. *Proc. Natl. Acad. Sci. U.S.A.* **79,** 7679–7683.

Weber, I. T., Shabb, J. B., and Corbin, J. D. (1989). Predicted structures of the cGMP binding domains of the cGMP-dependent protein kinase: A key alanine/threonine difference in evolutionary divergence of cAMP and cGMP binding sites. *Biochemistry* **28,** 6122–6127.

Wernet, W., Flockerzi, V., and Hofmann, F. (1989). The cDNA of the two isoforms of bovine cGMP-dependent protein kinase. *FEBS Lett.* **251,** 191–196.

Wolfe, L., Francis, S. H., Landiss, L. R., and Corbin, J. D. (1987). Interconvertible cGMP-free and cGMP-bound forms of cGMP-dependent protein kinase in mammalian tissues. *J. Biol. Chem.* **262,** 16906–16913.

Wolfe, L., Francis, S. H., and Corbin, J. D. (1989a). Properties of a cGMP-dependent monomeric protein kinase from bovine aorta. *J. Biol. Chem.* **264,** 4157–4162.

Wolfe, L., Corbin, J. D., and Francis, S. H. (1989b). Characterization of a novel isozyme of cGMP-dependent protein kinase from bovine aorta. *J. Biol. Chem.* **264,** 7734–7741.

Wyatt, T. A., Lincoln, T. M., and Pryzwansky, K. B. (1991). Vimentin is transiently co-localized with and phosphorylated by cyclic GMP-dependent protein kinase in formyl-peptide-stimulated neutrophils. *J. Biol. Chem.* **266,** 21274–21280.

Yeaman, S. J., Cohen, P., Watson, D. C., and Dixon, G. H. (1977). The substrate specificity of adenosine 3′,5′-monophosphate-dependent protein kinase of rabbit skeletal muscle. *Biochem. J.* **162,** 411–421.

Yoshida, Y., Sun, H.-T., Cai, J.-Q., and Imai, S. (1991). Cyclic GMP-dependent protein kinase stimulates the plasma membrane Ca^{2+} pump ATPase of vascular smooth muscle via phosphorylation of a 240-kDa protein. *J. Biol. Chem.* **266,** 19819–19825.

Yoshida, Y., Cai, J.-Q., and Imai, S. (1992). Plasma membrane Ca^{2+}-pump ATPase is not a substrate for cGMP-dependent protein kinase. *J. Biochem.* (*Tokyo*) **111,** 559–562.

Effects of Cyclic GMP on Smooth Muscle Relaxation

Timothy D. Warner,* Jane A. Mitchell,* Hong Sheng,† and Ferid Murad‡

**William Harvey Research Institute*
St. Bartholomew's Hospital Medical College
London EC1M 6BQ, United Kingdom

†Department of Pharmacology
University of California, Los Angeles
School of Medicine
Los Angeles, CA 90024

‡Molecular Geriatrics Corporation
Lake Bluff, Illinois 60044

I. Introduction

It is now well established that elevation of cyclic GMP within smooth muscle leads to relaxation. This elevation can be the result of the activity of two distinct guanylyl cyclase enzymes, cytoplasmic and membrane-bound, respectively, which convert GTP to cyclic GMP. Although both of these enzymes produce the same second messenger, they become stimulated to do this by clearly different agents and possess different molecular structures. The biochemical pathways involved in the stimulation and activation of these isoforms are discussed at greater length elsewhere (e.g., see Murad, 1986; Waldman and Murad, 1987; Rosenzweig and Seid-

Advances in Pharmacology, Volume 26

man, 1991) and in this volume. They will be mentioned here to illustrate the physiological and pharmacological routes by which guanylyl cyclase in smooth muscle may become activated, as this is the central theme of this review. We will also attempt to provide an outline of the current theories on the ways in which elevation of intracellular cyclic GMP leads to relaxation of smooth muscle. However, these biochemical details are reviewed at greater length both within this volume and elsewhere (e.g., see Lincoln, 1989). In addition, we will also provide information on neural pathways that may stimulate cyclic GMP formation in smooth muscle, variations in regional and tissue responses to agents that elevate cyclic GMP, and other evidence showing the diversity of cyclic GMP responses in smooth muscle.

II. Isoforms of Guanylyl Cyclase Present in Smooth Muscle

There are two isoforms of guanylyl cyclase present in smooth muscle. These are situated, respectively, in the soluble (cytosolic) and particulate (membrane-bound) fractions of the cells. They are stimulated by clearly different agents, although stimulation of either results in elevated levels of intracellular cyclic GMP and relaxation.

A. Particulate Guanylyl Cyclase

1. Atrial Natriuretic Peptide

Although the membrane-associated form of guanylyl cyclase was originally isolated from sea urchin spermatozoa (Garbers, 1976; Radany *et al.*, 1983), subsequent work has demonstrated its presence in almost all mammalian tissues, including many smooth muscles. It is now known that there are different forms of particulate guanylyl cyclase that act as selective receptors for a family of circulating peptides (see below). Indeed, it was the discovery of these endogenous activators of particulate guanylyl cyclase that propelled this field. This area of study grew out of work that demonstrated the presence of membrane-bound granules within the atria, but not the ventricles of guinea pigs (Kisch, 1956) and subsequently the demonstration by de Bold *et al.* (1981) that injection of a granule-enriched extract from rat atria provoked a rapid diuresis in rats. This led to the proposal of a natriuretic agent within these granules that was given the appropriate name, atrial natriuretic factor (ANF) or atrial natriuretic peptide (ANP). Within a very short time it was reported that extracts of atria also caused powerful relaxations of isolated smooth muscle and cardiac muscle preparations (Currie *et al.*, 1983; Kleinert *et al.*, 1984; Winquist *et al.*, 1984a) and that this relaxation was associated with the stimulation of particulate guanylyl cyclase (Winquist *et al.*, 1984b; Waldman *et al.*, 1984). So was born the clear idea of a circulating factor, derived from the atria, which, acting via particulate guanylyl cyclase, could elevate intracellular cyclic GMP and so relax smooth muscle. The effects of atrial natriuretic peptide are also discussed in Chapter 5.

It is now known that ANP is a 28 amino acid peptide with a 17 member ring formed by a disulfide bond between the amino acid residues at positions 7 and 23, and that its amino acid sequence is conserved across species, apart from variations at position 12 (Rosenzweig and Seidman, 1991). In addition to ANP, there are also other related peptides classified within groups as B natriuretic peptide (BNP) and C natriuretic peptide (CNP). Although BNP was originally isolated from the porcine brain, it was demonstrated subsequently that more was present within the cardiac atria than within the CNS (Saito *et al.*, 1989). BNP, like ANP, contains a 17 member ring structure but within this ring it differs from ANP at 7 positions. In addition, in the portions of the molecule outside the ring other and greater differences are shown at the amino and carboxyl extensions of the molecule. Various forms of BNP have been reported, including a 32 amino acid form in the circulations of humans and pigs (Kojima *et al.*, 1989); a 45 amino acid form, classified as iso-ANP, within the circulation of rodents (Flynn *et al.*, 1989; Kambayashi *et al.*, 1989); and a noncirculating 26 amino acid BNP from the pig brain (Sudoh *et al.*, 1989). Type C natriuretic peptide is a 22 amino acid peptide that also contains the 17 member ring but no carboxyl extension past this point (Sudoh *et al.*, 1990). Its physiological significance is currently unclear, although it may act on a specific receptor population (see below).

2. Enzyme Structure: Receptors for Atrial Natriuretic Peptide

The effects of ANP and related peptides are mediated via the activity of distinct particulate guanylyl cyclases, which are also ANP receptors (Kuno *et al.*, 1986). At this time two mammalian, membrane-bound guanylyl cyclase-linked ANF receptors, classified as GC-A and GC-B, that may be present on smooth muscle have been identified (Chinkers *et al.*, 1989; Lowe *et al.*, 1989; Thorpe and Garbers, 1989). These have a molecular mass of 120–140 kDa and are present as single-chain polypeptides. These two receptors may provide selectivity for the effects of circulating ANF peptides, as they have been shown to be activated by different isoforms: GC-A by ANP (Chinkers *et al.*, 1989; Lowe *et al.*, 1989) and GC-B by BNP more than ANP. However, the fact that relatively high concentrations of either ANP or BNP are required to stimulate the GC-B receptor suggests that neither is an endogenous ligand (Chang *et al.*, 1989); in fact, recent evidence has been presented which suggests that CNP may be the true ligand (Koller *et al.*, 1991). In addition, there is also one so-called clearance receptor (60–70kDa). Additional discussion about ANP receptors and guanylyl cyclase can be found in Chapter 5.

B. Cytosolic Guanylyl Cyclase

Agents that stimulate cytosolic, or soluble, guanylyl cyclase and so elevate cyclic GMP have been in therapeutic use for a large number of years. For instance, one of the best known of these agents, glyceryl trinitrate (GTN), was first synthesized about one and a half centuries ago. Very soon after this it was proposed as a homoeopathic remedy for a number of diseases including angina pectoris (Ahlner *et al.*, 1991a). However, it was not until about 15 years ago that it was clearly demonstrated that smooth muscle relaxing agents including sodium nitroprusside (SNP), sodium nitrite, isosorbide dinitrate (IDN), GTN, hydroxylamine, and sodium azide also increase tissue levels of cyclic GMP (Kimura *et al.*, 1975a,b; Arnold *et al.*, 1977; Katsuki and Murad, 1977; Katsuki *et al.*, 1977a,b). At this same time it was also noted that all the agents had in common the possibility to release, either spontaneously or via enzymatic degradation, the common factor nitric oxide (NO). So it was suggested that all these agents acted via the release of NO (Katsuki *et al.*, 1977a; Arnold *et al.*, 1977) and that this was the active moiety that stimulated soluble guanylyl cyclase, via a heme-dependent mechanism (Ignarro, 1990). The common term coined for this group of agents was therefore nitrovasodilators (Murad, 1986).

1. Metabolism of Nitrovasodilators: Tolerance

Although NO is capable of directly stimulating guanylyl cyclase (Arnold *et al.*, 1977; Katsuki *et al.*, 1977a) and so relaxing smooth muscle (Katsuki *et al.*, 1977b; Greutter *et al.*, 1979) organic nitrate and nitrate esters (e.g., GTN, IDN) require metabolism to release NO. Studies have shown that they may react with thiol groups of sulfhydryl compounds to generate NO via *S*-nitrosothiol intermediates (Needleman *et al.*, 1969; Ignarro *et al.*, 1980, 1981) and this process of denitration has been shown to correlate with the potency of agents to elevate cyclic GMP and relax smooth muscle (Wingren *et al.*, 1981; Yeates *et al.*, 1985; Kawamoto *et al.*, 1990). However, this picture may be complicated by the presence of multiple pathways leading to degradation of organic nitrate esters, not all of which may liberate NO (Feelisch *et al.*, 1988).

Many groups have reported that repeated administration of organic nitrates leads to tolerance to their effects, in both clinical and experimental studies, and this may well be linked to a reduction in the ability of the smooth muscle to liberate NO from these molecules (Bennett *et al.*, 1989; Ahlner *et al.*, 1991a). However, it must be noted that tolerant tissues have also been shown to have smaller responses to nitric oxide and SNP (which spontaneously releases NO) (e.g., Romanin and Kukovetz, 1989). Thus there may additionally be a frank change in the guanylyl cyclase enzyme associated with a reduction in activity, that is only reversed after the synthesis of new enzyme (Waldman *et al.*, 1986; Schröder *et al.*, 1988).

2. Endothelium-Derived Relaxing Factor

The finding that nitric oxide was such a powerful activator of guanylyl cyclase led directly to the question of what was the endogenous substance within tissues that stimulated this pathway. So came the suggestion that some hormones could increase cyclic GMP accumulation in tissues by altering the rate of formation of an "endogenous nitrovasodilator" from some precursor substance, and that this endogenous nitrovasodilator could be NO itself, or a related substance (Murad *et al.*, 1978a,b). This field of research then took an unexpected turn when Furchgott and Zawadzki (1980) reported that the relaxation of vascular smooth muscle induced by a variety of agents including acetylcholine, substance P, ATP and ADP, bradykinin, 5-hydroxytryptamine, and thrombin (Furchgott *et al.*, 1984) was dependent on the presence of an intact endothelial cell layer. They further suggested that this relaxation was mediated by a highly unstable endothelium-derived relaxing factor (EDRF), which was released from the endothelial cells and passed to the smooth muscle, where it caused relaxation. It was subsequently shown that EDRF-induced relaxations were associated with increased cyclic GMP accumulation and cyclic GMP-dependent protein kinase activation within the smooth muscle (Rapoport and Murad, 1983a; Rapoport *et al.*, 1983), and that EDRF could directly activate guanylyl cyclase (Förstermann *et al.*, 1986). So within the vascular system EDRF could act as the endogenous nitrovasodilator that stimulated soluble guanylyl cyclase within the smooth muscle.

The structure of EDRF was unknown for more than 6 years after its discovery but then it was reported that NO could account fully for the activity of EDRF, and that in a number of systems EDRF and NO were indistinguishable (Ignarro *et al.*, 1987; Palmer *et al.*, 1987; Furchgott, 1988). At the same time, studies on macrophages revealed that these cells released NO as a cytotoxic agent and the synthesis of this was inhibited by arginine analogues such as N^G-monomethyl-L-arginine (L-NMMA, Hibbs *et al.*, 1987). Extending this work to endothelial cells demonstrated that L-arginine was the precursor for EDRF/NO formation (Palmer *et al.*, 1988a), and that L-NMMA also inhibited the release of EDRF/NO (Sakuma *et al.*, 1988; Palmer *et al.*, 1988b). Continuing studies have now isolated the enzyme responsible for the formation of EDRF/NO in endothelial cells (Pollock *et al.*, 1991) and shown that it has an activity consistent with the concept that in endothelial cells L-arginine is converted to NO, or a related species, which is identical to EDRF.

Within blood vessels the complete pathway for activation of soluble guanylyl cyclase and relaxation of the smooth muscle has now been characterized. This consists of an EDRF/NO synthase present within the endothelial cells that releases NO following activation of the cells. This diffuses to the smooth muscle where it produces relaxation by activating the soluble form of guanylyl cyclase and so elevating cyclic GMP. Thus NO derived from the endothelium is important in control of blood vessel tone and, therefore, probably also in the physiological control of blood pressure (Rees *et al.*, 1989). In addition, it may also be involved in other blood vessel specific effects. For instance, it has been reported to mediate the rhythmic smooth muscle activity seen in hamster aortae, via a cyclic GMP-dependent mechanism (Jackson *et al.*, 1991).

3. Nonadrenergic Noncholinergic Nerves

It is also appropriate to discuss here the relaxation of smooth muscle, which results from the activity of certain nonadrenergic noncholinergic (NANC) nerves, as this too has been shown recently to be mediated through elevation of cyclic GMP.

Work in this area began with the observation that NANC inhibitory nerves were present within the rat anococcygeus (Gillespie, 1972) and bovine retractor penis (BRP) muscles (Klinge and Sjöstrand, 1974). Stimulation of these nerves was shown to be associated with stimulation of guanylyl cyclase within the smooth muscle (Bowman and Drummond, 1984), and the activity, of presumably the NANC transmitter, was inhibited by hemoglobin (Bowman *et al.*, 1982), or by hypoxia (Bowman and McGrath, 1985). In addition, electrical stimulation of inhibitory NANC nerves within the lower esophageal sphincter was shown to cause an elevation in cyclic GMP levels (Torphy *et al.*, 1986), as has now been demonstrated in the rat anococcygeus (Mirzazadeh *et al.*, 1991). These results were in parallel to what was known at the time about EDRF, for this too elevated cyclic GMP (Rapoport and Murad, 1983a; Rapoport *et al.*, 1983; Förstermann *et al.*, 1986) and was inhibited by hemoglobin (Martin *et al.*, 1985). Hemoglobin inhibition of nitrovasodilator activation of guanylyl cyclase was previously known (Murad *et al.*, 1978a). As research into EDRF/NO continued and inhibitors of NO synthase became available (see above) these were tested in the NANC model systems and found to be active inhibitors of the responses to nerve stimulation, at least within the rat or mouse anococcygeus (Gillespie *et al.*, 1989; Gibson *et al.*, 1990). Thus an increasing weight of evidence suggested that the NANC mediator within these tissues may be NO. This possibility was confirmed by experiments employing the canine ileocolonic junction. These experiments showed by bioassay, chemical instability, inactivation by superoxide anion and hemoglobin, inhibition by N^{G}-nitro-L-arginine, and potentiation by L-arginine that the activity of the biologically transferable NANC transmitter from this tissue could be accounted for by NO (Bult *et al.*, 1990). Many subsequent reports have now established functional evidence for NO as a NANC transmitter in a wide variety of other smooth muscle preparations including the guinea pig trachea (Tucker *et al.*, 1990), the human or rabbit penile corpus carvernosum (Ignarro *et al.*, 1990a; Sjöstrand *et al.*, 1990; Kim *et al.*, 1991), the rabbit urethra (Dokita *et al.*, 1991), the dog duodenum (Toda *et al.*, 1990a), the rat gastric fundus (Li and Rand, 1990; Boeckxstaens *et al.*, 1991), and the opossum lower esophogeal sphincter (Tøttrup *et al.*, 1991). However, these functional responses do not necessarily mean that NO is released from the NANC nerves, for an alternative explanation could be that another unknown transmitter is released from the nerves and in turn stimulates the underlying smooth muscle to generate NO (Ignarro *et al.*, 1990b). This concern may now be answered by studies that have demonstrated that the NO synthase within these tissues is of a type similar to that isolated from brain, consistent with a neural origin, and that the enzyme can be immunohistochemically located within the nerve fibers in the tissue (Mitchell *et al.*, 1991; Sheng *et al.*, 1992, 1993). It is also interesting to note that it has been reported that stimulation of NANC nerves within tissues in which ATP, rather than NO, is the putative neurotransmitter is associated with an elevation in cyclic GMP, as well as cyclic AMP levels (Baird and Muir, 1990).

Although the functional role of these nerves is still not clear, within the gastrointestinal system they may be associated with adaptive changes in smooth muscle tone in response to changes in the intraluminal volume or mediate the effects of parasympathetic nerve activity, as has been demonstrated in the stomach (Desai *et al.*, 1991a,b). In addition, NO-releasing nerves ("nitrinergic nerves") may be responsible for the vasodilatation following nerve stimulation in both cerebroarterial vessels (Toda *et al.*, 1990b) and bovine mesenteric arteries (Ahlner *et al.*, 1991a).

Evidence has now accumulated, therefore, for the widespread distribution of nerves releasing NO, which in its turn acts on guanylyl cyclase within target smooth muscle. One other process by which cyclic GMP modulates neurotransmission in smooth muscle may be by effects within the nerve ending, affecting the amount of transmitter released. For instance, the presence of a functional endothelium both inhibits the responses of arteries and veins to nerve stimulation (Tesafamariam *et al.*, 1987) and the efflux of labeled noradrenaline (Cohen and Weisbrodt, 1988). In addition, ANF decreases the release of labeled noradrenaline from sympathetic nerves innervating the rabbit vas deferens (Drewett *et al.*, 1989). Further studies have shown that agents that elevate cyclic GMP also act as inhibitors of the response to low levels of nerve stimulation and the release of transmitter (Greenberg *et al.*, 1990, 1991). Although these authors were not able to identify the origin of the cyclic GMP as being within the nerve endings, they did report that cyclic GMP within the smooth muscle preparation did effect the release of transmitter. It must, however, be borne in mind that some evidence suggests that the effects of ANF on the release of transmitter are not mediated via cyclic GMP (Drewett *et al.*, 1990).

4. NO-Generating Enzymes in Smooth Muscle: Induction by Cytokines

One other source of NO that may stimulate guanylyl cyclase in smooth muscle is from within the smooth muscle itself. It has been reported that perfused endothelium-denuded bovine pulmonary artery releases a factor indistinguishable from EDRF, that L-arginine causes a time-dependent endothelium-independent relaxation of smooth muscle rings, and that smooth muscle cells in culture produce and release NO (Wood *et al.*, 1990; Berhnardt *et al.*, 1991; Moritoki *et al.*, 1991; Mollace *et al.*, 1991). In addition, it has also been reported that bovine tracheal smooth muscle can generate an EDRF-like factor (Sheng *et al.*, 1991). However, as mentioned above, many smooth muscles are also known to contain NO-synthesizing nerves. Thus, some of these effects may be due to the release of NO from a neural source. In addition, these studies may also be complicated by the recent observation that very many cell types, including smooth muscle, may be induced by cytokines to express NO synthase. First reports showed that interleukin-1 or endotoxin inhibited contractions of rat aortic rings in a time-dependent manner. This inhibition was only seen after an incubation period of several hours and was inhibited by cycloheximide (Beasley *et al.*, 1989a,b). Further studies by the same and other groups showed that incubation with either agent was associated with activation of soluble guanylyl cyclase within vascular smooth muscle and led to the suggestion that IL-1 and endotoxin may activate a soluble form of guanylate cyclase by inducing the formation of nitric oxide (Beasley, 1990; Busse and Mülsch, 1990; Fleming *et al.*, 1990). Earlier work had already shown that aortae taken from rats treated with endotoxin had a decreased contractile response (Fink *et al.*, 1985; Schaller *et al.*, 1985; Wakabayishi *et al.*, 1987) and subsequent studies have shown that this is due to the induction of NO synthase within the smooth muscle (Julou-Schaeffer *et al.*, 1990). In addition, studies using cultured smooth muscle cells have shown that these too can be induced to produce NO (Beasley *et al.*, 1991; Schini *et al.*, 1991). These observations have suggested that induction of NO synthesis within vascular smooth muscle cells may underly the condition of septic shock (see Glauser *et al.*, 1991), which appears to be largely due to bacterial endotoxin (Natanson *et al.*, 1989). In this condition there is marked vascular hyporesponsiveness to pressor agonists and frequent mortality due to peripheral vascular failure. These recent results, therefore, suggest clearly that the effects of endotoxin on blood vessel tone and blood pressure are due to the induction of NO-synthase within the vascular tissue. This in turn generates nitric oxide, which stimulates guanylyl cyclase resulting in reduction in smooth muscle tone and hyporesponsiveness to contractile or pressor agonists.

5. Enzyme Structure

Cytosolic guanylyl cyclase exists as a heterodimer of 70- and 82-kDa subunits (Kamisaki *et al.*, 1986) and the presence of both subunits is required for the full catalytic and regulatory activity of the enzyme (Harteneck *et al.*, 1990; Buechler *et al.*, 1991). Although the two subunits have now been purified, cloned, and sequenced (Nakane *et al.*, 1988, 1990, this volume; Koesling *et al.*, 1988), there are still unanswered questions about the regulation of enzyme activity. As to the mechanism of stimulation by NO, this is most probably subsequent to it binding to a heme moiety within the protein structure (Ignarro, 1990; see also Chapter 4).

Thus there are a number of pathways that can lead to stimulation of guanylyl cyclase within smooth muscle. These can be divided into two primary groups: those that lead to stimulation of particulate guanylyl cyclase, i.e., ANF and related peptides, and those that cause stimulation of soluble guanylyl cyclase, e.g., NO. This latter stimulator could derive from endothelial cells, in the case of vascular smooth muscle, or from NANC nerves in a wide variety of other smooth muscles.

III. Mechanism of Cyclic GMP-Mediated Smooth Muscle Relaxation

A. Correlation between Cyclic GMP Levels and Relaxation of Smooth Muscle

The earliest reports that associated cyclic GMP with smooth muscle relaxation were from experiments on tracheal and gastrointestinal smooth muscle preparations examining the effects of various nitrovasodilators (Katsuki and Murad, 1977; Katsuki *et al.*, 1977b). Indeed, previous reports had concluded that cyclic GMP did not influence smooth muscle motility or that it caused contraction. Subsequently, similar studies were performed with various smooth muscle preparations including vascular smooth muscle. For example, both ANF and SNP produced elevations in intracellular cyclic GMP levels in, e.g., rabbit aorta or canine trachealis that correlated with both the magnitude and the time course of relaxation of the tissue (Winquist *et al.*, 1984b; Ohlstein and Berkowitz, 1985; Zhou and Torphy, 1991). Similarly, other nitrovasodilators have been shown to elevate cyclic GMP in smooth muscles as diverse as bovine and guinea pig trachea, guinea pig taenia coli, vas deferens, and aorta, and coronary, mesenteric, femoral, and umbilical arteries (Rapoport and Murad, 1983; Murad, 1986; Waldman and Murad, 1987). It has also been demonstrated *in vivo* that increase in blood flow in response to GTN is preceded by an elevation of cyclic GMP (Kobayishi *et al.*, 1980). In addition, treatment of tissues with agents such as methylene blue, ferricyanide, or hemoglobin that inhibit the activation of guanylyl cyclase by nitrovasodilators (Murad *et al.*, 1978a; Katsuki *et al.*, 1977a) also inhibits smooth muscle relaxation (Gruetter *et al.*, 1981a,b). Conversely, inhibitors of cyclic GMP phosphodiesterase, which metabolizes cyclic GMP, directly relax smooth muscle (Katsuki and Murad, 1977; Lorenz and Wells, 1983). Finally, cyclic GMP analogues can cause relaxation of a variety of smooth muscles in a manner that can be potentiated by phosphodiesterase inhibitors (Katsuki and Murad, 1977; Lincoln, 1983; Rapoport and Murad, 1983a).

It is also worth remembering that some reports have noted a separation in the abilities of, e.g., ANF analogs to elevate cyclic GMP and relax smooth muscle (Budzik *et al.*, 1987), which suggests that ANF may also relax smooth muscle by cyclic GMP-independent mechanisms. Similarly, studies using methylene blue, an inhibitor of soluble guanylyl cyclase, have shown that vascular tissues may still relax to SNP (Otsuka *et al.*, 1988) or EDRF (Vidal *et al.*, 1991) even when cyclic GMP accumulation is greatly decreased.

1. Variability in Smooth Muscle Responses

Differences in both species and vascular bed have been reported to effect the responses to a number of nitrovasodilators. For instance, it has been shown in dog vascular tissues that there are differences in responses between femoral vein (most sensitive) and renal artery (least sensitive) (Shibata *et al.*, 1986). Indeed many groups have shown that, in general, nitrovasodilators are more potent as relaxants of veins than arteries *in vitro* (see Ahlner *et al.*, 1991a), which in part may be associated with differences in the organic nitrate metabolism (Kawamoto *et al.*, 1990). Others have also found coronary arteries to be more sensitive than other arterial vessels (Gharaibeh and Gross, 1984; Miwa and Toda, 1985) and large arterial vessels more sensitive than small ones (Harder *et al.*, 1979; Tillmanns *et al.*, 1979). Vascular preparations are more sensitive to nitrovasodilators than nonvascular smooth muscle. Regional differences in the responses to ANF have also been reported. For instance, large arteries relax more fully to ANF than smaller or distal arteries (Faison *et al.*, 1985). This may be correlated with the finding that higher amounts of high affinity ANF receptors have been localized on large/central arteries (e.g., aorta) than on smaller/peripheral arteries (e.g., ear) (Winquist *et al.*, 1985). However, this cannot fully account for the differences for some tissues that did not respond well to ANF were found to have a good population of high affinity receptors. In addition, the ability of ANF to relax smooth muscle is dependent on both the tone of the tissue and the contractile agent. For instance, ANF has been reported to be more effective at relaxing tissues contracted with angiotensin II than those contracted with noradrenaline (Kleinert *et al.*, 1984). In addition, various groups have reported that most relaxants, including ANF, are more effective in relaxing tissues contracted by receptor agonists than those contracted by potassium depolarization (Winquist, 1985). This suggests that the difference in potency of ANF against different receptor-mediated agonists may be explained by the degree to which they also depolarize the membrane. Similar effects have been suggested for relaxations mediated by the activity of soluble guanylyl cyclase. For instance, rhythmic contractions of the myometrium are refractory to the effects of nonhydrolyzable analogs of cyclic GMP (Word *et al.*, 1991) possibly because in these types of contractions inositol phosphate turnover plays only a minor role, and this is the level at which cyclic GMP acts (see below).

B. Cyclic GMP-Dependent Protein Kinase

Many of the effects of cyclic GMP are believed to be mediated through the activity of cyclic GMP-dependent protein kinase, the activity of which is governed by cyclic GMP. Indeed, the relevance of cyclic GMP-dependent protein kinase within smooth muscle is strengthened by the observation that, although in most tissues the specific activity or amount of cAMP protein kinase is approximately 20 times higher than that of cyclic GMP protein kinase, within the rat aorta they are approximately equi-active (Lincoln, 1989). In many tissues, including, for instance, rat aorta (Fiscus *et al.*, 1985), an increase in cyclic GMP is associated with activation of this kinase by a mechanism that requires its binding to two distinct subunit sites (Corbin *et al.*, 1986; and Chapter 7, this volume). The subsequent activity of cyclic GMP kinase is sufficient to account for decreases in calcium within the cell, decreased phosphorylation of myosin light chains, and smooth muscle relaxation, following elevation of cyclic GMP (Murad, 1986; Cornwell and Lincoln, 1989; Lincoln and Cornwell, 1991).

Cyclic GMP-dependent protein kinase exists in two major forms, types I (including types Ia and Ib) and II. Type Ia has been purified from lung (Gill *et al.*, 1976; Lincoln *et al.*, 1977) and heart (Flockerzi *et al.*, 1978) and is a dimer consisting of two 80-kDa subunits (Lincoln *et al.*, 1977). Within the rabbit aorta it has been detected in both the soluble and the particulate fractions (Ives *et al.*, 1980) and can be separated from the membrane fraction by treatment with high salt concentrations. Type Ib has been isolated recently from bovine aorta (Lincoln *et al.*, 1988; Wolfe *et al.*, 1989), and it has been suggested that this may be the more important form in smooth muscle. Cyclic GMP-dependent protein kinase is discussed in greater detail in Chapter 7.

C. Cyclic GMP, Calcium, and Intracellular Signaling

A variety of effects of the cyclic GMP pathway on ion fluxes and intracellular signaling have been reported, most of which are probably secondary to the activity of cyclic GMP kinase. For instance, the elevation of cyclic GMP has been reported to stimulate a number of mechanisms that actively decrease calcium levels within the cell. These include activation of a calcium extrusion pump within the sarcolemma (Popescu *et al.*, 1985; Furukawa *et al.*, 1988), stimulation of a calcium ATPase that would extrude calcium from the cell (Fujii *et al.*, 1986), and activation of sodium/calcium exchange (Furukawa *et al.*, 1991) and sodium/potassium/chlorine cotransport (O'Donnell and Owen, 1986). It may also blockade calcium translocation across the plasma membrane (Taylor and Meisheri, 1986) or accelerate calcium uptake by the sarcoplasmic reticulum (Twort and van Breemen, 1988). All, or most, of these effects may well be attributed to the activity of a cyclic GMP kinase for protein phosphorylation by calcium-dependent protein kinase may activate sodium/calcium exchange in squid axons and heart cells (Caroni and Carafoli, 1983; DiPolo and Beaugé, 1987) and cyclic GMP kinase catalyzes the phosphorylation of phospholamban, a regulator of the sarcoplasmic reticulum calcium-ATPase (Raeymaekers *et al.*, 1988). Although cyclic GMP kinase also stimulates a sarcolemmal calcium/magnesium ATPase, this is not due to a direct effect on the enzyme but is probably mediated through phosphorylation of other membrane components (Imai *et al.*, 1990).

Thus, cGMP may cause a reduction in intracellular calcium by stimulating calcium extrusion and sequestration systems within vascular smooth muscle, and possibly other smooth muscles. Cyclic GMP may also inhibit calcium release from intracellular stores (Meisheri *et al.*, 1986; Fujii *et al.*, 1986) and decrease calcium mobilization following agonist activation by an inhibitory effect on IP_3 formation (Rapoport, 1986; Hirata *et al.*, 1990). This would correlate with studies in human myometrium which have demonstrated that nonhydrolyzable analogs of cyclic GMP cause a significant decrease in the resting and agonist-induced increase in intracellular calcium (Word *et al.*, 1991). This latter effect may be mediated by cyclic GMP altering the phosphorylation of proteins in the sarcoplasmic reticulum involved in IP_3-induced calcium release (Lincoln, 1989), or may be at the level of the interaction between a guanine nucleotide regulatory protein and phospholipase C. (Hirata *et al.*, 1990). However, the effect of cyclic GMP on inositol phosphate turnover is not clear, for it has been suggested that elevation of cyclic GMP is associated with both a decrease in inositol phosphate accumulation (Chuprun and Rapoport, 1987) and a stimulation of inositol phosphate production (Resink *et al.*, 1988).

The decrease in intracellular calcium in smooth muscle is associated with a decrease in myosin light-chain phosphorylation, due to a reduction in the activity of myosin light chain kinase, and so relaxation (see Murad, 1986). This is also observed in spontaneously contracting myometrium, where SNP has been shown to decrease both spontaneous contractions and myosin light-chain phosphorylation (Word *et al.*, 1991). However, although it has been suggested that cyclic GMP kinase may directly phosphorylate myosin light-chain kinase (Vrolix *et al.*, 1988), this is not supported by studies on the substrates of cyclic GMP kinase in vascular smooth muscle (Baltensperger *et al.*, 1990).

D. Other Possible Mechanisms

In addition to binding to specific protein kinases cyclic GMP binds with high affinity to type III phosphodiesterase present within intestinal smooth muscle (Lincoln *et al.*, 1976; Lincoln, 1989). This binding is not associated with the metabolism of cyclic GMP and it has not been demonstrated that type III phosphodiesterase mediates any actions of cyclic GMP (Lincoln, 1989).

Cyclic GMP may also act to potentiate the levels of cyclic AMP within smooth muscle, leading to relaxation (Kauffman *et al.*, 1987; Silver *et al.*, 1988; Lindgren *et al.*, 1991). This effect is mediated by an influence on cyclic GMP-inhibitable cyclic nucleotide phosphodiesterases (cyclic GMP-I-PDE), the presence of which has been demonstrated in aortic smooth muscle (Silver *et al.*, 1988; Lindgren *et al.*, 1991). Indeed, within the rat aorta this may be the principal pathway regulating hydrolysis of cyclic AMP, as can be demonstrated in experiments using either selective PDE inhibitors or guanylyl cyclase stimulators (Lindgren *et al.*, 1990, 1991; Maurice and Haslam, 1990). The complexities of cyclic nucleotide hydrolysis by cyclic nucleotide phosphodiesterases and their regulation are discussed in greater detail in Chapter 6.

It is also worth noting that cyclic GMP has been shown to bind to ion channels within the membrane. In particular it has been demonstrated that cyclic GMP binds to a membrane-associated sodium channel (Zimmerman *et al.*, 1985; Cook *et al.*, 1987) leading to entry of sodium and depolarization of the cells. Although such a channel has not been shown in smooth muscle it does suggest the possibility that such a mechanism could be involved in relaxation of smooth muscle induced by cyclic GMP.

IV. Summary

Cyclic GMP levels within smooth muscle are affected then by a number of different pathways. Physiologically NO and ANF are probably the two most important regulators for smooth muscle function, but a variety of other mediators and pharmacological agents may also influence this system. Because of the important role that cyclic GMP plays in the control of smooth muscle tone, which clearly includes vascular smooth muscle, it is now and will continue to be in the future an important physiological and biochemical target for research and a pharmacological target for therapeutic agents.

References

Ahlner, J., Andersson, R. G. G., Torfgård, and Axelsson, K. (1991a). Organic nitrate esters: Clinical use and mechanisms of actions. *Pharmacol. Rev.* **43,** 351–423.

Ahlner, J., Ljusegren, M. E., Grundström, N., and Axelsson, K. L. (1991b). Role of nitric oxide and cyclic GMP as mediators of endothelium-independent neurogenic relaxation in bovine mesenteric artery. *Circ. Res.* **68,** 756–762.

Arnold, W. P., Mittal, C., Katsuki, S., and Murad, F. (1977). Nitric oxide activates guanylate cyclase and increase guanosine 3′ : 5′-cyclic monophosphate levels in various tissue preparations. *Proc. Natl. Acad. Sci. U.S.A.* **74,** 3203–3207.

Baird, A. A., and Muir, T. C. (1990). Membrane hyperpolarization, cyclic nucleotide levels and relaxation in the guinea-pig internal anal sphincter. *Br. J. Pharmacol.* **100,** 329–335.

Baltensperger, K., Chiesi, M., and Carafoli, E. (1990). Substrates of cGMP kinase in vascular smooth muscle and their role in the relaxation process. *Biochemistry,* **29,** 9753–9760.

Beasley, D. (1990). Interleukin and endotoxin activate soluble guanylate cyclase in vascular smooth muscle. *Am. J. Physiol.* **259,** R38–R-44.

Beasley, D., Cohen, R. A., and Levinsky, N. G. (1989a). Interleukin-1 inhibits contraction of vascular smooth muscle. *J. Clin. Invest.* **83,** 331–335.

Beasley, D., Cohen, R. A., and Levinsky, N. G. (1989b). Endotoxin inhibits contraction of vascular smooth muscle in vitro. *Am. J. Physiol.* **258,** H1187–H1192.

Beasley, D., Schwartz, J. H., and Brenner, B. M. (1991). Interleukin 1 induces prolonged L-arginine-dependent cyclic guanosine monophosphate and nitrite production in rat vascular smooth muscle cells. *J. Clin. Invest.* **87,** 602–608.

Bennett, B., Leitman, D., Schroder, H., Kauramotto, J., Nakatsu, K., and Murad, F. (1989). Relationship between biotransformation of glyceryl trinitrate and cyclic GMP accumulation in various cultural cell lines. *J. Pharmacol. Exp. Ther.* **250,** 316–323.

Bernhardt, J., Tschudi, M. R., Dohi, Y., Gut, I., Urwyler, B., Bühler, F. R., and Lüscher, T. F. (1991). Release of nitric oxide from human vascular smooth muscle cells. *Biochem. Biophys. Res. Commun.* **180,** 907–912.

Boeckxtaens, G. E., Pelckmans, P. A., Borgers, J. J., Bult, H., De Man, J. G., Osterboch, L., Herman, A. G., and Van Maercke, Y. M. (1991). Release of nitric oxide upon stimulation of nonadrenergic noncholinergic nerves in rat gastric fundus. *J. Pharmacol. Exp. Ther.* **256,** 441–447.

Bowman, A., and Drummond, A. H. (1984). Cyclic GMP mediates neurogenic relaxation in the bovine retractor penis muscle. *Br. J. Pharmacol.* **81,** 665–674.

Bowman, A., and McGrath, J. C. (1985). The effect of hypoxia on neuroeffector transmission in the bovine retractor penis and rat anococcygeus muscles. *Br. J. Pharmacol.* **85,** 869–875.

Bowman, A., Gillespie, J. S., and Pollock, P. (1982). Oxyhaemoglobin blocks non-adrenergic non-cholindergic inhibition in the bovine retractor penis muscle. *Eur. J. Pharmacol.* **85,** 221–224.

Budzik, G. P., Firestone, S. L., Bush, E. N., Connolly, P. J., Rockway, T. W., Sarin, V. K., and Holleman, W. (1987). Divergence of ANF analogs in smooth muscle cell cGMP response and aorta vasorelaxation: Evidence for receptor subtypes. *Biochem. Biophys. Res. Commun.* **144,** 422–431.

Buechler, W. A., Nakane, M., and Murad, F. (1991). Expression of soluble guanylate cyclase activity requires both enzyme subunits. *Biochem. Biophys. Res. Commun.* **174,** 351–357.

Bult, H., Boeckxstaens, G. E., Pelckmans, P. A., Jordaens, F. H., Van Maercke, Y. M., and Herman, A. G. (1990). Nitric oxide as an inhibitory non-adrenergic non-cholinergic neurotransmitter. *Nature* (*London*) **345,** 346–347.

Busse, R., and Mülsch, A. (1990). Induction of nitric oxide synthase by cytokines on vascular smooth muscle cells. *FEBS Lett.* **275,** 87–90.

Caroni, P., and Carafoli, E. (1983). The regulation of the Na^+-Ca^{2+} exchanger of the heart sarcolemma. *Eur. J. Biochem.* **132,** 451–460.

Chang, M. S., Lowe, D. G., Lewis, M., Hellmiss, R., Chen, E., and Goeddel, D. V. (1989). Differential activation by atrial and brain natriuretic peptides of two different receptor guanylate cyclases. *Nature (London)* **341,** 68–72.

Chinkers, M., Garbers, D. L., Chang, M.-S., Lowe, D. G., Chin, H., Goeddel, D. V., and Schulz, S. (1989). A membrane form of guanylate cyclase is an atrial natriuretic peptide receptor. *Nature (London)* **338,** 78–83.

Chuprun, J. K., and Rapoport, R. M. (1987). Nitroglycerine-induced desensitization of vascular smooth muscle may be mediated through cyclic GMP-disinhibition of phosphatidylinositol hydrolysis. *Experientia* **43,** 316–318.

Cohen, R. A., and Weisbrodt, R. M. (1988). The endothelium inhibits release of norepinephrine from rabbit carotid artery during electrical stimulation. *Am. J. Physiol.* **255,** H871.

Cook, N. J., Hanke, W., and Kaupp, U. B. (1987). Identification, purification and functional reconstitution of the cyclic GMP-dependent channel from rod photoreceptors. *Proc. Natl. Acad. Sci. U.S.A.* **84,** 585–589.

Corbin, J. D., Ogreid, D., Miller, J. P., Suva, R. H., Jastorff, B., and Doskeland, S. O. (1986). Studies of cGMP analog specificity and function of two intrasubunit binding sites of cGMP-dependent protein kinase. *J. Biol Chem.* **261,** 1208–1214.

Cornwell, T. L., and Lincoln, T. M. (1989). Regulation of intracellular Ca^{2+} levels in cultured vascular smooth muscle cells: Reduction of Ca^{2+} by atriopeptin and 8-bromo-cyclic GMP is mediated by cyclic GMP-dependent protein kinase. *J. Biol. Chem.* **264,** 1146–1155.

Currie, M. G., Geller, D. M., Cole, B., Boylan, J. G., Shenz, W. Y., Holmberg, S. W., and Needleman, P. (1983). Bioactive cardiac substances: Potent vasorelaxant activity in mammalian atria. *Science* **221,** 71–73.

de Bold, A. J., Borenstein, H. B., Veress, A. T., and Sonnenberg, H. (1981). A rapid and potent natriuretic response to intravenous injection of atrial myocardial extract in rats. *Life Sci.* **28,** 89–94.

Desai, K. M., Sessa, W. C., and Vane, J. R. (1991a). Involvement of nitric oxide in the reflex relaxation of the stomach to accomodate food and fluid. *Nature (London)* **351,** 477–479.

Desai, K. M., Zemobowicz, A., Sessa, W. C., and Vane, J. R. (1991b). Nitroxergic nerves mediate vagally induced relaxation in the isolated stomach of the guinea pig. *Proc. Natl. Acad. Sci. U.S.A.* **88,** 11490–11494.

DiPolo, R., and Beaugé, L. (1987). In squid axons, ATP modulates Na^+-CA^{2+} exchange by a $Ca^{2+}{}_i$-dependent phosphorylation. *Biochim. Biophys. Acta* **897,** 347–354.

Dokita, S., Morgan, W. R., Wheeler, M. A., Yoshida, M., Latifpour, J., and Weiss, R. M. (1991). N^G-nitro-L-arginine inhibits non-adrenergic, non-cholinergic relaxation in rabbit urethral smooth muscle. *Life Sci.* **48,** 2429–2436.

Drewett, J. G., Trachte, G. J., and Marchand, G. R. (1989). Atrial natriuretic factor inhibits adrenergic and purinergic neurotransmission in the rabbit isolated vas deferens. *J. Pharmacol. Exp. Ther.* **248,** 135–142.

Drewett, J. G., Ziegler, R. J., and Trachte, G. J. (1990). Neuromodulatory effects of atrial natriuretic factor are independent of guanylate cyclase in adrenergic neuronal pheochromocytoma cells. *J. Pharmacol. Exp. Ther.* **255,** 497–503.

Faison, E. P., Siegl, P. K. S., Moran, G., and Winquist, R. J. (1985). Regional vasorelaxant selectivity of atrial natriuretic factor in isolated rabbit vessels. *Life Sci.* **37,** 1073–1079.

Feelisch, M., Noack, E., and Schröder, H. (1988). Explanation of the discrepancy between the degree of organic nitrate decomposition, nitrite formation and guanylate cyclase stimulation. *Eur. Heart J.* **9**Suppl. A, 57–62.

Fink, M. P., Homer, L. D., and Fletcher, J. R. (1985). Diminished pressor response to exogenous norepinephrine and angiotensin II in septic, unanesthetized rats: evidence for a prostaglandin mediated effect. *J. Surg. Res.* **38,** 335–342.

Fiscus, R. R., Rapoport, R. M., Waldman, S. A., and Murad, F. (1985). Atriopeptin II elevates cyclic GMP, activates cyclic GMP-dependent protein kinase, and causes relaxation in rat thoracic aorta. *Biochim. Biophys. Acta* **846,** 179–184.

Fleming, I., Gray, G. A., Julou-Schaeffer, G., Parratt, J. R., and Stoclet, J.-C. (1990). Incubation with endotoxin activates the L-arginine pathway in vascular tissue. *Biochem. Biophys. Res. Commun.* **171,** 562–568.

Flockerzi, V., Speichermann, N., and Hofmann, F. (1978). A guanosine 3′:5′-monophosphate protein kinase from bovine heart muscle: Purification and phosphorylation of histone I and IIb. *J. Biol. Chem.* **253,** 3395–3399.

Flynn, T. G., Brar, A., Tremblay, L., Sarda, I., Lyons, C., and Jennings, D. B. (1989). Isolation and characterization of ISO-rANP, a new natriuretic peptide. *Biochem. Biophys. Res. Commun.* **161,** 830–837.

Förstermann, U., Mülsch, A., Bohme, E., and Büsse, R. (1986). Stimulation of soluble guanylate cyclase by an acetylcholine-induced endothelium-dependent factor from rabbit and canine arteries. *Circ. Res.* **58,** 531–538.

Fujii, K., Ishimatsu, T., and Kuriyama, H. (1986). Mechanism of vasodilation induced by α-human atrial natriuretic polypeptide in rabbit and guinea pig renal arteries. *J. Physiol. (London)* **377,** 315–332.

Furchgott, R. F. (1988). Studies on relaxation of rabbit aorta by sodium nitrite: the basis for the proposal that the acid-activatable inhibitory factor from bovine retractor penis is organic nitrite and the endothelium-derived relaxing factor is nitric oxide. *In* "Vasodilatation: Vascular Smooth Muscle, Peptides, Autonomic Nerves and Endothelium" (P. M. Vanhoutte, ed.), pp. 401–414. Raven Press, New York.

Furchgott, R. F., and Zawadzki, J. V. (1980). The obligatory role of endothelial cells in the relaxation of arterial smooth muscle to acetylcholine. *Nature (London)* **288,** 373–376.

Furchgott, R. F., Cherry, P. D., Zawadzki, J. V., and Jothianandan, D. (1984). Endothelial cells as mediators of vasodilation of arteries. *J. Cardiovasc. Pharmacol.* **6,** S336–S343.

Furukawa, K.-I., Tawada, Y., and Shigekawa, M. (1988). Regulation of the plasma membrane Ca^{2+} pump by cyclic nucleotides in cultured vascular smooth muscle cells. *J. Biol. Chem.* **263,** 8058–8065.

Furukawa, K.-I., Ohshima, N., Tawada-Iwata, Y., and Shigekawa, M. (1991). Cyclic GMP stimulates Na^{+}/Ca^{2+} exchange in vascular smooth muscle cells in primary culture. *J. Biol. Chem.* **266,** 12337–12341.

Garbers, D. L. (1976). Sea urchin sperm guanylate cyclase: Purification and loss of cooperativity. *J. Biol. Chem.* **251,** 4071–4077.

Gharaibeh, M. N., and Gross, G. J. (1984). Comparative relaxant effects of nitroglycerin in isolated rings from various canine vascular beds. *Gen. Pharmacol.* **15,** 217–221.

Gibson, A., Mirzazadeh, S., Hobbs, A. J., and Moore, P. K., (1990). L-N^{G}-Monomethylarginine and L-N^{G}-nitro-arginine inhibit non-adrenergic non-cholinergic relaxation of the mouse anococcygeus muscle. *Br. J. Pharmacol.* **99,** 602–606.

Gill, G. N., Holdy, K. E., Walton, G. M., and Kanstein, C. B. (1976). Purification and characterization of 3′:5′-cyclic GMP-dependent protein kinase. *Proc. Natl. Acad. Sci. U.S.A.* **73,** 3918–3922.

Gillespie, J. S. (1972). The rat anococcygeus muscle and its response to nerve stimulation and to some drugs. *Br. J. Pharmacol.* **45,** 404–416.

Gillespie, J. S., Liu, X., and Martin, W. (1989). The effect of L-arginine and N^G-monomethyl-L-arginine on the response of the rat anococcygeus muscle to NANC stimulation. *Br. J. Pharmacol.* **98,** 1080–1082.

Glauser, M. P., Zanetti, G., Baumgartner, J.-D., and Cohen, J. (1991). Septic shock: Pathogenesis. *Lancet* **338,** 732–736.

Greenberg, S. S., Diecke, F. P. J., Cantor, E., Peevy, K., and Tanaka, T. P. (1990). Inhibition of sympathetic neurotransmitter release by modulators of cyclic GMP in canine vascular smooth muscle. *Eur. J. Pharmacol.* **187,** 409–423.

Greenberg, S. S., Cantor, E., Diecke, F. P. J., and Tanaka, T. P. (1991). Cyclic GMP modulates release of norepinephrine from adrenergic nerves innervating canine arteries. *Am. J. Hypertens.* **4,** 173–176.

Gruetter, C. A., Barry, B. K., McNamara, D. B., Gruetter, D. Y., Kadowitz, P. J., and Ignarro, L. J. (1979). Relaxation of bovine coronary artery and activation of coronary arterial guanylate cyclase by nitric oxide, nitroprusside and a carcinogenic nitrosamine. *J. Cyclic Nucleotide Res.* **5,** 211–224.

Gruetter, C. A., Gruetter, D. Y., Lyon, E. J., Kadowitz, P. J., and Ignarro, L. J. (1981a). Relation between cyclic guanosine 3′ : 5′-monophosphate formation and relaxation of coronary artery smooth muscle by gylceryltrinitrate, nitroprusside, nitrite, and nitric oxide: Effects of methylene blue and methemoglobin. *J. Pharmacol. Exp. Ther.* **219,** 181–186.

Gruetter, C. A., Kadowitz, P. J., and Ignarro, L. J. (1981b). Methylene blue inhibits coronary arterial relaxation and guanylate cyclase activation by nitroglycerin, sodium nitrite, and amylnitrate. *Can. J. Physiol. Pharmacol.* **59,** 150–165.

Harder, D. R., Belardinelli, L., Sprelakis, N., Rubio, R., and Berne, R. M. (1979). Differential effects of adenosine and nitroglycerin on the action potentials of large and small coronary arteries. *Circ. Res.* **44,** 176–182.

Harteneck, C., Koesling, D., Söling, A., Schultz, G., and Böhme, E. (1990). Expression of soluble guanylate cyclase. Catalytic activity requires two enzyme subunits. *FEBS Lett.* **272,** 221–223.

Hibbs, J. R., Jr., Taintor, R. R., and Vavrin, Z. (1987). Macrophage cytotoxicity: Role for L-arginine deiminase and imino nitrogen oxidation to nitrite. *Science* **235,** 473–476.

Hirata, M., Kohse, K. P., Chang, C.-H., Ikebe, T., and Murad, F. (1990). Mechanism of cyclic GMP inhibition of inositol phosphate formation in rat aorta segments and cultured bovine aortic smooth muscle cells. *J. Biol. Chem.* **265,** 1268–1273.

Ignarro, L. J. (1990). Haem-dependent activation of guanylate cyclase and cyclic GMP formation by endogenous nitric oxide: A unique transduction mechanism for transcellular signaling. *Pharmacol. Toxicol.* **67,** 1–7.

Ignarro, L. J., Edwards, J. C., Gruetter, D. Y., Barry, B. K., and Gruetter, C. A. (1980). Possible involvement of S-nitrosothiols in the activation of guanylate cyclase by nitroso compounds. *FEBS Lett.* **110,** 275–278.

Ignarro, L. J., Lippton, H., Edwards, J. C., Baricos, W. H., Hyman, A. L., Kadowitz, P. J., and Gruetter, C. A. (1981). Mechanism of vascular smooth muscle relaxation by organic nitrates, nitrites, nitroprusside and nitirc oxide: Evidence for the involvement of S-nitrosothiols as active intermediates. *J. Pharmacol. Exp. Ther.* **218,** 739–749.

Ignarro, L. J., Buga, G. M., Wood, K. S., Byrns, R. E., and Chaudhuri, G. (1987). Endothelium-derived relaxing factor produced and released from artery and vein is nitric oxide. *Proc. Natl. Acad. Sci. U.S.A.* **84,** 9265–9269.

Ignarro, L. J., Bush, P. A., Buga, G. M., Wood, K. S., Fukuto, J. M., and Rajfer, J. (1990a). Nitric oxide and cyclic GMP formation upon electrical field stimulation cause relaxation of corpus cavernosum smooth muscle. *Biochem. Biophys. Res. Commun.* **170,** 843–850.

Ignarro, L. J., Bush, P. A., Buga, G. M., and Rajfer, J. (1990b). Neurotransmitter identity doubt. *Nature (London)* **347,** 131–132.

Imai, S., Yoshida, Y., and Sun, H.-T. (1990). Sarcolemmal (Ca^{2+} + Mg^{2+})-ATPase of vascular smooth muscle and the effects of protein kinases thereupon. *J. Biochem. (Tokyo)* **107,** 755–761.

Ives, H. E., Casnellie, J. E., Greengard, P., and Jamieson, J. D. (1980). Subcellular localization of cyclic GMP-dependent protein kinase and its substrates in vascular smooth muscle. *J. Biol. Chem.* **255,** 3777–3785.

Jackson, W. F., Mülsch, A., and Busse, R. (1991). Rhythmic smooth muscle activity in hamster aortas is mediated by continuous release of NO from endothelium. *Am. J. Physiol.* **260,** H248–H253.

Julou-Schaeffer, G., Gray, G. A., Fleming, I., Schott, C., Parratt, J. R., and Stoclet, J.-C. (1990). Loss of vascular responsiveness induced by endotoxin involves L-arginine pathway. *Am. J. Physiol.* **259,** H1038–H1043.

Kambayashi, Y., Nakao, K., Itoh, H., Hosoda, K., Saito, Y. *et al.* (1989). Isolation and sequence determination of rat cardiac natriuretic peptide. *Biochem. Biophys. Res. Commun.* **163,** 233–240.

Kamisaki, Y., Saheki, S., Nakane, M., Palmieri, J. A., Kuno, T., Chang, B. Y., Waldman, S. A., and Murad, F. (1986). Soluble guanylate cyclase from rat lung exists as a heterodimer. *J. Biol. Chem.* **261,** 7236–7241.

Katsuki, S., and Murad, F. (1977). Regulation of adenosine cyclic 3′:5′-monophosphate and guanosine 3′:5′-monophosphate levels and contractility in bovine tracheal smooth muscle. *Mol. Pharmacol.* **13,** 330–341.

Katsuki, S., Arnold, W., Mittal, C., and Murad, F. (1977a). Stimulation of guanylate cyclase by sodium nitroprusside, nitroglycerin and nitric oxide in various tissue preparations and comparison to the effects of sodium azide and hydroxylamine. *J. Cyclic Nucleotide Res.* **3,** 23–35.

Katsuki, S., Arnold, W. P., and Murad, F. (1977b). Effect of sodium nitroprusside, nitroglycerin and sodium azide on levels of cyclic nucleotides and mechanical activity of various tissues. *J. Cyclic Nucleotide Res.* **3,** 239–247.

Kauffman, R. F., Schenck, K. W., Utterback, G., Crowe, V. G., and Cohen, M. L. (1987). In vitro vascular relaxation by new inotropic agents: Relationship to phosphodiesterase inhibition and cyclic nucleotides. *J. Pharmacol. Exp. Ther.* **242,** 864–872.

Kawamoto, J. H., McLaughlin, B. E., Brien, J. F., Marks, G. S., and Nakatsu, K. (1990). Biotransformation of glyceryl trinitrate and elevation of cyclic GMP precede glyceryl trinitrate-induced vasodilation. *J. Cardiovasc. Pharmacol.* **15,** 714–719.

Kim, N., Azadzoi, K. M., Goldstein, I., and de Tajeda, I. S. (1991). A nitric oxide-like factor mediates nonadrenergic-noncholinergic neurogenic relaxation of penile corpus cavernosum smooth muscle. *J. Clin. Invest.* **88,** 112–118.

Kimura, H., Mittal, C. K., and Murad, F. (1975a). Increases in cyclic GMP levels in brain and liver with sodium azide, an activator of guanylate cyclase. *Nature (London)* **257,** 700–702.

Kimura, H., Mittal, C. K., and Murad, F. (1975b). Activation of guanylate cyclase from rat liver and other tissues with sodium azide. *J. Biol. Chem.* **250,** 8016–8022.

Kisch, B. (1956). Electron microscopy of the atrium of the heart. I. Guinea pig. *Exp. Med. Surg.* **14,** 99–112.

Kleinert, H. D., Maack, T., and Atlas, S. A. (1984). Atrial natriuretic factor inhibits angiotensin-, norepinephrine- and potassium-induced vascular contractility. *Hypertension,* **6,**Suppl. I, I-143–I-147.

Klinge, E., and Sjöstrand, N. O. (1974). Contraction and relaxation of the retractor penis muscle and the penile artery of the bull. *Acta Physiol. Scand., Suppl.* **420,** 1–88.

Kobayishi, A., Suzuki, Y., Kamikawa, T., Hayashi, H., and Yamazaki, N. (1980). The effects of nitroglycerin on cyclic nucleotides in the coronary artery in vivo. *Life Sci.* **27,** 1679–1685.

Koesling, D., Herz, J., Gausepohl, H., Niroomand, F., Hinsch, K.-D., Mülsch, A., Böhme, E., Schultz, G., and Frank, R. (1988). The primary structure of the 70 kDa subunit of bovine soluble guanylate cyclase. *FEBS Lett.* **239,** 29–34.

Kojima, M., Minamino, N., Kangawa, K., and Matsuo, H. (1989). Cloning and sequence analysis of cDNA encoding a precursor for rat brain natriuretic peptide. *Biochem. Biophys. Res. Commun.* **159,** 1420–1426.

Koller, K. J., Lowe, D. G., Bennet, G. L., Minamino, N., Kangawa, K., Matsuo, H., and Goeddel, D. V. (1991). Selective activation of the B natriuretic peptide receptor by C-type natriuretic peptide (CNP). *Science* **252,** 120–123.

Kuno, T. Andresen, J., Komisaki, Y., Waldman, S. A., Chang, L., Soheki, S., Leitman, D., Nakone, M., and Murad, F. (1986). Co-purification of an atrial natriuretic factor receptor and particulate guanylate cyclose from rat lung. *J. Biol. Chem.* **261,** 5817–5823.

Li, C. G., and Rand, M. J. (1990). Nitric oxide and vasoactive intestinal polypeptide mediate non-adrenergic, non-cholinergic inhibitory transmission to smooth muscle of the rat gastric fundus. *Eur. J. Pharmacol.* **191,** 303–309.

Lincoln, T. M. (1983). Effects of nitroprusside and 8-bromo cyclic GMP on the contractile activity of the rat aorta. *J. Pharmacol. Exp. Ther.* **224,** 100–107.

Lincoln, T. M. (1989). Cyclic GMP and mechanisms of vasodilation. *Pharmacol. Ther.* **41,** 479–502.

Lincoln, T. M., and Cornwell, T. L. (1991). Towards an understanding of the mechanism of action of cyclic AMP and cyclic GMP in smooth muscle relaxation. *Blood Vessels* **28,** 129–137.

Lincoln, T. M., Hall, C. L., Park, C. R., and Corbin, J. D. (1976). Guanosine 3′ : 5′-monophosphate binding proteins in rat tissues. *Proc. Natl. Acad. Sci. U.S.A.* **73,** 2559–2563.

Lincoln, T. M., Dills, W. L., and Corbin, J. D. (1977). Purification and subunit composition of guanosine 3′ : 5′-monophosphate-dependent protein kinase from bovine lung. *J. Biol. Chem.* **252,** 4269–4275.

Lincoln, T. M., Thompson, M., and Cornwell, T. L. (1988). Purification and characterization of two forms of cyclic GMP-dependent protein kinase from bovine aorta. *J. Biol. Chem.* **263,** 17632–17637.

Lindgren, S. H. S., Andersson, T. L. G., Vinge, E., and Andersson, K.-E. (1990). Effects of isozyme-selective phosphodiesterase inhibitors on rat aorta and human platelets: Smooth muscle tone, platelet aggregation and cAMP levels. *Acta Physiol. Scand.* **140,** 209–219.

Lindgren, S., Rascón, A., Andersson, K.-E., Manganiello, V., and Degerman, E. (1991). Selective inhibition of cGMP-inhibited and cGMP-noninhibited cyclic nucleotide phosphodiesterases and relaxation of rat aorta. *Biochem. Pharmacol.* **42,** 545–552.

Lorenz, K. L., and Wells, J. N. (1983). Potentiation of the effects of sodium nitroprusside and of isoproterenol by selective phosphodiesterase inhibitors. *Mol. Pharmacol.* **23,** 424–430.

Lowe, D. G., Chang, M.-S., Hellmiss, R., Chen, E., Singh, S., Garbers, D. L., and Goeddel, D. V. (1989). Human atrial natriuretic peptide receptor defines a new paradigm for second messenger signal transduction. *EMBO J.* **8,** 1377–1384.

Martin, W., Villani, G. M., Jothianandan, D., and Furchgott, R. F. (1985). Selective blockade of endothelium-dependent and glyceryl trinitrate-induced relaxation by haemoglobin and methylene blue in the rabbit aorta. *J. Pharmacol. Exp. Ther.* **232,** 708–716.

Maurice, D. H., and Haslam, R. J. (1990). Nitroprusside enhances isoprenaline-induced increases in cAMP in rat aortic smooth muscle. *Eur. J. Pharmacol.* **191,** 471–475.

Meisheri, K. D., Taylor, C. J., and Saneii, H. H. (1986). Synthetic atrial peptide inhibits intracellular Ca^{++} release in smooth muscle. *Am. J. Physiol.* **250,** C171–C174.

Mirzazdeh, S., Hobbs, A. J., Tucker, J. F., and Gibson, A. (1991). Cyclic nucleotide content of the rat anococcygeus during relaxations induced by drugs or by non-adrenergic, non-cholinergic field stimulation. *J. Pharm. Pharmacol.* **43,** 247–257.

Mitchell, J. A., Sheng, H., Förstermann, U., and Murad, F. (1991). Characterization of nitric oxide synthases in non-adrenergic noncholinergic nerve containing tissue from the rat anococcygeus. *Br. J. Pharmacol.* **104,** 289–291.

Miwa, K., and Toda, N. (1985). The regional difference of relaxations induced by various vasodilators in isolated dog coronary and mesenterial arteries. *Jpn. J. Pharmacol.* **38,** 313–330.

Mollace, V., Salvemini, D., Anggard, E., and Vane, J. (1991). Nitric oxide from vascular smooth muscle cells: Regulation of platelet reactivity and smooth muscle cell guanylate cyclase. *Br. J. Pharmacol.* **104,** 633–638.

Moritiko, H., Ueda, H., Yamamoto, T., Hisayama, T., and Takeuchi, S. (1991). L-Arginine induces relaxation of rat aorta possibly through non-endothelial nitric oxide formation. *Br. J. Pharmacol.* **102,** 841–846.

Murad, F. (1986). Cyclic guanosine monophosphate as a mediator of vasodilation. *J. Clin. Invest.* **78,** 1–5.

Murad, F., Mittal, C. K., Arnold, W. P., Katsuki, S., and Kimura, H. (1978a). Guanylate cyclase: Activation by azide, nitro compounds, nitric oxide, and hydroxyl radical and inhibition by hemoglobin and myoglobin. *Adv. Cyclic Nucleotide Res.* **9,** 145–158.

Murad, F., Mittal, C. K., Arnold, W. P., and Braughler, J. M. (1978b). Effect of nitro-compound smooth muscle relaxants and other materials on cyclic GMP metabolism. *Adv. Pharmacol. Thera.* **3,** 123–132.

Nakane, M., Saheki, S., Kuno, T., Ishii, K., and Murad, F. (1988). Molecular cloning of a cDNA coding for 70 kilodalton subunit of soluble guanylate cyclase from rat lung. *Biochem. Biophys. Res. Commun.* **157,** 1139–1147.

Nakane, M., Arai, K., Saheki, S., Kuno, T., Buechler, W., and Murad, F. (1990). Molecular cloning and expression of cDNAs coding soluble guanylate cyclase from rat lung. *J. Biol. Chem.* **265,** 16841–16845.

Natanson, C., Eichenholz, P. W., Danner, R. L., Eichacker, P. Q., Hoffman, W. D., Kuo, G. C., Banks, S. M., MacVittie, T. J., and Parrillo, J. E. (1989). Endotoxin and tumor necrosis factor challenges in dogs simulate the cardiovascular profile of human septic shock. *J. Exp. Med.* **169,** 823–832.

Needleman, P. J., Blehm, D. J., and Rotskoff, K. S. (1969). Relationship between glutathione-dependent denitration and the vasodilator effectiveness of organic nitrates. *J. Pharmacol. Exp. Ther.* **237,** 286–288.

O'Donnell, M. E., and Owen, N. E. (1986). Atrial natriuretic factor stimulates Na/K/Cl cotransport in vascular smooth muscle cells. *Proc. Natl. Acad. Sci. U.S.A.* **83,** 6132–6136.

Ohlstein, E. H., and Berkowitz, B. A. (1985). Cyclic guanosine monophosphate mediates vascular relaxation induced by atrial natriuretic factor. *Hypertension* **7,** 306–310.

Otsuka, Y., DiPiero, A., Hirt, E., Brennaman, B., and Lockette, W. (1988). Vascular relaxation and cyclic GMP in hypertension. *Am. J. Physiol.* **254,** H163–H169.

Palmer, R. M. J., Ferridge, A. G., and Moncada, S. (1987). Nitric oxide release accounts for the biological activity of endothelium-derived relaxing factor. *Nature (London)* **327,** 524–526.

Palmer, R. M. J., Ashton, D. S., and Moncada, S. (1988a). Vascular endothelial cells synthesize nitric oxide from L-arginine. *Nature (London)* **333,** 664–666.

Palmer, R. M. J., Rees, D. D., Ashton, D. S., and Moncada, S. (1988b). L-arginine is the physiological precursor for the formation of nitric oxide in endothelium-dependent relaxation. *Biochem. Biophys. Res. Commun.* **153,** 1251–1256.

Pollock, J. S., Förstermann, U., Mitchell, J. A., Warner, T. D., Schmidt, H. H. H. W., Nakane, M., and Murad, F. (1991). Purification and characterization of particulate endothelium-derived relaxing factor synthase from cultured and native bovine aortic endothelial cells. *Proc. Natl. Acad. Sci. U.S.A.* **88,** 10480–10484.

Popescu, L. M., Panoiu, C., Hinescu, M., and Nutu, O. (1985). The mechanism of cGMP-induced relaxation in vascular smooth muscle. *Eur. J. Pharmacol.* **107,** 393–394.

Radany, E. W., Gerzer, R., and Garbers, D. L. (1983). Purification and characterization of particulate guanylate cyclase from sea urchin spermatozoa. *J. Biol. Chem.* **258,** 8346.

Raeymaekers, L., Hofmann, F., and Casteels, R. (1988). Cyclic GMP-dependent protein kinase phosphorylates phospholamban in isolated sarcoplasmic reticulum from cardiac and smooth muscle. *Biochem. J.* **252,** 269–273.

Rapoport, R. M. (1986). Cyclic guanosine monophosphate inhibition of contraction may be mediated through inhibition of phosphatidylinositol hydrolysis in rat aorta. *Circ. Res.* **58,** 407–410.

Rapoport, R. M., and Murad, F. (1983a). Agonist induced endothelium-dependent relaxation in rat thoracic aorta may be mediated through cGMP. *Circ. Res.* **52,** 352–357.

Rapoport, R. M., and Murad, F. (1983b). Endothelium-dependent and nitrovasodilation induced relaxation of vascular smooth muscle: Role for cyclic GMP *J. Cyclic Nucleotide Protein Phosphorylation Res.* **9,** 281–296.

Rapoport, R. M., Draznin, M. B., and Murad, F. (1983). Endothelium-dependent relaxation in rat aorta may be mediated through cyclic GMP-dependent protein phosphorylation. *Nature (London)* **306,** 174–176.

Rees, D. D., Palmer, R. M. J., and Moncada, S. (1989). Role of endothelium-derived nitric oxide in the regulation of blood pressure. *Proc. Natl. Acad. Sci. U.S.A.* **86,** 3375–3378.

Resink, T., Scott-Burden, T., Baur, U., Jones, C. T., and Buhler, F. R. (1988). Atrial natriuretic peptide induces breakdown of phosphatidylinositol phosphates in cultured vascular smooth muscle cells. *Eur. J. Biochem.* **172,** 499–505.

Romanin, C., and Kukovetz, W. R. (1989). Tolerance to nitroglycerin is caused by reduced guanylate cyclase activation. *J. Mol. Cell. Cardiol.* **21,** 41–48.

Rosenzweig, A., and Seidman, C. E. (1991). Atrial natriuretic factor and related peptide hormones. *Annu. Rev. Biochem.* **60,** 229–255.

Saito, Y., Nakao, K., Itoh, H., Yamada, T., Mukoyama, M., Arai, H., Hosoda, K., Shirakami, G., Suga, S.-I., Minamino, N., Kangawa, K., Matsuo, H., and Imura, H. (1989). Brain natriuretic peptide is a novel cardiac hormone. *Biochem. Biophys. Res. Commun.* **158,** 360–368.

Sakuma, I., Stuehr, D. J., Gross, S. S., Nathan, C., and Levi, R. (1988). Identification of L-arginine as a precursor of endothelium-derived relaxing factor. *Proc. Natl. Acad. Sci. U.S.A.* **85,** 8664–8667.

Schaller, M. D., Waeber, B., Nussberger, J., and Brunner, H. R. (1985). Angiotensin II, vasopressin and sympathetic activity in conscious rats with endotoxaemia. *Am. J. Physiol.* **249,** H1086–H1092.

Schini, V. B., Junquero, D. C., Scott-Burden, T., and Vanhoutte, P. M. (1991). Interleukin-1 β induces the production of an L-arginine-derived relaxing factor from cultured smooth muscle cells from rat aorta. *Biochem. Biophys. Res. Commun.* **176,** 114–121.

Schröder, H., Leitman, D. C., Bennet, B. M., Waldman, S. A., and Murad, F. (1988). Glyceryl trinitrate-induced desensitization of guanylate cyclase in cultured rat lung fibroblasts. *J. Pharmacol. Exp. Ther.* **245,** 413–418.

Sheng, H., Ishii, K., and Murad, F. (1991). Generation of endothelium-derived relaxing factor-like substance in bovine tracheal smooth muscle. *Am. J. Physiol.* **260,** L489–L493.

Sheng, H., Schmidt, H., Nakane, M., Mitchell, J., Pollock, J., Forstermann, U., and Murad, F. (1992). Characterization and localization of nitric oxide synthase in non-adrenergic non-cholinergic nerves from bovine retractor penis muscles. *Br. J. Pharmacol.* **106,** 768–773.

Sheng, H., Mitchell, J. A., Nakane, M., Schmidt, H. H. H. W., Pollock, J. S., Warner, T. D., Förstermann, U., and Murad, F. (1993). Characterization of NO synthase from non-adrenergic non-cholinergic nerves in rat anococcygeus and bovine retractor penis muscles. *In* "Biology of Nitric Oxide" (S. Moncada, M. A. Marletta, J. B. Hibbs, Jr., and E. A. Higgs, eds.) Vol. II, pp. 108–110. Springer-Verlag, Berlin.

Shibata, T., Ogawa, K., Ito, T., Hashimoto, H., Nakagawa, H., and Satake, T. (1986). Role of cyclic GMP in canine vascular smooth muscle relaxation by organic nitroesters. *Jpn. Circ. J.* **50,** 1091–1099.

Silver, P. J., Lepore, R. E., O'Connor, B., Lemp, B. M., Hamel, L. T., Bentley, R. G., and Harris, A. L. (1988). Inhibition of the low K_m cyclic AMP phosphodiesterase and activation of cyclic AMP system in vascular smooth muscle by milrinone. *J. Pharmacol. Exp. Ther.* **247,** 34–42.

Sjöstrand, N. O., Eldh, J., Samuelson, U. E., Alaranta, S., and Klinge, E. (1990). The effects of L-arginine and N^G-monomethyl L-arginine on the inhibitory neurotransmission of the human corpus cavernosum. *Acta Physiol. Scand.* **140,** 297–298.

Sudoh, T., Kangawa, K., Minamino, N., and Matsuo, H. (1989). A new natriuretic peptide in porcine brain. *Nature* (*London*) **332,** 78–81.

Sudoh, T., Minamino, N., Kangawa, K., and Matsuo, H. (1990). C type natriuretic peptide (CNP): A new member of natriuretic peptide family identified in porcine brain. *Biochem. Biophys. Res. Commun.* **168,** 863–870.

Taylor, C. J., and Meisheri, K. D. (1986). Inhibitory effects of a synthetic atrial peptide on contractions and 45Ca fluxes in vascular smooth muscle. *J. Pharmacol. Exp. Ther.* **237,** 803–808.

Tesafamariam, B., Weisbrod, R. M., and Cohen, R. A. (1987). Endothelium inhibits responses of rabbit carotid artery to adrenergic nerve stimulation. *Am. J. Physiol.* **253,** H792.

Thorpe, D. S., and Garbers, D. L. (1989). The membrane form of guanylate cyclase: homology with a subunit of the cytoplasmic form of the enzyme. *J. Biol. Chem.* **264,** 6545–6549.

Tillmanns, H., Steinhausen, M., Leinberger, H., and Thederan, H. (1979). Different response of the ventricular micro-circulation to coronary vasodilators. *Circulation* **60**Suppl. II, 142.

Toda, N., Baba, H., and Okamura, T. (1990a). Role of nitric oxide in nonadrenergic, noncholinergic nerve mediated relaxation in dog duodenal longitudinal muscle strips. *Jpn. J. Pharmacol.* **53,** 281–284.

Toda, N., Minami, Y., and Okamura, T. (1990b). Inhibitory effects of L-N^G-nitro-arginine on the synthesis of EDRF and the cerebroarterial response to vasodilator nerve stimulation. *Life Sci.* **47,** 345–351.

Torphy, T. J., Fine, C. F., Burman, M., Barnette, M. S., and Ormsbee, H. S. (1986). Lower esophageal sphincter relaxation is associated with increased cyclic nucleotide content. *Am. J. Physiol.* **251,** G786–G793.

Tøttrup, A., Svane, D., and Forman, A. (1991). Nitric oxide mediating NANC inhibition in opossum lower esophageal sphincter. *Am. J. Physiol.* **260,** G385–G389.

Tucker, J. F., Brave, S. R., Charalambous, L., Hobbs, A., and Gibson, A. (1990). L-N^G-nitro arginine inhibits non-adrenergic, non-cholinergic relaxations of guinea-pig isolated tracheal smooth muscle. *Br. J. Pharmacol.* **100,** 663–664.

Twort, C. H. C., and van Breemen, C. (1988). Cyclic guanosine monophosphate-enhanced sequestration of Ca^{2+} by sarcoplasmic reticulum in smooth muscle. *Circ. Res.* **62,** 961–964.

Vidal, M., Vanhoutte, P. M., and Miller, V. M. (1991). Dissociation between endothelium-dependent relaxations and increases in cGMP in systemic veins. *Am. J. Physiol.* **260,** H1531–H1537.

Vrolix, M., Raeymaekers, L., Wuytack, F., Hofmann, F., and Casteels, R. (1988). Cyclic GMP-dependent protein kinase stimulates the plasmalemmal Ca^{2+} pump of smooth muscle via phosphorylation of phosphatidylinositol. *Biochem. J* **255,** 855–863.

Wakabayishi, I., Hatake, K., Kakishita, E., and Nagai, K. (1987). Diminuition of contractile response of the aorta from endotoxin-injected rats. *Eur. J. Pharmacol.* **141,** 117–122.

Waldman, S. A., and Murad, F. (1987). Cyclic GMP synthesis and function. *Pharmacol. Rev.* **39,** 163–196.

Waldman, S. A., Rapoport, R., Ginsburg, R., and Murad, F. (1986). Desensitization to nitroglycerin in vascular smooth muscle from rat and human. *Biochem. Pharmacol.* **35,** 3525–3531.

Waldman, S. A., Rapoport, R., and Murad, F. (1984). Atrial natriuretic factor selectively activates particulate guanylate cyclase and elevates cyclic GMP in rat tissue. *J. Biol. Chem.* **259,** 14332–14334.

Wingren, A.-K., Axelsson, K. L., and Andersson, R. G. G. (1981). Enzymatic denitration of organic nitro-esters as a possible requirement for causing cGMP elevation of vascular smooth muscle. *Acta Pharmacol. Toxicol.* **49,**Suppl. III, 49.

Winquist, R. J. (1985). The relaxant effects of atrial natriuretic factor on vascular smooth muscle. *Life Sci.* **37,** 1081–1087.

Winquist, R. J., Faison, E. P., and Nutt, R. F. (1984a). Vasodilator profile of synthetic atrial natriuretic factor. *Eur. J. Pharmacol.* **102,** 169–173.

Winquist, R. J., Faison, E. P., Waldman, S. A., Schwartz, K., Murad, F., and Rapoport, R. A. (1984b). Atrial natriuretic factor elicits an endothelium-independent relaxation and activates particulate cyclase in vascular smooth muscle. *Proc. Natl. Acad. Sci. U.S.A.* **81,** 7661–7665.

Winquist, R. J., Napier, M. A., Vandlen, R. L., Arcuri, K., Keegan, M. E., Faison, E. P., and Baskin, E. P. (1985). Pharmacology and receptor binding of atrial natriuretic factor in vascular smooth muscle. *Clin. Exp. Hypertens.* **A7,** 869–884.

Wolfe, L., Corbin, J. D., and Francis, S. H. (1989). Characterization of a novel isoenzyme of cGMP-dependent protein kinase from bovine aorta. *J. Biol. Chem.* **264,** 7734–7741.

Wood, K. S., Buga, G. M., Byrns, R. E., and Ignarro, L. J. (1990). Vascular smooth muscle-derived relaxing factor (MDRF) and its close similarity to nitric oxide. *Biochem. Biophys. Res. Commun.* **170,** 80–88.

Word, R. A., Casey, M. L., Kamm, K. E., and Stull, J. T. (1991). Effects of cGMP on $[Ca^{2+}]_i$, myosin light chain phosphorylation, and contraction in human myometrium. *Am. J. Physiol.* **260,** C861–C867.

Yeates, R. A., Lauffen, H., and Leithold, M. (1985). The reaction between organic nitrates and sulfhydryl compounds. A possible model system for the activation of organic nitrates. *Mol. Pharmacol.* **28,** 555–559.

Zhou, H.-L., and Torphy, T. J. (1991). Relationship between cyclic guanosine monophosphate accumulation and relaxation of canine trachealis induced by nitrovasodilators. *J. Pharmacol. Exp. Ther.* **258,** 972–978.

Zimmerman, A. L., Yamanaka, G., Eckstein, F., Baylor, D. A., and Stryer, L. (1985). Interaction of hydrolysis-resistant analogs of cyclic GMP with the phosphodiesterase and light-sensitive channel of retinal rod outer segments. *Proc. Natl. Acad. Sci. U.S.A.* **82,** 8813–8817.

Interrelationships of Cyclic GMP, Inositol Phosphates, and Calcium

Masato Hirata* and Ferid Murad†

**Department of Biochemistry*
Faculty of Dentistry
Kyushu University
Fukuoka 812, Japan

†Molecular Geriatrics Corporation
Lake Bluff, Illinois 60044

I. Introduction

Cellular functions are regulated by neurotransmitters, hormones, and a wide variety of regulatory and growth-promoting factors. The agonists produce a host of physiological responses in their target tissues as a result of their interactions with specific cell surface receptors. Despite the numerous agonists and their specific receptors, the transmembrane signaling mechanisms can be categorized into relatively few pathways. Thus, interactions between these signaling pathways must occur to explain a remarkable diversity of cellular responses that are evoked by a variety of agonists through a simple signaling pathway. For example, stimulation of cells with 12-*O*-tetradecanoyl phorbol-13-acetate (TPA), a phorbol ester that activates protein kinase C (C-kinase), can influence hormone-sensitive adenylate cyclase; in some cells TPA induces desensitization of receptor-mediated stimulation of adenylate cyclase (Sibly *et al.*, 1984; Kelleher *et al.*, 1984), whereas in others, such as frog erythrocytes, phorbol ester treatment results in increased agonist-stimulated adenylate cyclase activities (Sibly *et al.*, 1986; Bell *et al.*, 1985). In another example, TPA attenuates the agonist-induced hydrolysis of polyphosphoinositides (PPI), and

Advances in Pharmacology, Volume 26

thus messengers may also influence their own signaling pathway. These interactions between transmembrane signaling pathways are called "cross-talk," and a large body of evidence for various types of cross-talk has been accumulating.

The cyclic GMP signal transduction system also regulates the Ca^{2+}/inositol phosphates pathways. Regulation of cyclic GMP synthesis and the interactions with Ca^{2+} have been reviewed previously (Murad *et al.*, 1986; Waldman and Murad, 1987). Therefore, we have chosen to concentrate on the cyclic GMP effects on the Ca^{2+}/inositol phosphate pathways.

II. Cyclic GMP Effects on Calcium

A. Plasmalemmal Ca^{2+} Pump

The plasma membrane Ca^{2+} pump is an enzyme that removes Ca^{2+} from all eukaryotic cells studied so far (Fig. 1). It is an ATPase of the P type; i.e., it forms an acylphosphate intermediate during the reaction cycle and is inhibited by low concentration of vanadate. The pump is the largest of all known P-type ion-motiveATPases, its molecular weight being 138,000 and is of very low abundance, constituting less than 0.05% of the total plasma membrane protein in erythrocytes. The large size and low abundance of this enzyme have greatly complicated studies of its structural aspects and attempts to elucidate its primary structure by conventional protein chemistry or by molecular biology approaches. Despite these difficulties, progress has been made using both approaches and thus the mechanisms underlying the ATP-driven Ca^{2+} transport and its modification by agents such as calmodulin have been elucidated (Verma *et al.*, 1988; Strehler *et al.*, 1989; Shull and Greeb, 1988; James *et al.*, 1988).

The first suggestion of a cyclic GMP effect on the plasmalemmal Ca^{2+} pump was made in 1977 by Schultz *et al.* Thereafter, agents, including nitroprusside, nitroglycerine, hydroxylamine, sodium nitrite, and nitric oxide, were demonstrated to activate guanylate cyclase and increase cyclic GMP levels in a wide variety of tissues. These agents also elevated cGMP in a variety of vascular and nonvascular smooth muscle preparations, including bovine tracheal smooth muscle, guinea pig tracheal smooth muscle, and guinea pig taenia coli, ductus deferens, aorta, coronary, mesenteric, femoral, and umbilical arteries (Waldman and Murad, 1987). Elevation of cyclic GMP with these agonists was associated in a dose- and time-dependent fashion with relaxation of many of these smooth muscle preparations. The elevations in cyclic GMP stimulated by these agents preceded the associated relaxation, consistent with a cause-and-effect

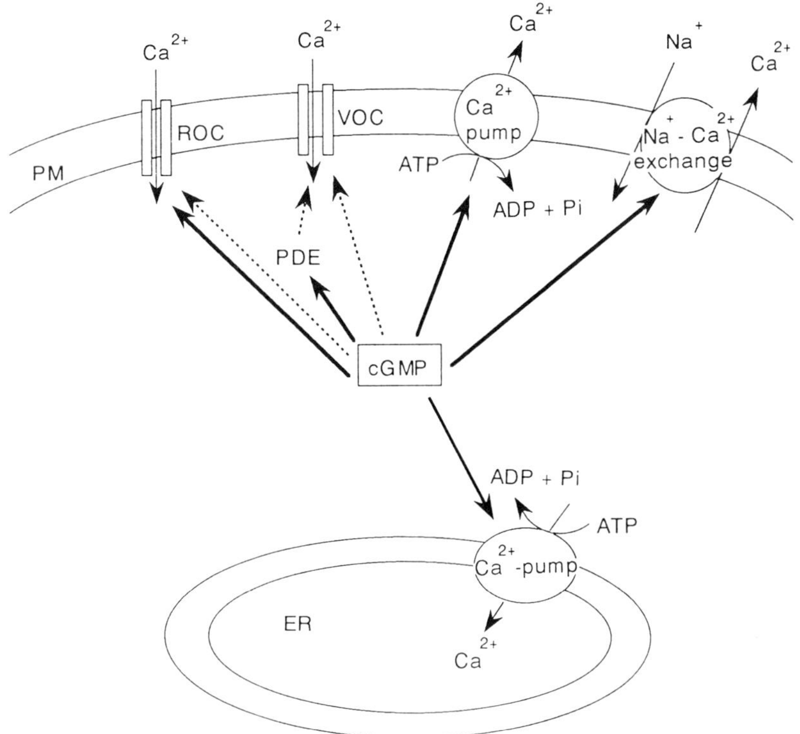

Fig. 1 Proposed effect of cyclic GMP/G-kinase on components that regulate intracellular free Ca^{2+} concentration. Solid or dotted arrow represents stimulation or inhibition, respectively.

relationship between these events. Since the main determinant of the contraction–relaxation cycle of smooth muscle is the change in intracellular free Ca^{2+} concentration, it was reasonable to assume that cyclic GMP acts on the plasmalemmal Ca^{2+} pump to reduce intracellular free Ca^{2+} concentrations.

Suematsu *et al.* (1984) studied the effect of cyclic GMP with G-kinase using membrane preparations from vascular smooth muscle. Cyclic GMP alone did not enhance the Ca^{2+} uptake of the plasmalemma-rich fraction. With addition of both G-kinase and cyclic GMP, there was a 20% enhancement in the maximal uptake with a concomitant phosphorylation of a membrane protein with a molecular mass of 35 kDa. However, in other reports, 250-, 130-, 85-, and 75-kDa proteins were phosphorylated by cyclic GMP in the sarcolemma of rabbit aortic smooth muscle (Ives *et al.*, 1980); 130- and 85-kDa proteins were phosphorylated in microsomes of canine tracheal smooth muscle (Hogaboom *et al.*, 1982); and lower molecular weight substrates were phosphorylated in intact rat aortae (Rapoport *et al.*, 1983). Thus, which phosphorylation in the sarcolemmal membrane proteins by cyclic GMP is correlated with the increased uptake of Ca^{2+} remains obscure (see below).

Itoh *et al.* (1983, 1985) recorded mechanical responses induced by 10 μM norepinephrine in rabbit mesenteric artery. Repeated application of the agent in the absence of external Ca^{2+} and Na^+ caused the gradual decrease in the tension development, due to Ca^{2+} extrusion to the external medium. Because of the absence of external Na^+, Na^+–Ca^{2+} exchange mechanisms were not involved and the Ca^{2+} pump would probably be functioning in this process. By adding 10 μM nitroglycerin during the repeated application of norepinephrine, the gradual decrease in tension development was markedly enhanced, indicating that cyclic GMP produced by nitroglycerin enhanced Ca^{2+} efflux through activation of the Ca^{2+} pump. Kobayashi *et al.* (1985) directly measured the concentration of intracellular free Ca^{2+} using a microfluorimetric recording method in cultured rat aortic smooth muscle cells loaded with quin 2. They found that nitroglycerin decreased Ca^{2+} in both the presence and the absence of external Ca^{2+}, and strongly and progressively decreased the extent of the transient increase in Ca^{2+} induced by repeated applications of caffeine in the absence of external Ca^{2+}. Popescu *et al.* (1985) observed that monospecific antibodies against G-kinase prevent cyclic GMP-dependent activation of the Ca^{2+}–ATPase in the sarcolemmal membrane isolated from the smooth muscle. All of these results indirectly suggested that cyclic GMP via activation of G-kinase stimulates the plasmalemmal Ca^{2+} pump of smooth muscle cells.

More direct evidence for cyclic GMP actions on the Ca^{2+} pump in smooth muscle has been obtained using purified preparations of the plasmalemmal Ca^{2+} pump. As described earlier in this section, the Ca^{2+} pump is stimulated by calmodulin through a direct interaction. Casteels and colleagues (Wuytack *et al.*, 1981a,b; Raeymaekers *et al.*, 1983; DeSchutter *et al.*, 1984) utilized this interaction to purify the Ca^{2+}-transport ATPase from smooth muscle preparations, as had previously been done for the erythrocytes (Niggli *et al.*, 1979; Gietzell *et al.*, 1980) and the cardiac sarcolemmal Ca^{2+} pump (Caroni and Carafoli, 1981). The Triton X-100-soluble fraction from porcine stomach smooth muscle microsomes was incubated with a calmodulin–Sepharose gel, which had been equilibrated with a Ca^{2+}-containing buffer. The retained proteins were eluted with a Ca^{2+}-free, EGTA-containing buffer. When the eluate was analyzed with polyacrylamide gel electrophoresis, it contained a prominent double band of molecular weight of 140,000, which is similar to the apparent molecular weight of the erythrocyte Ca^{2+}-transport ATPase. Furthermore, a Ca^{2+}–ATPase activity could be measured in this eluate. To prove that the calmodulin-affinity-purified Ca^{2+}–ATPase activity actually was a Ca^{2+}-transport enzyme, the purified ATPase was incorporated into liposomes. On reconstitution in asolectin vesicles, an ATP-energized Ca^{2+} uptake into the liposomes was found to be tightly coupled to the Ca^{2+}–ATPase activity. The Ca^{2+}/ATP stoichiometry of the reconstituted system was approximately 1, suggesting that the transport of 1 Ca^{2+} ion is driven by the hydrolysis of 1 ATP molecule. Furukawa and Nakamura (1987) examined the effect of cyclic GMP on purified plasma membrane Ca^{2+}–ATPase prepared from bovine aortic smooth muscle. The purification procedure for the enzyme was the same as that by Casteels and his group. The purified Ca^{2+}–ATPase was incorporated into soybean phospholipid liposomes through the removal of Triton X-100 on a Bio-Beads SM-2 column. The reconstituted enzyme was phosphorylated by cyclic GMP plus G-kinase with 100 μM [γ-^{32}P]ATP in a medium containing 0.13 M KCl, 10 mM $MgCl_2$, 0.5 mM EGTA, and 20 mM Hepes buffer (pH 7.4). The analysis of the mixture on SDS–polyacrylamide gel electrophoresis revealed that only the 135-kDa protein was phosphorylated in the presence of either Ca^{2+} or cyclic GMP plus G-kinase. In the former condition, all the ^{32}P label incorporated into this protein was released with hydroxylamine treatment, indicating that the phosphoprotein formed was of the acylphosphate type. On the other hand, ^{32}P label incorporated in the latter case was found to be resistant to hydroxylaminolysis. The results, therefore, suggest that it was the Ca^{2+}–ATPase that was phosphorylated by G-kinase. The $^{45}Ca^{2+}$ uptake activity into vesicles reconstituted with purified Ca^{2+}–ATPase was enhanced upon the phosphorylation by cyclic GMP plus G-kinase, and the stimulation correlated with the extent of phosphorylation. From these observations, they concluded that the activity of the plasma membrane Ca^{2+}-transport ATPase of vascular smooth muscle is directly stimulated by cyclic GMP-dependent phosphorylation of the same enzyme. Baltensberger *et al.* (1988) examined the suggestions that the major substrate of the G-kinase might be identical to the Ca^{2+} pump of the plasma membrane obtained from pig aorta. The separation of the plasma membrane vesicles on a gradient (5–10%) acidic gel system after radioactive labeling discriminated the phosphoprotein by G-kinase from the Ca^{2+}–ATPase, identified by the acylphosphate intermediate, indicating that the substrate phosphorylated by G-kinase and the Ca^{2+}–ATPase in the plasma membrane vesicles are not identical. Different approaches using a calmodulin-overlay technique or anti-Ca^{2+}–ATPase antibodies have also been used to separate and identify on various types of gels the Ca^{2+}-pumping ATPase. The results have consistently shown that the protein phosphorylated by G-kinase had a smaller apparent molecular weight than the Ca^{2+}-pump. Furthermore, Vrolix *et al.* (1988) studied the effect of phosphorylation by G-kinase on the activity of the plasma membrane Ca^{2+}-transport ATPase in isolated plasma membranes vesicles and in the ATPase purified from pig erythrocytes and from the smooth muscle of pig stomach and pig aorta. Incubation with G-kinase resulted in both smooth muscle preparations, but not in the erythrocyte ATPase, in an increase in the maximal rate of Ca^{2+} uptake. The stimulation of the Ca^{2+}–ATPase activity of the purified Ca^{2+} pump reconstituted in liposomes was dependent on the phospholipid used for reconstitution; i.e., the activation by G-kinase was only observed in the presence of phosphatidylinositol (PI). G-kinase, but not A-kinase, stimulated the phosphorylation of PI to phosphatidylinositol phosphate (PIP) in a preparation of Ca^{2+}–ATPase obtained by calmodulin-affinity chromatography from smooth muscle, but not in a similar preparation from erythrocytes. Adenosine, which is known to inhibit the phosphorylation of PI catalyzed by a PI-kinase, inhibited the stimulation of the Ca^{2+}–ATPase by G-kinase. Taken together with the previous reports that acidic phospholipids such as PIP, PIP_2, or PS (PIP is most potent) are capable of stimulating the enzyme activity (Niggli *et al.*, 1981a,b; Choquette *et al.*, 1984; Nelson and Hanahan, 1985; Enyadi *et al.*, 1987), they proposed that G-kinase stimulates the plasma membrane Ca^{2+} pump of smooth muscle cells indirectly via the phosphorylation of an associated PI-kinase. This is an attractive hypothesis, but there has been no report demonstrating the activation of PI-kinase by the phosphorylation with G-kinase. Furthermore, PI-kinase (specific to produce PI-4-P) has been purified and shown to have molecular masses of 55 or 76 kDa for bovine uterus (Porter *et al.*, 1988) and rat brain (Yamakawa and Takenawa, 1988), respectively. Another PI-kinase-specific phosphorylation at 3-position (to produce PI-3-P) has also been purified and proven to be a heterodimer with molecular masses of 110 and 85 kDa for both rat liver (Carpenter *et al.*, 1990) and bovine thymus (Shibasaki *et al.*, 1991). Thus, the proteins with the same molecular weights as these PI-kinases have not been reported to be phosphorylated by G-kinase, as described above.

B. Endoplasmic Reticulum Ca^{2+} Pump

The endoplasmic reticulum Ca^{2+} pump is also an ATPase that forms an acylphosphate intermediate, and is also responsible for regulating intracellular free Ca^{2+} concentration.

Phospholamban was first identified as the major substrate for A-kinase in isolated sarcoplasmic reticulum (SR) of cardiac muscle (LaRaia and

Morkin, 1974; Tada *et al.*, 1975). Since the Ca^{2+} uptake by the SR vesicles is stimulated, concomitantly with the phosphorylation of phospholamban, this event may mediate part of the mechanical responses of the heart to catecholamines and other agonists that raise cellular cyclic AMP concentrations. Phospholamban has been shown to be a substrate for C-kinase (Movsenian *et al.*, 1984) and calmodulin-dependent kinase (LePeuch *et al.*, 1979), as well as A-kinase, and, furthermore, identified in the SR vesicles of smooth muscle (Raeymaekers and Jones, 1986).

Raeymaekers *et al.* (1988) examined the phosphorylation of phospholamban catalyzed by G-kinase and its effect on the Ca^{2+} uptake in isolated SR vesicles from dog and pig cardiac muscle and bovine pulmonary arterial smooth muscle. Phospholamban could be phosphorylated by G-kinase at about the same rate as by A-kinase and the amount of ^{32}P incorporation by both kinases was not additive, indicating that both kinases phosphorylate the same site of phospholamban. The rate of Ca^{2+} uptake was increased with the amount of ^{32}P incorporation into phospholamban by both kinases. Twort and van Breemen (1988) investigated the effects of cyclic GMP on sequestration of Ca^{2+} into SR of vascular smooth muscle. Using saponin-skinned primary cultures of rat aortic smooth muscle, $^{45}Ca^{2+}$ uptake by these cells was measured in the presence or absence of cyclic GMP. Addition of cyclic GMP increased both the initial Ca^{2+} uptake at 2 min and the final steady-state uptake reached at 20 min. Thus, it is possible that this process acts synergistically with the activation of other systems, such as the stimulation of cyclic GMP of the Ca^{2+} extrusion via the plasmalemmal Ca^{2+}–ATPase (see previous section) and the Na^+–Ca^{2+} exchange (see next section), to reduce the intracellular free Ca^{2+} concentration.

C. Na^+–Ca^{2+} Exchange Mechanisms

Na^+–Ca^{2+} exchange through the plasma membrane of cells was first described in heart muscle by Reuter and Seitz (1968), and in squid axon by Baker *et al.* (1967). Thereafter, many studies have been reported on the Na^+–Ca^{2+} exchange mechanisms in a variety of cells. Although Na^+–Ca^{2+} exchange is present in a variety of cells, its role in the physiological regulation of the intracellular free Ca^{2+} concentration has not well been defined, and may be dependent on the cell types examined. For instance, in heart muscle the Na^+–Ca^{2+} exchange is generally presumed to be primarily responsible for extrusion of Ca^{2+}, which was carried through the action potential. Hirata *et al.* (1981) measured $^{45}Ca^{2+}$ efflux from single-cell suspensions prepared from guinea pig taenia coli and porcine coronary artery, and the efflux was enhanced by the presence of external Na^+ with cells from taenia coli but not those from coronary artery, indicating that taenia coli but not coronary artery possesses the Na^+–Ca^{2+} exchange mechanisms. Furukawa *et al.* (1988) first quantitatively characterized Ca^{2+} extrusion systems in intact smooth muscle cells obtained from rat aorta by measuring $^{45}Ca^{2+}$ efflux from cells as a function of intracellular free Ca^{2+} concentration in the presence or absence of added extracellular Na^+. They found that Na^+–Ca^{2+} exchange extrudes Ca^{2+} in an amount comparable to that by the sarcolemmal Ca^{2+} pump at high cytosolic Ca^{2+} concentration (~1 μM) but that the former extrudes less Ca^{2+} than the latter at lower cytosolic Ca^{2+} concentrations.

Regulation of Na^+–Ca^{2+} exchange mechanisms has recently begun to be studied. Na^+–Ca^{2+} exchange in rat aortic smooth muscle cells has been shown to be stimulated by the treatment of cells with phorbol ester (Vigne *et al.,* 1988), suggesting that the phosphorylation catalyzed by C-kinase may be involved in the regulation of Na^+–Ca^{2+} exchange. More recently, Na^+–Ca^{2+} exchange in cultured rat aortic smooth muscle cells has been shown to be inhibited by mitochondrial poisons (Smith and Smith, 1990), suggesting that ATP modulates the exchange mechanism. Furukawa *et al.* (1991) examined the effect of cyclic GMP on Na^+–Ca^{2+} exchange in rat aortic smooth muscle cells in primary culture. The cells loaded with $^{45}Ca^{2+}$ were pretreated with 8-bromo-cyclic GMP or atrial natriuretic peptide, and were then stimulated with the Ca^{2+}-ionophore ionomycin to cause $^{45}Ca^{2+}$ release from the SR and thus evoke a transient increase in cytosolic free Ca^{2+} concentration. $^{45}Ca^{2+}$ efflux was measured in the medium containing high Mg^{2+} concentration (20 mM) and at high pH (8.8), conditions, which inhibit activity of the sarcolemmal Ca^{2+} pump. $^{45}Ca^{2+}$ efflux under these conditions was presumably driven from the mechanism of Na^+–Ca^{2+} exchange, because the $^{45}Ca^{2+}$ efflux was primarily dependent on the extracellular Na^+ and was totally inhibited when the extracellular Na^+ was replaced with choline$^+$. Treatment of the cells with 8-bromo-cyclic GMP or atrial natriuretic peptide caused the enhancement of the extracellular Na^+-dependent $^{45}Ca^{2+}$ efflux by up to 60%. Conversely, when the cells were loaded with Na^+ by treating them with various concentrations of extracellular Na^+ in the presence of ouabain, monensin, and sodium bicarbonate and then were exposed to a Na^+-free medium, the rate of $^{45}Ca^{2+}$ uptake into the cells increased as the intracellular Na^+ increased. Prior treatment of cells with 8-bromo-cyclic GMP or atrial natriuretic peptide accelerated $^{45}Ca^{2+}$ uptake by up to 60% without influencing Na^+ loading itself. These results indicate that 8-bromo-cyclic GMP or atrial natriuretic peptide enhances the extracellular Na^+-dependent $^{45}Ca^{2+}$ efflux and/or the intracellular Na^+-dependent $^{45}Ca^{2+}$ influx, thereby suggesting that elevation of cytosolic cyclic GMP stimulates both forward and reverse modes of Na^+–Ca^{2+} exchange in these cells. The underlying mechanisms by which cyclic GMP stimulates the Na^+–Ca^{2+} exchange remain to be elucidated, but presumably protein phosphorylation by G-kinase is involved, since protein phosphorylation by Ca^{2+}-dependent protein kinase has already been suggested to activate Na^+–Ca^{2+} exchange in squid axons and heart cells (Caroni and Carafoli, 1983; Dipolo and Beaugé, 1987).

D. Voltage-Dependent Ca^{2+} Channels

In excitable cells, voltage-dependent Ca^{2+} channels, which are now classified into three types, T-, L-, and N-type according to the electrophysiological and pharmacological properties, mediate Ca^{2+}-dependent depolarization and translate changes in membrane potential into an intracellular Ca^{2+} signal (Hagiwara and Byerly, 1981; Tsien, 1983). Among these three types of voltage-dependent Ca^{2+} channels, only the L-type channel is either blocked (Triggle, 1981; Cauvin *et al.*, 1983) or activated (Schramm *et al.*, 1983), by a group of compounds that is derived from dihydropyridine and now termed "calcium antagonists" (Lee and Tsien, 1983). The presence of these specific drugs and the successful synthesis of radiolabeled nitrendipine, one of the "calcium antagonists," facilitated the research on the molecular characterization of the Ca^{2+} channel. Using these compounds, Catterall and colleagues and other groups have purified dihydropyridine receptors to near homogeniety from skeletal muscle T-tubular membranes (Curtis and Catterall, 1984; Borsotto *et al.*, 1984, 1985). The purified receptor consists of a noncovalently associated complex of α, β, and γ subunits having apparent molecular weights of 162,000, 50,000 and 33,000, respectively (Curtis and Catterall, 1984). If disulfide bonds are reduced, α subunits are separated into two protein bands with apparent molecular weights of 165,000 and 135,000 (Curtis and Catterall, 1984; Flockerzi *et al.*, 1986). The purified dihydropyridine receptor has been incorporated into phospholipid membranes and shown to mediate dihydropyridine-sensitive Ca^{2+} conductance, providing evidence that these three subunits are sufficient to mediate the physiological function of the Ca^{2+} channel (Curtis and Catterall, 1984; Flockerzi *et al.*, 1986). Molecular cloning of cDNA for α subunit of skeletal muscle (Tanabe *et al.*, 1987), rat aorta (Koch *et al.*, 1990), and brain (Mori *et al.*, 1991), for β subunit of skeletal muscle (Ruth *et al.*, 1989), and for γ subunit of skeletal muscle (Jay *et al.*, 1990; Bosse *et al.*, 1990), has been performed. Thus, the deduced primary structures and predicted secondary structures have been clarified.

The effect of cyclic GMP on the Ca^{2+} current carried through the L-type Ca^{2+} channel has been best studied with cardiac tissues since the first hormonal effect on cyclic GMP levels to be discovered was that produced by acetylcholine, an agent that exerts negative inotropic effects in rat ventricle (George *et al.*, 1970). It has been proposed that the negative inotropic effect of cyclic GMP, like that of acetylcholine, is mediated by a decrease in the Ca^{2+} current because cyclic GMP decreases $^{45}Ca^{2+}$ flux (Nawrath, 1977), shortens action potential duration (Trautwein *et al.*, 1982), and inhibits Ca^{2+}-dependent action potentials (Kohlhardt and Haap, 1978; Wahler and Sperelakis, 1985). But, it has also been reported that low concentrations of acetylcholine depress contractility without changing cyclic GMP levels (Watanabe and Besch, 1975), and that increases in cyclic GMP produced by chemicals other than acetylcholine do not correlate with changes in contractility (Diamond *et al.*, 1977). Furthermore, some investigators have failed to find a significant effect of exogenous cyclic GMP on contractile force (Linden and Brooker, 1979), and elevation of cytosolic level of cyclic GMP by a photoactivated derivative of cyclic GMP had no effect on the Ca^{2+} current (Nargeot *et al.*, 1983). The negative inotropic effect of acetylcholine is not blocked by drugs that prevent the increase in cyclic GMP levels (Diamond and Chu, 1985). Thus, the effect of cyclic GMP on the Ca^{2+} current is still controversial.

To clarify the possible role and mechanism of action of cyclic GMP, the effect of intracellular perfusion with cyclic GMP and/or cyclic AMP on Ca^{2+} current carried through the L-type Ca^{2+} channel has been studied in single cells isolated from frog ventricle using the whole-cell patch-clamp technique, and a perfusion pipet (Hartzell and Fischmeister, 1986; Fischmeister and Hartzell, 1987). Intracellular perfusion with cyclic GMP had no effect on the basal Ca^{2+} current. However, when the Ca^{2+} current was increased by isoproterenol or by intracellular perfusion of cyclic AMP, perfusion with cyclic GMP reduced the Ca^{2+} current by an average of 67%. This effect of cyclic GMP was apparently not mediated by stimulation of G-kinase because 8-bromo-cyclic GMP, a very potent activator of the protein kinase, was without effect. Cyclic GMP had no effect on the Ca^{2+} current elevated by the nonhydrolyzable 8-bromo-cyclic AMP. The effect of cyclic GMP on cyclic AMP-elevated Ca^{2+} current was partially blocked by the phosphodiesterase inhibitor methylisobutylxanthine. From these observations, the authors hypothesized that the effect of cyclic GMP on the Ca^{2+} current is indirectly exerted through cyclic AMP, which increases the Ca^{2+} current; as a result of a stimulation of a cyclic nucleotide phosphodiesterase by cyclic GMP the hydrolysis of cyclic AMP is enhanced, thus resulting in the decrease of cellular level of cyclic AMP.

An alternate explanation for the inhibitory effect of cyclic GMP on the Ca^{2+} current has recently been made with cardiac myocytes of guinea pig (Levi *et al.*, 1989), embryonic chicken (Wahler *et al.*, 1990), and rat (Mery *et al.*, 1991). Mery *et al.* (1991) performed patch-clamp measurement of Ca^{2+} current on isolated rat myocytes internally perfused with a catalytically active fragment of G-kinase, which was prepared by limited proteolysis of the holoenzyme with trypsin. This proteolysis removes the amino-terminal end of G-kinase, which has been shown to contain domains responsible for both the dimerization and the regulation of G-kinase, the latter of which inhibits enzyme activity in the absence of cyclic GMP (Heil *et al.*, 1987). Use of this active fragment of G-kinase allowed the authors to examine the direct effect of G-kinase on Ca^{2+} current without being influenced by cyclic GMP. The proteolyzed and thus active G-kinase by itself exerted no significant effect on the Ca^{2+} current. However, when added after the Ca^{2+} current stimulation by intracellular perfusion of cyclic AMP, the kinase exerted a strong inhibitory effect on the current, thereby indicating that G-kinase-catalyzed phosphorylation is responsible for the inhibition of Ca^{2+} current stimulated with cyclic AMP.

A membrane-permeable derivative of cyclic GMP, 8-bromo-cyclic GMP, was also reported to be effective in smooth muscle preparations; 8-bromo-cyclic GMP markedly inhibited the channels mediating tetraethylammonium-induced action potential in canine tracheal smooth muscle (Richards *et al.*, 1986). On the other hand, Ohya *et al.* (1987) reported that the Ca^{2+} channel in smooth muscle cells of rabbit portal vein was not enhanced for cyclic AMP applied intracellularly, indicating that mechanisms by cyclic nucleotide regulation of Ca^{2+} channels will be variable, depending on cell types.

E. Receptor-Operated Ca^{2+} Entry

The elevation of intracellular free Ca^{2+} concentration in response to Ca^{2+}-mobilizing agonists generally occurs in two phases, the first of which is a transient mobilization of Ca^{2+} from intracellular stores, most likely mediated by inositol 1,4,5-trisphosphate ($Ins(1,4,5)P_3$) (Berridge and Irvine, 1984; Berridge, 1987). The second phase is a sustained increase in Ca^{2+} entry from the extracellular space and continues as long as the agonists stimulate. Bolton is apparently the first to call this Ca^{2+} entry accompanying no changes in membrane potential "receptor-operated Ca^{2+} entry (channel)" (Bolton, 1979). There are no specific inhibitors for this continued entry of Ca^{2+}, like dihydripyridine Ca^{2+} antagonists against voltage-dependent Ca^{2+} channels as described in the previous section, and there has been no attempt to clarify the structural basis of the Ca^{2+} entry mechanisms.

The agonists that give rise to the sustained increase in Ca^{2+} entry generally cause the hydrolysis of a minor membrane phospholipid, PIP_2

(Michell, 1975), and it is thought that PIP_2 metabolites are involved in the proposed mechanisms for activation of the channels. Irvine has proposed that inositol 1,3,4,5-tetrakisphosphate ($Ins(1,3,4,5)P_4$), a product of a specific phosphorylation of $Ins(1,4,5)P_3$, acts as a mediator for activating the Ca^{2+} channels or $Ins(1,3,4,5)P_4$ together with $Ins(1,4,5)P_3$ causes a sustained increase in Ca^{2+} entry from the extracellular space (Irvine and Moor, 1985; Irvine, 1990; Morris *et al.*, 1987; Changya *et al.*, 1989; Lückhoff and Clapham, 1992). $Ins(1,4,5)P_3$ by itself has recently been proposed as a messenger for Ca^{2+} entry through its specific receptor located in the plasma membrane of T lymphocytes (Khan *et al.*, 1992a,b). Putney and his group have proposed a "capacitative Ca^{2+} entry" mechanism which argues that depletion of intracellular Ca^{2+} stores, most likely mediated by $Ins(1,4,5)P_3$ under physiological conditions, provides a signal to allow Ca^{2+} entry from the extracellular space (Putney *et al.*, 1989; Takemura *et al.*, 1989). Very recently, Huang *et al.* (1991) examined the possible mechanisms for the prolonged Ca^{2+} entry in platelet-derived growth factor (PDGF)-stimulated rat vascular smooth muscle cells, microinjecting with heparin and/or monoclonal anti-PIP_2 antibodies. Microinjection of monoclonal anti-PIP_2 antibodies totally abolished both mobilization of intracellular Ca^{2+} stores and the sustained entry of extracellular Ca^{2+}, whereas microinjection of heparin prevented the initial mobilization of intracellular Ca^{2+} but did not affect extracellular Ca^{2+} entry, thus providing evidence that the sustained increases in Ca^{2+} entry involves PIP_2 or PIP_2 metabolites.

Another possibility is that receptors might directly open plasma membrane Ca^{2+} channels, the putative receptor-operated channel, since Benham and Tsien (1987) have found the presence of such channels.

The mechanisms for receptor-operated Ca^{2+} entry are not known, but if the Ca^{2+} entry is mediated through the action of PIP_2 and its metabolism as originally proposed by Michell (1975), it would be probable that cyclic GMP inhibits this process, since cyclic GMP has been reported to inhibit the hydrolysis of PIP_2 as described in the next section. The inhibitory effect of cyclic GMP on receptor-operated Ca^{2+} entry was reported by Godfraind (1986). He measured $^{45}Ca^{2+}$ entry into rat aortic smooth muscle segments treated with norepinephrine. Inclusion of methylene blue, an inhibitor of guanylate cyclase markedly augmented the $^{45}Ca^{2+}$ entry induced by norepinephrine, but not by high K^+ medium, which stimulates the voltage-operated Ca^{2+} chanel, indicating that receptor (for norepinephrine)-operated Ca^{2+} entry is inhibited by cyclic GMP.

There is also a report indicating that the increase in cytosolic cyclic GMP is necessary to mediate the effect of the agonists on the plasma membrane Ca^{2+} entry mechanisms (Pandol and Schoeffield-Payne, 1990).

In guinea pig pancreatic acinar cells carbachol stimulated a transient 20- to 40-fold rise in cytosolic cyclic GMP followed by a sustained 3- to 4-fold rise in cytosolic cyclic GMP. The nonspecific guanylate cyclase inhibitor 6-anilino-5,8-quinolinedione (LY83583) caused a dose-dependent inhibition of carbachol-stimulated increases in cytosolic cyclic GMP. LY83583 also inhibited cellular Ca^{2+} influx during carbachol stimulation, and this was restored by the addition of dibutyryl cyclic GMP. Nitroprusside by itself increased both cellular cyclic GMP and the rate of Ca^{2+} influx. From these results, they suggested that cyclic GMP would be responsible for the increased rate of Ca^{2+} entry.

III. Cyclic GMP Effect on Ins(1,4,5)P_3 Production

A. Mechanisms of Ins(1,4,5)P_3 Production

The stimulation of PIP_2 hydrolysis by a wide variety of hormones, neurotransmitters, and growth factors to yield the second messengers Ins(1,4,5)P_3 and diacylglycerol is well documented (Berridge, 1987) (Fig. 2). The mechanism by which binding of ligands to their specific receptors triggers activation of intracellular phospholipase C (PLC) activity is not yet fully understood. A substantial body of evidence supports the view that a GTP-binding regulatory protein (G-protein) is involved in the coupling of receptors to PLC (Gilman, 1987). Briefly, agonist-induced hydrolysis of PIP_2 in isolated membrane vesicles requires the presence of GTP or nonhydrolyzable analogs of GTP, and GTP by itself at a relatively high concentration stimulates PIP_2 hydrolysis. GTP analogs also alter the affinity for

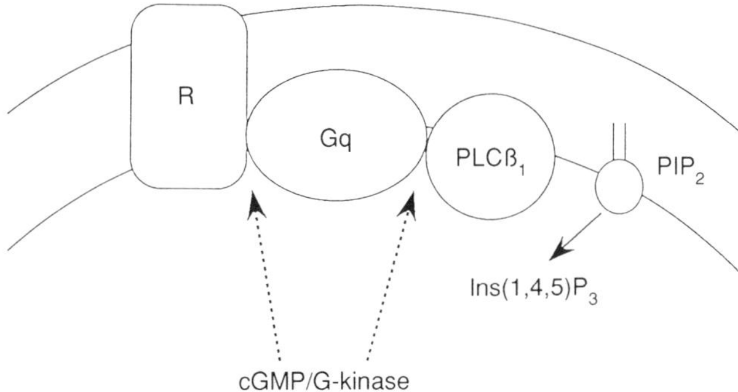

Fig. 2 Proposed mechanism for inhibition of cyclic GMP/G-kinase on Ins(1,4,5)P_3 production.

agonists receptors linking to PIP_2 hydrolysis. By analogy with the best understood pathway of adenylate cyclase and retinal cyclic GMP phosphodiesterase systems, it has been suggested that a heterotrimeric G-protein(s) transduces the signal from receptors to PLC. Such G-proteins are classified, depending on the sensitivity to a bacterial toxin, pertussis toxin. Pertussis toxin, which catalyzes ADP-ribosylation of the α subunit of G_i, G_o, and transducin and attenuates the interaction with their respective receptors, inhibits agonist-stimulated PIP_2 hydrolysis in some cell types, but not others. In the toxin-sensitive systems, G_o or a G_o-like protein has been reported to represent the toxin-sensitive stimulatory G-protein of PIP_2 hydrolysis (Boyer *et al.*, 1989a,b; Moriarty *et al.*, 1988, 1989). However, in the majority of tissues and cells the activation of PLC is insensitive to the toxin, but also in these cases the involvement of a G-protein(s) is required, indicating that another class of G protein may be responsible for such an activation. A novel G-protein (Gq) has been identified with $\beta\gamma$-agarose (Pang and Sternweis, 1989; Strathmann and Simon, 1990), and this novel protein is also identified as a PLC activator, for an isozyme, PLCβ1, but not for other isozymes (Taylor *et al.*, 1990; Rhee and Choi, 1992).

Several PLC isoenzymes have been purified to apparent homogeniety from a variety of tissues by many investigators. Rhee's group carried out extensive studies using antibodies against bovine brain enzymes and rat seminal vesicle enzymes and classified PLC into four isozymes with the name of PLC-α, β, γ, and δ (Rhee *et al.*, 1989). Of these isozymes, the γ isozyme has been implicated in a signal transduction mechanism (Ullrich and Schlessinger, 1990), in which the tyrosine phosphorylation of the isozyme by EGF and/or PDGF is involved but a role for G-protein is not involved. Very recently, the β1 isozyme was found to be regulated by the α subunit of the above-mentioned novel G protein, Gq and G_{11}, and is expected to be regulated by a wide variety of agonists (Smarcka *et al.*, 1991; Taylor *et al.*, 1991; Rhee and Choi, 1992). Furthermore, the $\beta\gamma$-subunits of trimeric G protein have recently been clarified to activate specifically β2 isozyme of PLC (Camps *et al.*, 1992; Katz *et al.*, 1992).

B. Mode of Cyclic GMP Inhibition of Ins(1,4,5)P_3 Production

8-Bromo-cyclic GMP, and cyclic GMP-elevating agents such as nitroprusside or atrial natriuretic peptide attenuate the hydrolysis of PIP_2 in platelets (Takai *et al.*, 1981, 1982; Nakashima *et al.*, 1986) or vascular tissues (Rapoport, 1986; Fujii *et al.*, 1986), when the preparations are stimulated with thrombin or norepinephrine, respectively. Furthermore, cyclic GMP inhibits the secretion of amylase in rat pancreatic acinar cells stimulated with cholecystokinin; this effect is thought to be mediated through the activation of C-kinase by diacylglecerol, the other product of PIP_2 hydrolysis (Rogers *et al.*, 1988). Thus, cyclic GMP is reported to inhibit the PIP_2 hydrolysis.

The receptor–G protein–PLC signal transduction pathway provides several potential sites for the regulation by cyclic GMP. For example, Ins(1,4,5)P_3 production could be regulated by an alteration of ligand-binding, receptor-mediated activation of the G protein, G protein-mediated activation of PLC, or direct alteration of PLC activity. Hirata *et al.* (1990) have attempted to determine the mechanism of inhibition involved in cyclic GMP inhibition of Ins(1,4,5)P_3 production in cultured bovine aortic smooth muscle cells. For this purpose, phosphoinositide hydrolysis and GTPase activity in homogenates and membrane preparations of cultured bovine aortic smooth muscle cells were studied. Pretreatment of homogenate preparations with cyclic GMP plus ATP did not inhibit 8-arginine-[^{3}H]vasopressin binding, but resulted in a marked suppression of the agonist-induced GTPase activation. The pretreatment with cyclic GMP and ATP also inhibited the formation of inositol phosphates induced by 8-arginine[^{3}H]vasopressin in the presence of low concentrations of guanosine 5′-(γ-thio)triphosphate (GTPγS), or by high concentrations of GTPγS alone. However, the formation of inositol phosphates by high concentrations of Ca^{2+} alone was not blocked. These results suggest that the ability of cyclic GMP to inhibit phosphoinositide hydrolysis results from an inhibition of G protein activation, and the interaction between G proteins and PLC. Although the precise site of this inhibition is not currently known, the inhibition by cyclic GMP is dependent on the addition of ATP and probably entails a phosphorylation event, presumably cyclic GMP dependent, since adenylylimidodiphosphate could not substitute for the ATP requirement.

References

Baker, P. F., Blaustein, M. P., Hodgkin, A. L., and Steinhardt, R. A. (1967). The effect of sodium concentration on calcium movements in giant axon of *Loligo forbesi*. *J. Physiol. (London)* **192,** 43P–44P.

Baltensberger, K., Carafoli, E., and Chiesi, M. (1988). The Ca^{2+}-pumping ATPase and the major substrates of the cGMP-dependent protein kinase in smooth muscle sarcolemma are distinct entities. *Eur. J. Biochem.* **172,** 7–16.

Bell, J. D., Buxton, I. L., and Brunton, L. L. (1985). Enhancement of adenylate cyclase activity in S49 lymphoma cells by phorbol esters. *J. Biol. Chem.* **260,** 2625–2628.

Benham, C. D., and Tsien, R. W. (1987). A novel receptor-operated Ca^{2+} channel activated by ATP in smooth muscle. *Nature (London)* **328,** 275–278.

Berridge, M. J. (1987). Inositol trisphosphate and diacylglycerol: Two interacting second messengers. *Annu. Rev. Biochem.* **56,** 159–193.

Berridge, M. J., and Irvine, R. F. (1984). Inositol trisphosphate, a novel second messenger in cellular signal transduction. *Nature (London)* **312,** 315–321.

Bolton, T. B. (1979). Mechanisms of action of transmitters and other substances on smooth muscle. *Physiol. Rev.* **59,** 606–718.

Borsotto, M., Norman, R. I., Fosset, M., and Lazdunski, M. (1984). Solubilization of the nitrendipine receptor from skeletal muscle transverse tubular membranes. *Eur. J. Biochem.* **142,** 449–455.

Borsotto, M., Barhanin, J., Fosset, M., and Lazdunski, M. (1985). The 1,4-dihydropyridine receptor associated with the skeletal muscle voltage-dependent Ca^{2+} channel. *J. Biol. Chem.* **260,** 14255–14263.

Bosse, E., Regulla, S., Biel, M., Ruth, P., Meyer, H. E., Flockerzi, V., and Hofmann, F. (1990). The cDNA and deduced amino acid sequence of the γ subunit of the L-type calcium channel from rabbit skeletal muscle. *FEBS Lett.* **267,** 153–156.

Boyer, J. L., Waldo, G. L., Evans, T., Northup, J. K., Downes, C. P., and Harden, T. K. (1989a). Modification of AlF_4- and receptor-stimulated phospholipase C activity by G-protein $\beta\gamma$ subunits. *J. Biol. Chem.* **264,** 13917–13922.

Boyer, J. L., Hepler, J. R., and Harden, T. K. (1989b). Hormone and growth factor receptor-mediated regulation of phospholipase C activity. *Trends Pharmacol. Sci.* **10,** 360–364.

Camps, M., Carozzi, A., Schnabel, P., Scheer, A., Parker, P. J., and Gierschik, P. (1992). Isozyme-selective stimulation of phospholipase C-β2 by G protein $\beta\gamma$-subunits. *Nature (London)* **360,** 684–686.

Caroni, P., and Carafoli, E. (1981). The Ca^{2+}-pumping ATPase of heart sarcolemma. *J. Biol. Chem.* **256,** 3263–3270.

Caroni, P., and Carafoli, E. (1983). The regulation of the Na^+-Ca^{2+} exchange of heart sarcolemma. *Eur. J. Biochem.* **132,** 451–460.

Carpenter, C. L., Duckworth, B. C., Auger, K. R., Cohen, B., Schaffhausen, B. S., and Cantley, L. C. (1990). Purification and characterization of phosphoinositide 3-kinase from rat liver. *J. Biol. Chem.* **265,** 19704–19711.

Cauvin, C., Loutzenhister, R., and van Breemen, C. (1983). Mechanisms of calcium antagonist-induced vasodilation. *Annu. Rev. Pharmacol. Toxicol.* **23,** 373–396.

Changya, L., Gallacher, D. V., Irvine, R. F., and Petersen, O. H. (1989). Inositol 1,3,4,5-tetrakisphosphate and inositol 1,4,5-trisphosphate act by different mechanisms when controlling Ca^{2+} in mouse lacrimal acinar cells. *FEBS Lett.* **251,** 43–48.

Choquette, D., Hakim, G., Filoteo, A. G., Plishker, G. A., Bostwich, J. R., and Penniston, J. T. (1984). Regulation of plasma membrane Ca^{2+} ATPase by lipids of the phosphatidylinositol cycle. *Biochem. Biophys. Res. Commun.* **125,** 908–915.

Curtis, B. M., and Catterall, W. A. (1984). Purification of the calcium antagonist receptor of the voltage-sensitive calcium channel from skeletal muscle transverse tubules. *Biochemistry* **23,** 2113–2118.

DeSchutter, G., Wuytack, F., Verbist, J., and Casteels, R. (1984). Tissue levels and purification by affinity chromatography of the calmodulin-stimulated Ca^{2+}-transport ATPase in pig antrum smooth muscle. *Biochim. Biophys. Acta* **773,** 1–10.

Diamond, J., and Chu, E. B. (1985). A novel cyclic GMP-lowering agent, LY83583, blocks carbachol-induced cyclic GMP elevation in rabbit atrial strips without blocking the negative inotropic effects of carbachol. *Can. J. Physiol. Pharmacol.* **63,** 908–911.

Diamond, J., Ten Eick, R. E., and Trapani, A. J. (1977). Are increases in cyclic GMP levels responsible for the negative inotropic effects of acetylcholine in the heart? *Biochem. Biophys. Res. Commun.* **79,** 912–917.

Dipolo, R., and Beaugé, L. (1987). In squid axon, ATP modulates Na^+-Ca^{2+} exchange by a Ca^{2+}-dependent phosphorylation. *Biochim. Biophys. Acta* **897,** 347–354.

Enyadi, A., Flura, M., Sarkadi, B., Gardos, G., and Carafoli, E. (1987). The maximal velocity and the calcium affinity of the red cell calcium pump may be regulated independently. *J. Biol. Chem.* **262,** 6425–6430.

Fischmeister, R., and Hartzell, H. C. (1987). Cyclic guanosine 3′,5′-monophosphate regulates the calcium current in the single cells from frog ventricle. *J. Physiol. (London)* **387,** 453–472.

Flockerzi, V., Oeken, H.-J., Hofmann, F., Pelzer, D., Cavalie, A., and Trautwein, W. (1986). Purified dihydropyridine-binding site from skeletal muscle t-tubules is a functional calcium channel. *Nature (London)* **323,** 66–68.

Fujii, K., Ishimatsu, T., and Kuriyama, H. (1986). Mechanism of vasodilation induced by a-human atrial natriuretic polypeptide in rabbit and guinea pig renal arteries. *J. Physiol. (London)* **377,** 315–332.

Furukawa, K.-I., and Nakamura, H. (1987). Cyclic GMP regulation of the plasma membrane (Ca^{2+}-Mg^{2+})ATPase in vascular smooth muscle. *J. Biochem. (Tokyo)* **101,** 287–290.

Furukawa, K.-I., Tawada, Y., and Shigekawa, M. (1988). Regulation of the plasma membrane Ca^{2+} pump by cyclic nucleotides in cultured vascular smooth muscle cells. *J. Biol. Chem.* **263,** 8058–8065.

Furukawa, K.-I., Ohshima, N., Tawada-Iwata, Y., and Shigekawa, M. (1991). Cyclic GMP stimulates Na^{+}/Ca^{2+} exchange in vascular smooth muscle cells in primary culture. *J. Biol. Chem.* **266,** 12337–12341.

George, W. J., Polson, J. B., O'Toole, A. G., and George, N. (1970). Elevation of guanosine 3′,5′-cyclic phosphate in rat heart after perfusion with acetylcholine. *Proc. Natl. Acad. Sci. U.S.A.* **66,** 398–403.

Gietzell, K., Tejcka, M., and Wolf, H. U. (1980). Calmodulin affinity chromatography yields a functional purified erythrocyte ($Ca^{2+} + Mg^{2+}$)-dependent adenosine triphosphatase. *Biochem. J.* **189,** 81–88.

Gilman, A. G., (1987). G proteins:transducers of receptor-generated signals. *Annu. Rev. Biochem.* **56,** 615–649.

Godfraind, T. (1986). EDRF and cyclic GMP control gating of receptor-operated calcium channels in vascular smooth muscle. *Eur. J. Pharmacol.* **126,** 341–343.

Hagiwara, S., and Byerly, L. (1981). Calcium channel. *Annu. Rev. Neurosci.* **4,** 69–125.

Hartzell, H. C., and Fischmeister, R. (1986). Opposite effects of cyclic GMP and cyclic AMP on Ca^{2+} current in single heart cells. *Nature (London)* **323,** 273–275.

Heil, W. G., Landgraf, W., and Hofmann, F. (1987). A catalytically active fragment of cGMP-dependent protein kinase. *Eur. J. Biochem.* **168,** 117–121.

Hirata, M., Itoh, T., and Kuriyama, H. (1981). Effects of external cations on calcium efflux from single cells of the guinea pig taenia coli and procine coronary artery. *J. Physiol. (London)* **310,** 321–336.

Hirata, M., Kohse, K. P., Chang, C.-H., Ikebe, T., and Murad, F. (1990). Mechanism of cyclic GMP inhibition of inositol phosphate formation in rat aorta segments and cultured bovine aortic smooth muscle cells. *J. Biol. Chem.* **265,** 1268–1273.

Hogaboom, G. K., Emler, C. A., Butcher, F. R., and Fedan, J. S. (1982). Concerted phosphorylation of endogenous tracheal smooth muscle of membrane proteins by Ca^{2+}-clamodulin-, cyclicGMP- and cyclic AMP-dependent protein kinases. *FEBS Lett.* **139,** 309–312.

Huang, C.-L., Takenawa, T., and Ives, H. E. (1991). Platelet-derived growth factor-mediated Ca^{2+} entry is blocked by antibodies to phosphatidylinositol 4,5-bisphosphate but does not involve heparin-sensitive inositol 1,4,5-trisphosphate receptors. *J. Biol. Chem.* **266,** 4045–4048.

Irvine, R. F. (1990). 'Quantal' Ca^{2+} release and the control of Ca^{2+} entry by inositol phosphates-a possible mechanism. *FEBS Lett.* **263,** 1–5.

Irvine, R. F., and Moor, R. M. (1985). Micro-injection of inositol 1,3,4,5-tetrakisphosphate activates sea urchin eggs by a mechanism dependent on external Ca^{2+}. *Biochem. J.* **240,** 917–920.

Itoh, T., Kuriyama, H., and Ueno, H. (1983). Mechanisms of the nitroglycerine-induced vasodilation in vascular smooth muscles of the rabbit and pig. *J. Physiol. (London)* **343,** 233–252.

Itoh, T., Kanmura, Y., Kuriyama, H., and Sasaguri, T. (1985). Nitroglycerine- and isoprenaline-induced vasodilatation:assessment from the actions of cyclic nucleotides. *Br. J. Pharmacol.* **84,** 393–406.

Ives, H. E., Casnellie, J. E., Greengard, P., and Jamieson, J. D. (1980). Subcellular localization of cyclic GMP-dependent protein kinases and its substrates in vascular smooth muscle. *J. Biol. Chem.* **255,** 3777–3785.

James, P., Maeda, M., Fischer, R., Verma, A., Krebs, J., Penniston, J. T., and Carafoli, E. (1988). Identification and primary structure of a calmodulin-binding domain of the Ca^{2+} pump of human erythrocytes. *J. Biol. Chem.* **263,** 2905–2910.

Jay, S. D., Ellis, S. B., McCue, A. F., Williams, M. E., Vedvick, T. S., Harpold, M. M., and Campbell, K. P. (1990). Primary structure of the γ subunit of the DHP-sensitive calcium channel from skeletal muscle. *Science.* **248,** 490–492.

Katz, A., Wu, D., and Simon, M. I. (1992). Subunits $\beta\gamma$ of heterotrimeric G protein activate $\beta 2$ isoform of phospholipase C. *Nature (London)* **360,** 686–689.

Kelleher, D. J., Pessin, J. E., Ruoho, A., and Johnson, G. L. (1984). Phorbol ester induces desensitization of adenylate cyclase and phosphorylation of the b-adrenergic receptor in turkey erythrocytes. *Proc. Natl. Acad. Sci. U.S.A.* **81,** 4316–4320.

Khan, A. D., Steiner, J. P., and Snyder, S. H. (1992a). Plasma membrane inositol 1,4,5-trisphosphate receptor of lymphocytes: Selective enrichment in sialic acid and unique binding specificity. *Proc. Natl. Acad. Sci. U.S.A.* **89,** 2849–2853.

Khan, A. D., Steiner, J. P., Klein, M. G., Schneider, M. F., and Snyder, S. H. (1992b). IP_3 receptor: Localization to plasma membrane of T cells and cocapping with the T cell receptor. *Science* **257,** 815–818.

Kobayashi, S., Kanaide, H., and Nakamura, M. (1985). Cytosolic free calcium transient in cultured vascular smooth muscle cells. *Science* **229,** 553–556.

Koch, W. J., Ellinor, P. T., and Schwartz, A. (1990). cDNA cloning of a dihydropyridine-sensitive calcium channel from rat aorta. *J. Biol. Chem.* **265,** 17786–17791.

Kohlhardt, M., and Haap, K. (1978). 8-bromo-guanosine-3′,5′-monophosphate mimics the effect of acetylcholine on slow response action potential and contractile force in mammalian atrial myocardium. *J. Mol. Cell. Cardiol.* **10,** 573–586.

LaRaia, J. J., and Morkin, E. (1974). Adenosine 3′,5′-monophosphate-dependent membrane phosphorylation. *Circ. Res.* **35,** 298–306.

Lee, K. S., and Tsien, R. W. (1983). Mechanism of calcium channel blockade by verapamil, D600, diltiazem and nitrendipine in single dialysed heart cells. *Nature (London)* **302,** 790–794.

LePeuch, C. J., Haiech, J., and Demaille, J. G. (1979). Concerted regulation of cardiac sarcoplasmic reticulum calcium transport by cyclic adenosine monophosphate-dependent and calcium-calmodulin-dependent phosphorylations. *Biochemistry* **18,** 5150–5157.

Levi, R. C., Alloatti, G., and Fischmeister, R. (1989). Cyclic GMP regulates the Ca^{2+}-channel current in guinea pig ventricular myocytes. *Pfluegers Arch.* **413,** 685–687.

Linden, J. M., and Brooker, G. (1979). The questionable role of cyclic guanosine 3′:5′-monophosphate in heart. *Biochem. Pharmacol.* **28,** 3351–3360.

Lückhoff, A., and Clapham, D. E. (1992). Inositol 1,3,4,5-tetrakisphosphate activates an endothelial Ca^{2+}-permeable channel. *Nature (London)* **355,** 356–358.

Mery, P.-F., Lohmann, S. M., Walter, U., and Fischmeister, R. (1991). Ca^{2+} current is regulated by cyclic GMP-dependent protein kinase in mammalian cardiac myocytes. *Proc. Natl. Acad. Sci. U.S.A.* **88,** 1197–1201.

Michell, R. H. (1975). Inositol phospholipids and cell surface receptor function. *Biochim. Biophys. Acta* **415,** 81–147.

Mori, Y., Friedrich, T., Kim, M.-S., Mikami, A., Nakai, J., Ruth, P., Bosse, E., Hofmann, F., Flockerzi, V., Furuichi, T., Mikoshiba, K., Imoto, K., Tanabe, T., and Numa, S. (1991). Primary structure and functional expression from complementary DNA of a brain calcium channel. *Nature (London)* **350,** 398–402.

Moriarty, T. M., Gillo, B., Carty, D. J., Premont, R. T., Landau, E. M., and Iyengar, R. (1988). bg-subunits of GTP-binding proteins inhibit muscarinic receptor stimulation of phospholipase C. *Proc. Natl. Acad. Sci. U.S.A.* **85,** 8865–8869.

Moriarty, T. M., Sealfon, S. C., Carty, D. J., Roberts, J. L., Iyengar, R., and Landau, E. M. (1989). Coupling of exogenous receptors to phospholipase C in xenopus oocytes through pertussis toxin-sensitive and-insensitive pathways. *J. Biol. Chem.* **264,** 13524–13530.

Morris, A. P., Gallacher, D. V., Irvine, R. F., and Petersen, O. H. (1987). Synergism of inositol trisphosphate and tetrakisphosphate in activating Ca^{2+}-dependent K^+ channels. *Nature (London)* **330,**653–655.

Movsenian, M. A., Nishikawa, M., and Adelstein, R. S. (1984). Phosphorylation of phospholamban by calcium-activated, phospholipid-dependent protein kinase. *J. Biol. Chem.* **259,** 8029–8032.

Murad, F., Waldman, S. A., Fiscus, R. R., and Rapoport, R. M. (1986). Regulation of cyclic GMP synthesis and the interactions with calcium. *J. Cardiovasc. Pharmacol.* **8,** S57-S60.

Nakashima, S., Tohmatsu, T., Hattori, H., Okano, Y., and Nozawa, Y. (1986). Inhibitory action of cyclic GMP on secretion, polyphosphoinositide hydrolysis and calcium mobilization in thrombin-stimulated human platelets. *Biochem. Biophys. Res. Commun.* **135,** 1099–1104.

Nargeot, J., Nerbonne, J. M., Engels, J., and Lester, H. A. (1983). Time cource of the increase in the myocardial slow inward current after a photochemically generated concentration jump of intracellular cAMP. *Proc. Natl. Acad. Sci. U.S.A.* **80,** 2395–2399.

Nawrath, H. (1977). Does cyclic GMP mediate the negative inotropic effect of acetylcholine in the heart? *Nature (London)* **267,** 72–74.

Nelson, D. R., and Hanahan, D. J. (1985). Phospholipid and detergent effects on (Ca^{2+}-Mg^{2+})ATPase purified from human erythrocytes. *Arch. Biochem. Biophys.* **236,** 720–730.

Niggli, V., Penniston, J. T., and Carafoli, E. (1979). Purification of the (Ca^{2+}-Mg^{2+})-ATPase from human erythrocyte membranes using a calmodulin affinity column. *J. Biol. Chem.* **254,** 9955–9958.

Niggli, V., Adunyah, E. S., Penniston, J. T., and Carafoli, E. (1981a). Purified (Ca^{2+}-Mg^{2+})-ATPase of the erythrocyte membrane. *J. Biol. Chem.* **256,** 395–401.

Niggli, V., Adunyah, E. S., and Carafoli, E. (1981b). Acidic phospholipids, unsaturated fatty acid, and limited proteolysis mimic the effect of calmodulin on the purified erythrocyte Ca^{2+}-ATPase. *J. Biol. Chem.* **256,** 8588–8592.

Ohya, Y., Kitamura, K., and Kuriyama, H. (1987). Modulation of ionic currents in smooth muscle balls of the rabbit intestine by intracellulary perfused ATP and cyclic AMP. *Pfluegers Arch.* **408,** 465–473.

Pandol, S. J., and Schoeffield-Payne, M. S. (1990). Cyclic GMP mediates the agonist-stimulated increase in plasma membrane calcium entry in the pancreatic acinar cell. *J. Biol. Chem.* **265,** 12846–12853.

Pang, I.-H., and Sternweis, P. C. (1989). Isolation of the a subunits of GTP-binding regulatory proteins by affinity chromatography with immobilized bg subunits. *Proc. Natl. Acad. Sci. U.S.A.* **86,** 7814–7818.

Popescu, L. M., Panoiu, C., Hinescu, M., and Nutu, O. (1985). The mechanism of cGMP-induced relaxation in vascular smooth muscle. *Eur. J. Pharmacol.* **107,** 393–394.

Porter, F. D., Li, Y.-S., and Deuel, T. F. (1988). Purification and characterization of a phosphatidylinositol 4-kinase from bovine uteri. *J. Biol. Chem.* **263,** 8989–8995.

Putney, J. W., Jr., Takemura, H., Hughes, A. R., Horstman, D. A., and Thastrup, O. (1989). How do inositol phosphates regulate calcium signaling? *FASEB J.* **3,** 1899–1905.

Raeymaekers, L., and Jones, L. R. (1986). Evidence for the presence of phospholamban in the endoplasmic reticulum of smooth muscle. *Biochim. Biophys. Acta* **882,** 258–265.

Raeymaekers, L., Wuytack, F., Eggermont, J., DeSchutter, G., and Casteels, R. (1983). Isolation of a plasma-membrane fraction from gastric smooth muscle. Comparison of the calcium uptake with that in endoplasmic reticulum. *Biochem. J.* **210,** 315–322.

Raeymaekers, L., Hofmann, F., and Casteels, R. (1988). Cyclic GMP-dependent protein kinase phosphorylates phospholamban in isolated sarcoplasmic reticulum from cardiac and smooth muscle. *Biochem. J.* **252,** 269–273.

Rapoport, R. M. (1986). Cyclic guanosine monophosphate inhibition of contraction may be mediated through inhibition of phosphatidylinositol hydrolysis in rat aorta. *Circ. Res.* **58,** 407–410.

Rapoport, R. M., Draznin, M. B., and Murad, F. (1983). Endothelium-dependent relaxation in rat aorta may be mediated through cyclic GMP-dependent protein phosphorylation. *Nature* (*London*) **306,** 174–176.

Reuter, H., and Seitz, H. (1968). The dependence of calcium efflux from cardiac muscle on temperature and external ion composition. *J. Physiol.* (*London*) **195,** 451–470.

Rhee, S. G., and Choi, K. D. (1992). Regulation of inositol phospholipid-specific phospholipase C isozymes. *J. Biol. Chem.* **267,** 12393–12396.

Rhee, S. G., Suh, P.-G., Ryu, S.-H., and Lee, S. Y. (1989). Studies of inositol phospholipid-specific phospholipase C. *Science* **244,** 546–550.

Richards, I. S., Murlas, C., Ousterhout, J. M., and Sperelakis, N. (1986). 8-bromo-cyclic GMP abolishes TEA-induced slow action potentials in canine trachealis muscle. *Eur. J. Pharmacol.* **128,** 299–302.

Rogers, J., Hughes, R. G., and Matthews, E. K. (1988). Cyclic GMP inhibits protein kinase C-mediated secretion in rat pancreatic acini. *J. Biol. Chem.* **263,** 3713–3719.

Ruth, P., Rohrkasten, A., Biel, M., Bosse, E., Regulla, S., Meyer, H. E., Flockerzi, V., and Hofmann, F. (1989). Primary structure of the β subunit of the DHP-sensitive calcium channel from skeletal muscle. *Science* **245,** 1115–1118.

Schramm, M., Thomas, G., Towart, R., and Franckowiak, G. (1983). Novel dihydropyridines with positive inotropic action through activation of Ca^{2+} channels. *Nature* (*London*) **303,** 535–537.

Schultz, K.-D., Schultz, K., and Schultz, G. (1977). Sodium nitroprusside and other smooth muscle-relaxants increase cyclic GMP levels in rat ductus deference. *Nature* (*London*) **265,** 750–751.

Shibasaki, F., Hamma, Y., and Takenawa, T. (1991). Two types of phosphatidylinositol 3-kinase from bovine thymus. *J. Biol. Chem.* **266,** 8108–8114.

Shull, G., and Greeb, J. (1988). Molecular cloning of two isoforms of the plasma membrane Ca^{2+}-transporting ATPase from rat brain. *J. Biol. Chem.* **263,** 8646–8657.

Sibly, D. R., Nambi, P., Peters, J. R., and Lefkowitz, R. J. (1984). Phorbol diesters promote β-adrenergic receptor phosphorylation and adenylate cyclase desensitization in dick erythrocytes. *Biochem. Biophys. Res. Commun.* **121,** 973–979.

Sibly, D. R., Jeffs, R. A., Daniel, K., Nambi, P., and Lefkowitz, R. J. (1986). Phorbol diester treatment promotes enhanced adenylate cyclase activity in frog erythrocytes. *Arch. Biochem. Biophys.* **244,** 373–381.

Smrcka, A. V., Hepler, J. R., Brown, K. O., and Sternweis, P. C. (1991). Regulation of polyphosphoinositide-specific phospholipase C activity by purified Gq. *Science* **251,** 804–807.

Smith, J. B., and Smith, L. (1990). Energy dependence of sodium-calcium exchange in vascular smooth muscle cell. *Am. J. Physiol.* **259,** C302–C309.

Strathmann, M., and Simon, M. I. (1990). G protein diversity: A distinct class of a subunits is present in vertebrates and invertebrates. *Proc. Natl. Acad. Sci. U.S.A.* **87,** 9113–9117.

Strehler, E. E., Strehler, M. A., Vogel, G., and Carafoli, E. (1989). mRNAs for plasma membrane calcium pump isoforms differing in their regulatory domain are generated by alternative splicing that involves two internal donor sites in a single exon. *Proc. Natl. Acad. Sci. U.S.A.* **86,** 6908–6912.

Suematsu, E., Hirata, M., and Kuriyama, H. (1984). Effects of cAMP- and cGMP-dependent protein kinase, and calmodulin on Ca^{2+} uptake by highly purified sarcolemmal vesicles of vascular smooth muscle. *Biochim. Biophys. Acta* **773,** 83–90.

Tada, M., Kirchberger, M. A., and Katz, A. M. (1975). Phosphorylation of a 22,000-dalton component of the cardiac sarcoplasmic reticulum by adenosine 3′:5′-monophosphate-dependent protein kinase. *J. Biol. Chem.* **250,** 2640–2647.

Takai, Y., Kaibuchi, K., Matsubara, T., and Nishizuka, Y. (1981). Inhibitory action of guanosine 3′,5′-monophosphate on thrombin-induced phosphatidylinositol turnover and protein phosphorylation in human platelets. *Biochem. Biophys. Res. Commun.* **101,** 61–67.

Takai, Y., Kaibuchi, K., Sano, K., and Nishizuka, Y. (1982). Counteraction of calcium-activated, phospholipid-dependent protein kinase activation by adenosine 3′,5′-monophosphate and guanosine 3′,5′-monophosphate in platelets. *J. Biochem. (Tokyo)* **93,** 403–406.

Takemura, H., Hughes, A. R., Thastrup, O., and Putney, J. W., Jr. (1989). Activation of calcium entry by the tumor promotor thapsigargin in parotid acinar cells. *J. Biol. Chem.* **264,** 12266–12271.

Tanabe, T., Takeshima, H., Mikami, A., Flockerzi, V., Takahashi, H., Kangawa, K., Kojima, M., Matsuo, H., Hirose, T., and Numa, S. (1987). Primary structure of the receptor for calcium channel blockers from skeletal muscle. *Nature (London)* **328,** 313–318.

Taylor, S. J., Smith, J. A., and Exton, J. H. (1990). Purification from bovine liver membranes of a guanine nucleotide-dependent activator of phosphoinositide-specific phospholipase C. *J. Biol. chem.* **265,** 17150–17156.

Taylor, S. J., Chae, H. Z., Rhee, S. G., and Exton, J. H. (1991). Activation of the β1 isozyme of phospholipase C by a subunits of the Gq class of G proteins. *Nature (London)* **350,** 516–518.

Trautwein, W., Taniguchi, J., and Noma, A. (1982). The effect of intracellular cyclic nucleotides and calcium on the acetylcholine response of isolated cardiac cells. *Pfluegers Arch.* **392,** 307–314.

Triggle, D. J. (1981). Calcium antagonists: Basic chemical and pharmacological aspects. *In* "New Perspectives of Calcium Antagonists" (G. B. Weiss, ed.), pp. 1–18. Am. Physiol. Soc., Washington, DC.

Tsien, R. W. (1983). Calcium channels in excitable cell membranes. *Annu. Rev. Physiol.* **45,** 341–358.

Twort, C. H. C., and van Breemen, C. (1988). Cyclic guanosine monophosphate-enhanced sequestration of Ca^{2+} by sarcoplasmic reticulum in vascular smooth muscle. *Circ. Res.* **62,** 961–964.

Ullrich, A., and Schlessinger, J. (1990). Signal transduction by receptors with tyrosine kinase activity. *Cell (Cambridge, Mass.)* **61,** 203–212.

Verma, A. K., Filoteo, A. G., Stanford, D. R., Wieben, E. D., Penniston, J. T., Strehler, E. E., Fischer, R., Heim, R., Vogel, G., Matthews, S., James, P., Voherr, T., Krebs, J., and Carafoli, E. (1988). Complete primary structure of a human plasma membrane Ca^{2+} pump. *J. Biol. Chem.* **263,** 14152–14159.

Vigne, P., Breittmayer, J.-P., Duval, D., Frelin, C., and Lazdunski, M. (1988). The Na^{+}/Ca^{2+} antiporter in aortic smooth muscle cells. *J. Biol. Chem.* **263,** 8078–8083.

Vrolix, M., Raeymaekers, L., Wuytack, F., Hofmann, F., and Casteels, R. (1988). Cyclic GMP-dependent protein kinase stimulates the plasmalemmal Ca^{2+} pump of smooth muscle via phosphorylation of phosphatidylinositol. *Biochem. J.* **255,** 855–863.

Wahler, G. M., and Sperelakis, N. (1985). Intracellular injection of cyclic GMP depresses cardiac slow action potentials. *J. Cyclic Nucleotide Protein Phosphorylation Res.* **10,** 83–95.

Wahler, G. M., Rusch, N. J., and Sperelakis, N. (1990). 8-bromo-cyclic GMP inhibits the calcium channel current in embryonic chick ventricular myocytes. *Can. J. Physiol. Pharmacol.* **68,** 531–534.

Waldman, S. A., and Murad, F. (1987). Cyclic GMP synthesis and function. *Pharmacol. Rev.* **39,** 163–196.

Watanabe, A. M., and Besch, H. R. (1975). Interaction between cyclic adenosine monophosphate and cyclic guanosine monophosphate in guinea pig ventricular myocardium. *Circ. Res.* **37,** 309–317.

Wuytack, F., DeSchutter, G., and Casteels, R. (1981a). The effect of calmodulin on the active calcium-ion transport and ($Ca^{2+}+Mg^{2+}$)-dependent ATPase in microsomal fractions of smooth muscle compared with that in erythrocytes and cardiac muscle. *Biochem. J.* **190,** 827–831.

Wuytack, F., DeSchutter, G., and Casteels, R. (1981b). Partial purification of ($Ca^{2+}+Mg^{2+}$)-dependent ATPase from pig smooth msucle and reconstitution of an ATP-dependent Ca^{2+}-transport system. *Biochem. J.* **198,** 265–271.

Yamakawa, A., and Takenawa, T. (1988). Purification and characterization of membrane-bound phosphatidylinositol kinase from rat brain. *J. Biol. Chem.* **263,** 17555–17560.

Cyclic GMP Regulation of Calcium Slow Channels in Cardiac Muscle and Vascular Smooth Muscle Cells

Nicholas Sperelakis,* Noritsugu Tohse,† Yusuke Ohya,‡ and Hiroshi Masuda#

**Department of Physiology and Biophysics
College of Medicine
University of Cincinnati
Cincinnati, Ohio 45267*

*†Department of Pharmacology
School of Medicine
Hokkaido University
Sapporo 060, Japan*

*‡Second Department of Internal Medicine
School of Medicine
Kyushu University
Fukuoka 812, Japan*

*#Department of Pediatrics
School of Medicine
University of Hiroshima
Minami-ku, Hiroshima
Hiroshima 734, Japan*

I. Introduction and Overview

Considerable attention during the past few years has been given to phosphorylation of ion channels as a means whereby the activity of the ion channels can be regulated. This chapter will cover the evidence that cyclic nucleotides regulate the Ca^{2+} influx into the myocardial cells during each

Advances in Pharmacology, Volume 26

cardiac cycle and into vascular smooth muscle (VSM) cells. This regulation is presumably mediated by phosphorylation(s) of the Ca^{2+} slow channel protein (L-type) and/or of an associated regulatory protein(s). In myocardial cells, phosphorylation of the slow Ca^{2+} channels (or of an associated regulatory protein) by cAMP-PK (Fig. 1) presumably (*a*) increases the number of Ca^{2+} slow channels available for voltage activation during the action potential (AP), (*b*) increases the probability of their opening, and (*c*) increases their mean open time. A greater density of available Ca^{2+} channels increases Ca^{2+} influx and inward Ca^{2+} slow current (I_{Ca}) during the AP, and so increases the force of contraction of the heart. Phosphorylation by cGMP-PK depresses the activity of the slow Ca^{2+} channels (Bkaily and Sperelakis, 1985; Wahler and Sperelakis, 1985; Wahler *et al.*, 1990).

The Ca^{2+} slow channels in young (3-day-old) embryonic chick heart cells exhibited a high incidence of long openings, and the incidence was diminished by 17 days (Tohse and Sperelakis, 1990; Tohse *et al.*, 1992a). Cyclic GMP inhibited these long openings (Tohse and Sperelakis, 1990).

In some VSM cells, phosphorylation by cGMP-PK or cAMP-PK inhibits the Ca^{2+} slow channel activity and thereby produces vasodilation, whereas phosphorylation by protein kinase C (PK-C) stimulates the Ca^{2+} slow channel activity and produces vasoconstriction.

Besides the slow Ca^{2+} channel, a fast-type of Ca^{2+} channel (T-type) has been found in cardiac muscle and VSM cells on the basis of kinetics

Table I

Summary of Major Differences between the Slow (L-Type) and Fast (T-Type) Ca^{2+} Channels

Properties	Ca^{2+} Channels	
	Slow (L-Type)	Fast (T-Type)
Duration of current	Long-lasting (sustained)	Transient
Inactivation kinetics	Slower	Faster
Activation kinetics	Slower	Faster
Threshold	High (ca. −30 mV)	Low (ca. −50 mV)
Half-inactivation potential	ca. −20 mV	ca. −50 mV
Single-channel conductance	High (18–26 pS)	Low (8–10 pS)
Regulated by cAMP and cGMP	Yes	No
Regulated by phosphorylation	Yes	No
Blocked by Ca^{2+} antagonist drugs	Yes	No (slight)
Opened by Ca^{2+} agonist drugs	Yes	No
Permeation by Me^{2+}	Ba > Ca	Ba ≈ Ca
Inactivation by $[Ca]_i$	Yes	Slight (?)
Recordings in isolated patches	Runs down	Rel. stable

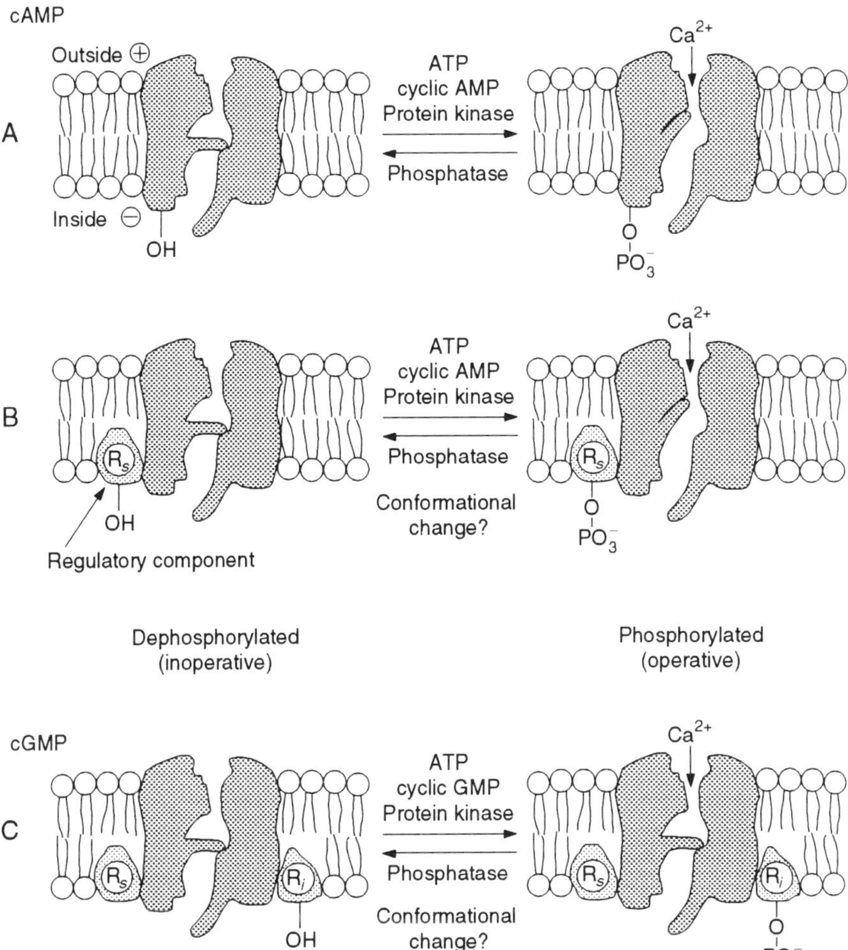

Fig. 1 Schematic model for a Ca^{2+} slow channel in myocardial cell membrane in two hypothetical forms: dephosphorylated (or electrically silent) form (left) and phosphorylated form (right). The two gates associated with the channel are an activation gate and an inactivation gate. The phosphorylation hypothesis states that a protein constituent of the slow channel itself (A) or a regulatory protein associated with the slow channel (B) must be phosphorylated in order for the channel to be in a state available for voltage activation. Phosphorylation of a serine or threonine residue occurs by a cAMP-dependent protein kinase (PK-A) in the presence of ATP. Phosphorylation may produce a conformational change that effectively allows the channel gates to operate. The slow channel (or an associated regulatory protein) may also be phosphorylated by a cGMP-PK (C), thus mediating the inhibitory effects of cGMP on the slow Ca^{2+} channel. Modified from Sperelakis and Schneider (1976).

(e.g., Bean, 1985; Nilius *et al.*, 1985). These fast Ca^{2+} channels are much more rapidly inactivated than the slow Ca^{2+} channels, are active over a more negative voltage range, and are little affected by cAMP or Ca^{2+} antagonists (Table I). Their function may be to trigger Ca^{2+} release from the sarcoplasmic reticulum (SR) (Ca-induced Ca release).

In addition, a new type of Ca^{2+} channel was discovered in 18-day-old fetal rat ventricular cells (Tohse *et al.*, 1992b). A residual I_{Ca} remaining in the presence of a high concentration (3 μM) of nifedipine (nifedipine-resistant I_{Ca}) was not blocked by diltiazem, tetramethrine (T-type channel blocker), or ω-conotoxin (N-type channel blocker) and had a half-inactivation potential about 20 mV more negative than the nifedipine-sensitive (L-type channel) I_{Ca}.

II. Cardiac Muscle

A. Cyclic AMP Stimulation of Slow Ca^{2+} Channels

The voltage- and time-dependent Ca^{2+} slow channels in the myocardial cell membrane are the major pathway by which Ca^{2+} ions enter the cell during excitation for initiation and regulation of the force of contraction of cardiac muscle. The slow channels have some special properties, including functional dependence on metabolic energy, selective blockade by acidosis, and regulation by the intracellular cyclic nucleotide levels. Because of these special properties of the slow channels, Ca^{2+} influx into the myocardial cell can be controlled by extrinsic factors (such as autonomic nerve stimulation or circulating hormones) and by intrinsic factors (such as cellular pH or ATP level).

Cyclic AMP (cAMP) modulates the functioning of the Ca^{2+} slow channels (Shigenobu and Sperelakis, 1972; Tsien *et al.*, 1972; Sperelakis and Schneider, 1976; Schneider *et al.*, 1976; Reuter and Scholz, 1977). Histamine and β-adrenergic agonists, after binding to their specific receptors, lead to rapid stimulation of adenylate cyclase with resultant elevation of cAMP levels. Methylxanthines enter the myocardial cells and inhibit the phosphodiesterase, thus causing an elevation of cAMP. These positive inotropic agents also concomitantly induce Ca^{2+}-dependent slow APs by increasing I_{Ca}.

Additional evidence for the regulatory role of cAMP has been obtained. (*a*) The GTP analogue GPP(NH)P, which directly activates adenylate cyclase, induced Ca^{2+}-dependent slow APs in heart cells (Josephson and Sperelakis, 1978). (*b*) Forskolin, another highly potent activator of adenylate cyclase activity, exerted a strong positive inotropic effect and induced and potentiated slow APs (Spah, 1984; Wahler and Sperelakis, 1986). (*c*) cAMP iontophoretically microinjected into Purkinje fibers and ventricular muscle cells induced slow APs in the injected cells within seconds (Vogel and Sperelakis, 1981). (*d*) Pressure injection of cAMP, GPP(NH)P, and cholera toxin (which irreversibly activates adenylate cyclase) rapidly induced and potentiated slow APs (Li and Sperelakis, 1983) (Fig. 2). (*e*) Liposome injection of cAMP into heart cells also induced slow APs (Bkaily and Sperelakis, 1985).

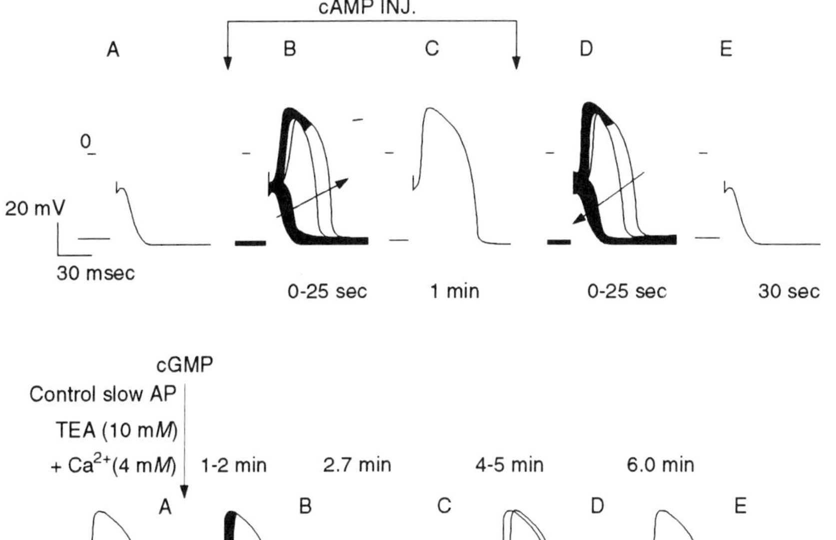

Fig. 2 (Top) Induction of Ca^{2+}-dependent slow action potentials (APs) in guinea pig papillary muscle by intracellular pressure injection of cyclic AMP. The muscle was depolarized in 22 m*M* $[K]_o$ to voltage inactivate fast Na$^+$ channels. (A) Small graded response (stimulation rate 30/min). (B) Superimposed records showing the gradual appearance of slow APs on cAMP injection over a 25-s period. (C) Presence of stable slow APs after injection for 1 min. (D) Gradual depression of slow APs over a period of 25 s after stopping injection. (E) Complete decay of slow APs 30 s after cessation of cAMP injection. All records are from one impaled cell. Data taken from Li and Sperelakis (1983). (Bottom) Transient abolition of Ca^{2+}-dependent slow APs by pressure injection of cGMP. (A) Control slow AP induced by 10 m*M* TEA plus 4.0 m*M* $[Ca]_o$ in 25 m*M* K$^+$ to inactivate fast Na$^+$ channels. (B–C) 1–2 min following the onset of cGMP injection (10-s duration), the slow APs were depressed and then abolished. (D–E) At 4–6 min, the slow APs recovered spontaneously to control levels. All records from the same cell. Taken from Wahler and Sperelakis (1985).

Table II

Comparison of Properties of Ca^{2+} Slow Channels in Myocardial Cells and Vascular Smooth Muscle Cells

	Myocardial cells	VSM cells
ATP	stim.	stim.
cAMP	stim.	inhib.[a]
cGMP	inhib.	inhib.[a]
acidosis	inhib.	?
Ca antagonist drugs	inhib.	inhib.
Ca agonist drugs	stim.	stim.

[a] In some VSM cells, e.g., rat aortic.

Other results also support a role for cAMP in stimulating the slow inward Ca^{2+} current in myocardial cells (Table II). (*a*) Injection of cAMP enhanced I_{Ca} in isolated single cardiac cells (Irisawa and Kokobun, 1983). (*b*) A photochemical activation method for suddenly increasing the intracellular cAMP level enhanced I_{Ca} in bullfrog atrial cells (Nargeot *et al.*, 1983). (*c*) Noise analysis and patch-clamp analysis suggest that cAMP increases the number of functional slow channels available in the sarcolemma and/or the probability of opening of a given channel (Cachelin *et al.*, 1983; Trautwein and Hoffman, 1983; Bean *et al.*, 1984). Both actions would increase the number of slow channels open at any instant of time. Isoproterenol increased the mean open time of single Ca^{2+} channels and decreased the intervals between bursts; the conductance of the single channel was not increased (Reuter *et al.*, 1982). Therefore, the increase in the slow Ca^{2+} current produced by isoproterenol could be produced by the observed increase in mean open time of each channel, as well as by an increase in the number of available channels.

B. Phosphorylation Hypothesis

Because of the relationship between cAMP and the number of available slow Ca^{2+} channels, and because of the dependence of the functioning of these channels on metabolic energy, it was postulated that the slow channel protein must be phosphorylated in order for it to become available for voltage activation (Shigenobu and Sperelakis, 1972; Tsien *et al.*, 1972; Sperelakis and Schneider, 1976; Trautwein and Hoffman, 1983; Sperelakis, 1988). Elevation of cAMP by a positive inotropic agent activates a cAMP-dependent protein kinase (cA-PK), which phosphorylates a variety of proteins in the presence of ATP. One protein that is phosphorylated might be the slow channel protein itself or a contiguous regulatory type of protein (Fig. 1).

Phosphorylation could make the slow channel available for activation by a conformational change that allowed the activation gate to be opened upon depolarization. In this model, the phosphorylated form of the slow channel is the active (operational) form, and the dephosphorylated form is the inactive (inoperative) form. The dephosphorylated channels are electrically silent. Thus, phosphorylation increases the probability of channel opening with depolarization. An equilibrium would exist between the phosphorylated and the dephosphorylated forms of the slow channels for a given set of conditions. Agents that elevate cAMP increase the fraction of the channels that are in the phosphorylated form, and hence readily available for voltage activation.

To test whether the regulatory effect of cAMP is exerted by means of the cA-PK and phosphorylation, intracellular injection of the catalytic subunit of the cA-PK was done. Such injections induced and enhanced the slow APs (Bkaily and Sperelakis, 1984) and potentiated I_{Ca} (Trautwein and Hoffman, 1983; Osterrieder *et al.*, 1982; Trautwein *et al.*, 1982). Another test of the phosphorylation hypothesis was done by liposome injection of an inhibitor (protein) of the cA-PK into heart cells, and showing that it inhibited the spontaneous slow APs (Bkaily and Sperelakis, 1984). This protein kinase inhibitor also was shown to inhibit I_{Ca} of cardiac cells (Kameyama *et al.*, 1986).

Based on the rapid decay of the response to injected cAMP (Fig. 2, top), the mean life span of a phosphorylated channel is likely to be only a few seconds at most, and it is possible that the channels are phosphorylated and dephosphorylated with every cardiac cycle (Li and Sperelakis, 1983). Hence, agents that affect or regulate the phosphatase would affect the life span of the phosphorylated channel. Thus, channel stimulation can be produced either by increasing the rate of phosphorylation (by cA-PK) or by decreasing the rate of dephosphorylation (inhibition of the phosphatase) (Vogel *et al.*, 1977). For example, the Ca^{2+}-dependent phosphatase calcineurin inhibits slow APs in 3-day-old embryonic chick hearts (Tripathi and Sperelakis, 1991). Phosphatases have been shown to decrease the Ca^{2+} current in neurons (Chad and Eckert, 1986) and ventricular myocardial cells (Hescheler *et al.*, 1987a). The catalytic subunit of the protein phosphatases 1 and 2A inhibited the Ca^{2+} channel, and okadaic acid, a protein phosphatase inhibitor, enhanced the amplitude of the I_{Ca} prestimulated by β-adrenergic agents (Hescheler *et al.*, 1988).

Consistent with the phosphorylation hypothesis, it has been found that the slow Ca^{2+} channel activity disappears within 90 s in isolated membrane inside-out patches (Reuter, 1983), and was restored by applying catalytic subunit of cA-PK and ATP-Mg (Armstrong and Eckert, 1987). This is consistent with the washing away of regulatory components of the slow channels or of the enzymes necessary to phosphorylate the channel. Even in whole-cell voltage clamp, there is a progressive rundown of the slow Ca^{2+} current, which is slowed or partially reversed by conditions that enhance cA-PK phosphorylation (Chad and Eckert, 1986).

Some agents that affect the force of contraction of the heart may do so without increasing the level of cyclic AMP. For example, fluoride ion (<1 m*M*) is a potent positive inotropic agent and potentiates the Ca^{2+}-dependent slow APs and Ca^{2+} influx (I_{si}), but yet does not elevate cAMP (Vogel *et al.*, 1977). Fluoride may act by inhibiting the phosphatase, which dephosphorylates the slow channel protein (or associated regulatory protein). This would prolong the life span of the phosphorylated channel, resulting in potentiation of I_{si} and contraction. In contrast, it is possible that some negative inotropic drugs depress the rate of phosphorylation. It might be difficult to distinguish between a drug that depressed phosphorylation of the slow Ca^{2+} channel and one that physically blocked the channel.

cAMP also has effects on other types of ion channels. For example, the following channels of heart are stimulated by cAMP: (*a*) delayed rectifier K^+ channel (Trautwein *et al.*, 1982; Yazawa and Kameyama, 1990), (*b*) hyperpolarization-activated Na-K I_f (I_h) channel (DiFrancesco and Tromba, 1988), and (*c*) catecholamine-activated Cl^- channel (Ehara and Ishihara, 1990). The fast Na^+ channel was reported to be inhibited by cAMP (Ono *et al.*, 1989). The delayed rectifier K^+ current was also stimulated by cGMP, presumably by PDE inhibition (Ono and Trautwein, 1991), and the catecholamine-activated Cl^- current was also stimulated by cGMP (Tareen *et al.*, 1991).

C. Cyclic GMP Inhibition of Slow Ca^{2+} Current

The physiological role of cyclic GMP on cardiac function is still controversial. 8-Br-cGMP (10^{-4} *M*) shortened the action potential duration in rat atria accompanied by a negative inotropic effect, and it was suggested that cyclic GMP might decrease the Ca^{2+} conductance (Nawrath, 1977). Acetylcholine (ACh) and 8-Br-cGMP reduced upstroke velocity and duration of the Ca-dependent slow action potential in guinea pig atria (Kohlhardt and Haap, 1978). The abbreviation of action potential duration was also observed following pressure injection of cGMP into isolated guinea pig cardiomyocytes (Trautwein *et al.*, 1982).

It has been proposed that cGMP plays an antagonistic role to that of cAMP, namely that there was a "Yin-Yang" relationship between cAMP and cGMP (Goldberg *et al.*, 1975). Superfusion of isolated ventricular muscle with 8-Br-cGMP abolished the Ca^{2+}-dependent slow APs and accompanying contractions within 7–20 min (Wahler and Sperelakis, 1985). A similar inhibition by cGMP was shown for the slow APs of atrial muscle and Purkinje fibers (Kohlhardt and Haap, 1978; Mehegan *et al.*, 1985). Intracellular pressure injection of cGMP into ventricular cells transiently depressed or abolished slow APs much more quickly (e.g., 1–2 min) (Wahler and Sperelakis, 1985) (Fig. 2, bottom). Injection of cGMP into heart cells by the liposome method also abolished the slow APs (Bkaily and Sperelakis, 1985). It was also demonstrated that 8-Br-cGMP inhibits the basal I_{Ca} (unstimulated by cAMP) in voltage-clamped ventricular myocytes (Wahler *et al.*, 1990) (Fig. 3).

We recently demonstrated cGMP inhibition of Ca^{2+} slow channel activity at the single-channel level (Tohse and Sperelakis, 1991a). Cyclic GMP

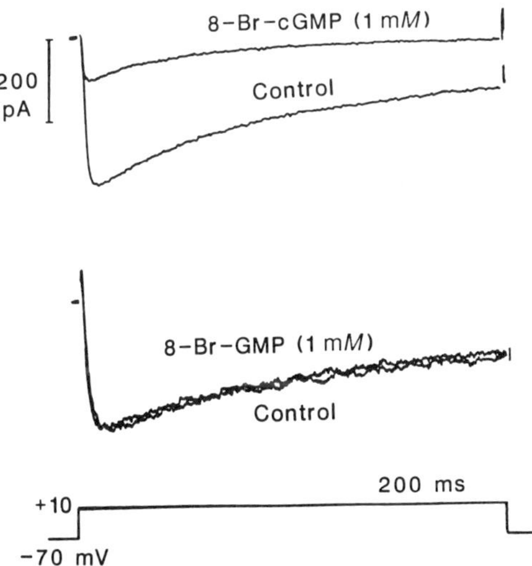

Fig. 3 Effect of 8-Br-cGMP on the slow inward Ca^{2+} current in two cultured embryonic chick ventricular myocytes. (Upper traces) Currents elicited by depolarizing pulses from −70 to +10 mV in the control bath solution and following 10 min superfusion with a solution containing 1 m*M* 8-Br-cGMP. Note the large inhibition of $I_{Ca(s)}$. (Lower traces) Currents elicited by depolarizing pulses in the control bath solution and following 10 min superfusion with a solution containing 1 m*M* 8-Br-GMP, the noncyclic analog of 8-Br-cGMP. The bath solution (20–22°C) included (in m*M*) 10 $BaCl_2$ and 135 TEA-Cl; the pipet solution included 150 Cs-glutamate, 5 MgATP, 1 EGTA. Reproduced from Wahler *et al.* (1990).

did not change unit amplitude and slope conductance of the Ca^{2+} channel, but prolonged the closed times and shortened the open times. Because 8-Br-cGMP is a potent activator of G-kinase and does not stimulate cAMP hydrolysis, cGMP-induced inhibition of the basal activity of the Ca^{2+} channels (not prestimulated by cAMP) may be mediated by G-kinase. Similar observations were made by Tohse *et al.* (1992c) on isolated rabbit ventricular myocytes.

In 3-day-old embryonic chick heart cells, the Ca^{2+} slow channels often exhibited long-lasting openings (e.g., for 300 ms) under normal conditions, especially at the more positive command potentials (Tohse and Sperelakis, 1990). That is, the Ca^{2+} slow channels naturally possessed mode 2 behavior, in the absence of any added Ca^{2+} channel agonist such as the dihydropyridine Bay-K-8644. Addition of Bay-K-8644 did not further prolong the open times, but appeared to recruit silent Ca^{2+} channels (Tohse *et al.*, 1991). Long-lasting openings were much less frequently observed in 17-day-old embryonic cells (Tohse *et al.*, 1992a). Addition of 8-Br-cGMP to the bath of 3-day cells exhibiting long openings completely inhibited Ca^{2+} slow channel activity (Fig. 4).

Therefore, cGMP regulates the functioning of the myocardial Ca^{2+} slow channels in a manner that is antagonistic to that of cAMP (Fig. 5). It is possible that the Ca^{2+} slow channel protein has a second site that can be phosphorylated by cG-PK and which, when phosphorylated, inhibits the slow channel. Another possibility is that there is a second type of regulatory protein that is inhibitory when phosphorylated (Fig. 1).

A single protein has been found to be specifically phosphorylated by cG-PK in guinea pig sarcolemmal preparations (Cuppoletti *et al.*, 1988). In the presence of the purified kinase plus 10^{-5} *M* cGMP or 8-Br-cGMP, a protein of approximately 47 kDa was phosphorylated. Thus, this substrate, the identity of which is unknown, may be a possible mediator of cGMP-mediated control of cardiac function.

Another mechanism proposed for cGMP inhibition is based on cGMP depression of the cAMP level. Intracellular application of cGMP inhibited I_{Ca} of frog ventricular myocytes, but only after the cAMP levels had been increased; i.e., there was no effect of cGMP on the basal I_{Ca} (Hartzell and Fischmeister, 1986; Fischmeister and Hartzell, 1987). It was concluded that cGMP inhibited I_{Ca} by stimulating a phosphodiesterase, resulting in increased degradation of cAMP. However, in a later study on guinea pig and rat cardiomyocytes, this same group reported a direct inhibition of I_{Ca} by cGMP and a direct cG-PK (Levi *et al.*, 1989; Mery *et al.*, 1991). In addition, 8-Br-cAMP inhibition of slow APs in mammalian cardiac muscle occurs without a decrease in cAMP levels (Thakkar *et al.*, 1988). Thus, it appears that in mammalian ventricular muscle, cGMP

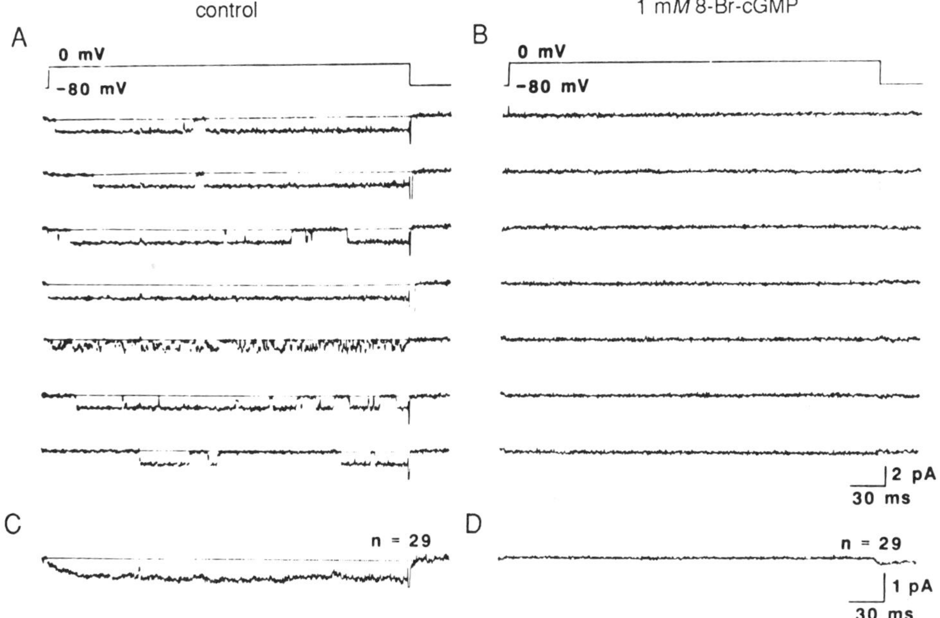

Fig. 4 Current recordings from a cell-attached patch showing effect of 8-Br-cGMP on the Ca^{2+} slow channel activity in a single myocardial cell isolated from a 3-day-old embryonic chick heart. Single-channel currents were evoked by depolarizing voltage pulses to 0 mV from a holding potential of −80 mV, at a duration of 300 ms and repetition rate of 0.5 Hz. (A, B) Examples of original current recordings from the same patch, before (A) and after (B) superfusion with 1.0 m*M* 8-Br-cGMP. (C, D) Ensemble-averaged currents calculated from the current recordings ($n = 29$). The current tracings were low-pass filtered at 1 kHz and corrected for leakage and capacitive currents. Data taken from Tohse and Sperelakis (1991a).

inhibits I_{Ca} directly through a cGMP-mediated phosphorylation (8-Br-cGMP is a potent activator of cG-PK) of some protein involved in the functioning of the slow Ca^{2+} channels (Fig. 1).

It has been proposed that muscarinic agonists also act to inhibit Ca^{2+} slow channel by cG-PK stimulation of a phosphatase that dephosphorylates the channel (Watanabe *et al.,* 1989) (Fig. 5). This would have the effect of lowering the fraction of channels in the phosphorylated form, and therefore the Ca^{2+} influx. In this mechanism, the rate of phosphorylation is unaffected, but the rate of dephosphorylation is increased. Muscarinic agonists are known to elevate cGMP (George *et al.,* 1970). However, it was reported that ACh was ineffective in reducing the basal I_{Ca} (Hescheler *et al.,* 1986), and that cGMP actually potentiated the stimulating effect of

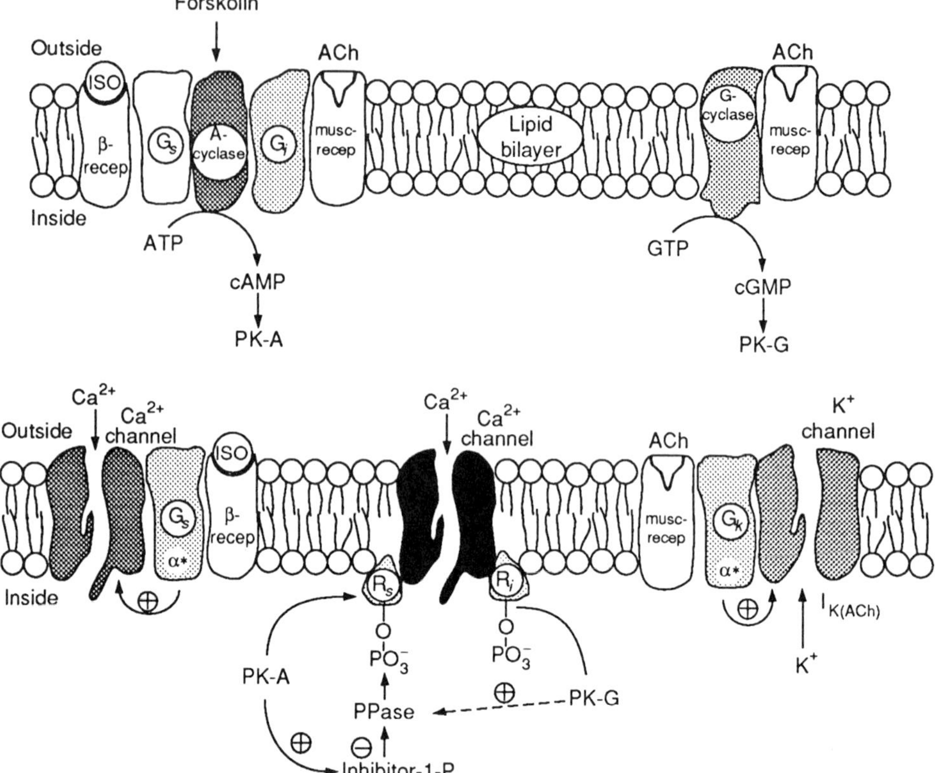

Fig. 5 Diagrammatic summary of the regulation of the Ca^{2+} slow channels in the myocardial cell membrane and the mechanisms of action of some inotropic agents. The β-adrenergic agonists and histamine H_2 agonists act via their receptors on a GTP-binding protein (G_S) to stimulate adenylate cyclase and cAMP production. The voltage-dependent myocardial slow Ca^{2+} channels are stimulated by cAMP, presumably because the channel (or an associated regulatory protein) must be phosphorylated in order for it to be in a form that is available for voltage activation. cGMP-dependent phosphorylation also regulates the slow channel in a manner antagonistic to cAMP, namely producing inhibition. Thus, the muscarinic receptor activated by ACh can produce inhibition of Ca^{2+} influx by at least the four mechanisms depicted: (*a*) reversal of adenylate cyclase stimulation produced by β-agonists or H_2 agonists; (*b*) stimulation of guanylate cyclase and production of cGMP; (*c*) activation of a K^+ channel ($I_{K(ACh)}$) that produces an outward K^+ current which depresses excitability and terminates the AP earlier, and thereby voltage deactivates the Ca^{2+} slow channels earlier; and (*d*) stimulation of a phosphatase (PPase) by cGMP and PK-G. Mechanism (*c*) may be absent in ventricular myocardial cells.

ISO (Isoproterenol) on I_{Ca}, perhaps mediated by PDE inhibition (Ono and Trautwein, 1991). Elevation of intracellular cGMP by photoactivation of a derivative had no effect on the L-type Ca^{2+} current (I_{Ca}) in isolated rat ventricular cells (Nargeot *et al.*, 1983).

The parasympathetic neurotransmitter ACh exerts a negative inotropic effect on ventricular myocardium prestimulated by β-adrenergic agonists. Activation of the muscarinic receptor by ACh exerts an inhibitory effect on adenylate cyclase and cAMP levels, via the G_i (inhibitory) coupling protein, to reverse the stimulation of adenylate cyclase produced by means of the G_s coupling protein due to, for example, activation of the β-adrenoceptor (Fig. 5). Thus, ACh may depress Ca^{2+} influx and contraction by reversing cAMP elevation produced by various agonists. For example, in cultured chick ventricular cells, ACh depressed I_{Ca} that had been stimulated by isoproterenol (Josephson and Sperelakis, 1978). ACh also reverses the electrophysiological effects of direct adenylate cyclase stimulation by forskolin (Wahler and Sperelakis, 1986). Additionally, in ventricular cells, ACh may inhibit the slow Ca^{2+} channels due, in part, to elevation of cGMP levels (MacLeod and Diamond, 1986). Adenosine exerts effects similar to those of ACh.

D. Calmodulin–Protein Kinase and Protein Kinase C

Inhibitors of calmodulin (trifluoperazine and calmidazolium) inhibit the slow APs of heart cells (Johnson *et al.*, 1983; Bkaily *et al.*, 1984; Bkaily and Sperelakis, 1986). Subsequent injection of calmodulin reverses the inhibition produced by calmidazolium. It appears that maximal activation of the slow channels requires two separate phosphorylation steps (calmodulin-dependent and cAMP-dependent). These may be on the same protein or on two separate proteins.

High concentration of the α-adrenergic agonist phenylephrine causes a positive inotropic effect in cardiac muscle (Bruckner and Scholz, 1984). The α-adrenoceptor agonists stimulate the phosphatidylinositol cycle and generation of inositol trisphosphate (IP_3) and diacyl glycerol (DAG) (Brown *et al.*, 1985). IP_3 acts as a second messenger to release stored Ca^{2+} from the SR. DAG and Ca^{2+} activate PK-C, which phosphorylates a number of proteins. It is currently controversial whether PK-C is involved in regulation of the myocardial slow Ca^{2+} channels. One group reported that phorbol ester and angiotensin-II (Ang-II) stimulated I_{Ca} (Dosemeci *et al.*, 1988), whereas another group did not observe such stimulation (Tohse *et al.*, 1990). There have been variable findings with respect to the effect of activation of α-adrenoceptor agonists on elevation of cAMP.

E. Summary and Conclusions

The slow Ca^{2+} channels of the heart are stimulated by cAMP. Elevation of cAMP produces a very rapid increase in number of slow channels available for voltage activation during excitation. The probability of a slow channel opening and the mean open time of the channel are increased. Therefore, any agent that increases the cAMP level of the myocardial cell will tend to potentiate I_{si}, Ca^{2+} influx, and contraction.

The myocardial slow Ca^{2+} channels are also regulated by cGMP, in a manner that is opposite to that of cAMP. The effect of cGMP may be mediated by means of phosphorylation of a protein, as, for example, a regulatory protein (inhibitory-type) associated with the slow channel. In addition, cGMP may act by stimulating a phosphatase that dephosphorylates the Ca^{2+} channel.

Preliminary data suggest that calmodulin also may play a role in regulation of the myocardial slow Ca^{2+} channels, possibly mediated by the Ca^{2+}-calmodulin protein kinase and phosphorylation of some regulatory-type of protein.

Thus, it appears that the slow Ca^{2+} channel is a complex structure, including perhaps several associated regulatory proteins, which can be regulated by a number of extrinsic and intrinsic factors (Fig. 5).

III. Vascular Smooth Muscle Cells

A. Inhibition of Ca^{2+} Slow Channels by Cyclic AMP and Cyclic GMP

Figure 6 depicts some of the types of ion channels in VSM cells, including the two types of Ca^{2+} channels: slow (L-type) and fast (T-type). The properties of these two types of Ca^{2+} channels are rather similar to those in myocardial cells described above, with the exception of the regulatory mechanisms (Table II). As depicted, both cAMP and cGMP markedly potentiate neurotransmitter release (NE and ATP) from the nerve terminals (Sperelakis *et al.*, 1991).

Both cAMP and cGMP elevations have been implicated in the relaxation of smooth muscle tissue in response to some vasodilators, such as EDRF, atrial natriuretic factor (ANF), and nitroprusside as discussed in earlier chapters. The effects of cyclic nucleotide analogs and related agents on the electrical properties of cultured rat aortic reaggregates were examined (Ousterhout and Sperelakis, 1987). Agents that activate adenylate cyclase, such as isoproterenol and forskolin, depressed or abolished the Ca^{2+}-dependent APs and hyperpolarized the membrane. These effects were

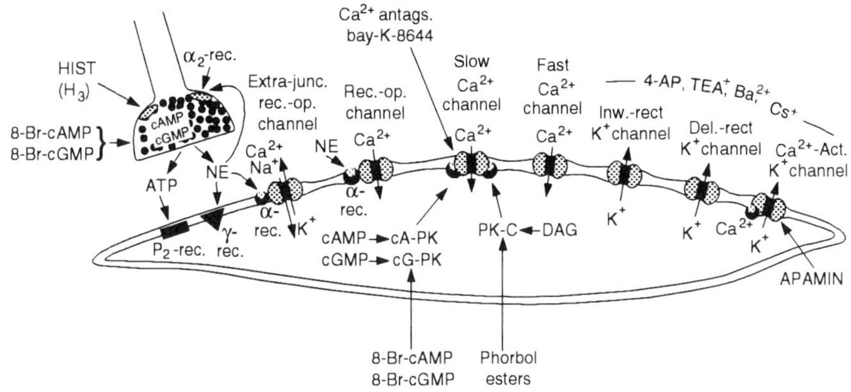

Fig. 6 Diagrammatic representation of various ion channels in a vascular smooth muscle (VSM) cell and a number of agents that may either activate or inhibit these channels. Depicted are three different types of Ca^{2+} channels (fast, slow, and receptor-operated) and a nonselective ion channel (extrajunctional, receptor-operated), which allows Ca^{2+}, Na^+, and K^+ to pass through. The voltage-dependent Ca^{2+} slow channels are blocked by Ca^{2+} antagonists and enhanced by Ca^{2+} agonists (Bay-K-8644). The three different K^+ channels (inward rectifier, delayed rectifier, and Ca^{2+} activated) are blocked by TEA^+, Ba^{2+}, and Cs^+. Also shown is an adrenergic nerve terminal from which norepinephrine (NE) and ATP are released to activate α- or γ-receptors and purinergic (P_2) receptors, respectively, on the postsynaptic membrane. Release of neurotransmitters is modulated by cyclic nucleotides and phorbol esters.

mimicked by membrane-permeable analogs of cAMP (dibutyryl or 8-bromo cAMP) (Fig. 7). 8-Bromo-cGMP also depressed or abolished the APs. Synthetic ANF, which activates the membrane-bound guanylate cyclase, had inhibitory effects, whereas nitroprusside, which stimulates only the cytosolic G-cyclase, had little effect (Table III).

In voltage-clamp experiments on cultured single cells from rabbit aorta, the dibutyryl analogs of cAMP and cGMP suppressed the inward Ba^{2+} current carried through the Ca^{2+} slow channels (Bkaily *et al.*, 1988b). The 8-bromo analogs of cAMP and cGMP and nitroprusside also enhanced the delayed-rectifier outward K^+ current (Bkaily *et al.*, 1988b). These results indicate that cAMP and cGMP decrease the inward Ca^{2+} current and increase an outward K^+ current in VSM cells. Thus, these effects of cyclic nucleotides on the Ca^{2+} and K^+ channels can explain their inhibitory effects on the APs.

We propose a phosphorylation model for regulation of Ca^{2+} slow channels in VSM cells, similar to that for myocardial cells, except in the case

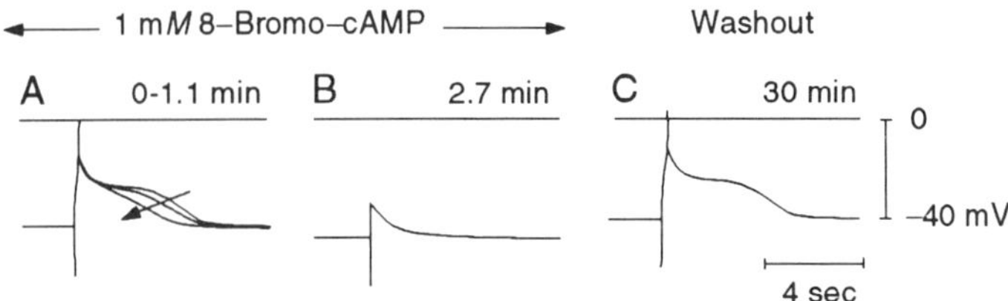

Fig. 7 Depression and abolition of TEA-induced APs in cultured rat aortic smooth muscle cells by 8-bromo cAMP. (A) Superimposed traces of a control AP (in 15 m*M* TEA) and depression of the AP within 1.1 min. after addition of 1 m*M* 8-bomo-cAMP. (B) Abolition of the AP after 2.7 min. (C) Recovery of the AP on washout for 30 min. (Records A–C were from the same cell.) The stimulation frequency was 0.04 Hz. Data taken from Ousterhout and Sperelakis (1987).

of VSM cells cAMP and cGMP have the same effect, namely both inhibit the Ca^{2+} slow channels (Fig. 8). A second regulatory protein, when phosphorylated by PK-C, stimulates the Ca^{2+} slow channels, as depicted in Fig. 8.

Figures 9 and 10 summarize the roles played by cyclic nucleotides and PK-C in regulating $[Ca]_i$ in VSM cells. Activation of A-cyclase or G-cyclase by appropriate membrane receptors (e.g., β-adrenergic, prostacyclin, ANP) and G coupling proteins or directly (by agents like nitroprusside, nitric oxide free radical, or forskolin) produces elevation of cAMP and cGMP and activation of cA-PK and cG-PK. These kinases can phosphorylate the Ca^{2+} slow channel (to inhibit channel opening) and the K^+ channels (to stimulate channel opening). Both mechanisms would inhibit

Table III

Summary of the Effects of Cyclic Nucleotides and Related Agents on the Action Potentials of Cultured Rat Aortic Cells

Compound	Effect on APs
8-Bromo cyclic AMP (10^{-3} *M*)	Abolished
Dibutyryl cyclic AMP (10^{-3} *M*)	Abolished
Isoproterenol (10^{-6} *M*)	Abolished
Forskolin (10^{-6} *M*)	Abolished
8-Bromo cGMP (10^{-4} – 10^{-3} *M*)	Abolished or depressed
Nitroprusside (10^{-6} – 10^{-5} *M*)	Depressed slightly
Atrial natriuretic factor (10^{-8} – 10^{-7} *M*)	Abolished

Note. Taken from Ousterhout and Sperelakis (1987).

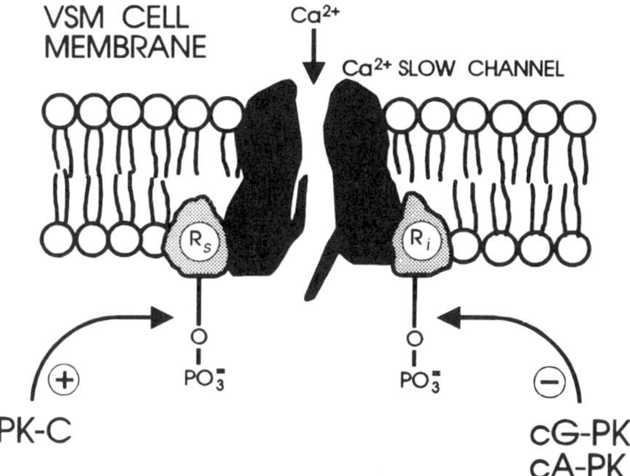

Fig. 8 Cartoon model of a Ca^{2+} slow channel protein imbedded in the lipid bilayer of the cell membrane of a vascular smooth muscle cell. Depicted are two hypothetical regulatory proteins, inhibitory and stimulatory, associated with the channel protein. As shown, both regulatory proteins have phosphorylation sites that affect the function of the regulatory proteins. Phosphorylation of the inhibitory protein by cAMP–protein kinase (PK-A) or by cGMP-PK (PK-G) inhibits the function of the ion channel, e.g., decreases probability of channel opening at depolarized potentials. Phosphorylation of the stimulatory protein by PK-C stimulates the Ca^{2+} slow channel, e.g., increases probability of opening. Thus, channel function is regulated by cyclic nucleotides and by diacylglycerol.

the Ca^{2+}-dependent APs, and thereby diminish Ca^{2+} influx and lower $[Ca]_i$. cG-PK and cA-PK also phosphorylate the sarcolemmal Ca^{2+}–ATPase and stimulate this Ca^{2+} pump, thus lowering $[Ca]_i$. Lowering $[Ca]_i$ inhibits contraction and produces vasodilation.

Activation of phospholipase C (PL-C) by appropriate membrane receptors (e.g., Ang-II receptor) and G coupling proteins stimulates PI turnover with IP_3 and DAG production. DAG activates PK-C to phosphorylate the Ca^{2+} slow channels and stimulate Ca^{2+} influx and raise $[Ca]_i$. Phorbol esters act, like DAG, to directly activate PK-C. IP_3 acts on the Ca-release channels of the SR to release Ca^{2+} and elevate $[Ca]_i$. Raising $[Ca]_i$ potentiates contraction and produces vasoconstriction.

The cA-PK also phosphorylates the myosin light-chain kinase (MLCK), which diminishes its Ca^{2+} sensitivity. In addition, the cyclic nucleotides have been reported to inhibit PL-C, and therefore production of IP_3 and diacylglycerol (DAG). Both of these factors would contribute to the vasodilation.

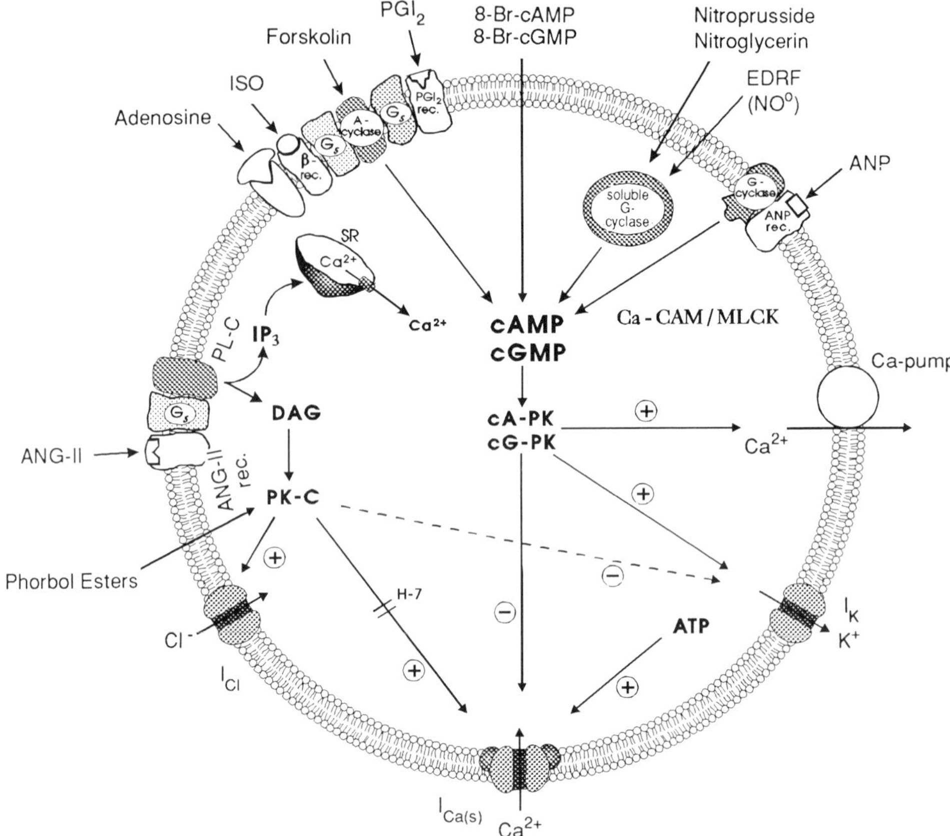

Fig. 9 Diagram of a vascular smooth muscle cell to illustrate one mechanism whereby elevation of cyclic nucleotides can lead to vasodilation. Depicted are membrane receptors for stimulation of adenylate cyclase and guanylate cyclase, resulting in elevation of either cAMP or cGMP. As shown, cAMP activates PK-A and cGMP activates PK-G, which phosphorylate at least two types of ion channels and a Ca pump in the sarcolemma. Phosphorylation of the Ca^{2+} slow channel (or an associated inhibitory-type regulatory protein) produces inhibition of the channel, e.g., decreases probability of opening at depolarized potentials. Also depicted is the fact that phosphorylation by PK-C stimulates the Ca^{2+} slow channel. The delayed rectifier K^+ channel is stimulated by phosphorylation with cA-PK and cG-PK, as illustrated. Also depicted is stimulation of the sarcolemmal Ca^{2+} pump by cA-PK and cG-PK.

B. Regulation of Ca^{2+} Channels by ATP

Two distinct types of voltage-sensitive Ca^{2+} channels are found in VSM cells (Bean *et al.*, 1986; Yatani *et al.*, 1987b; Pacaud *et al.*, 1987; Friedman *et al.*, 1986), as in myocardial cells. The two types of Ca^{2+} currents

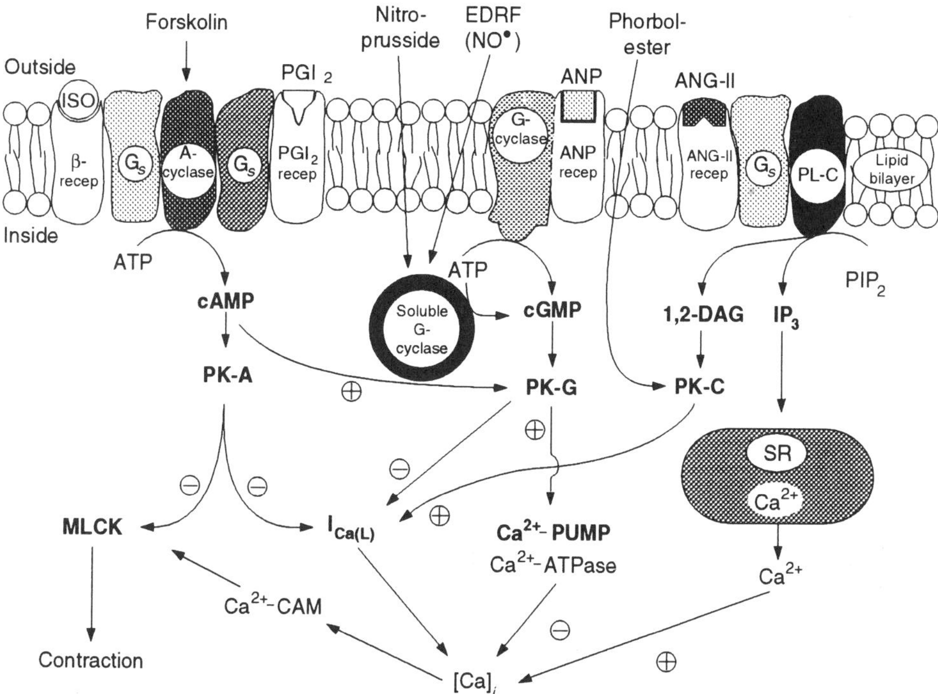

Fig. 10 Diagram of expanded region of the sarcolemma of a VSM cell to better illustrate the relationship between various membrane receptors and the cyclases and phospholipase C (PL-C). A receptor for atrial natriuretic peptide (ANP) is depicted, along with the guanylate (G) cyclase. Stimulation of membrane-bound G-cyclase by ANP, or the soluble G-cyclase directly by nitroprusside or EDRF (presumably nitric oxide free radical) elevates cGMP and activates cG-PK. cG-PK phosphorylates several proteins including the Ca^{2+} slow channel and the Ca^{2+}-ATPase. Receptors for prostacyclin (PGI_2) and isoproterenol (β-adrenergic agonist) are depicted, along with the adenylate (A) cyclase. The β-receptor is coupled to the A-cyclase by a GTP-binding protein (G_S). Activation of A-cyclase by agonists, or directly by forskolin, elevates cAMP level and thereby activates cA-PK, which phosphorylates several proteins including the Ca^{2+} slow channel. Also depicted is the angiotensin-II receptor, coupled by a G-protein to PL-C. Stimulation of PL-C stimulates PI turnover and leads to production of IP_3 and DAG. DAG activates PK-C, which phosphorylates several proteins including the Ca^{2+} slow channel. Phorbol esters directly stimulate PK-C (like DAG). Inositol trisphosphate (IP_3) causes Ca^{2+} release from the SR by activation of Ca release channels.

can be distinguished by differences in their kinetics, voltage ranges of activation and inactivation, and sensitivities to pharmacological agents (Table I). The slow (sustained) Ca^{2+} channel is sensitive to dihydropyridine Ca^{2+} agonists and antagonists, prefers Ba^{2+} over Ca^{2+} as charge carrier, and activates at more positive membrane potentials (high threshold) than the fast (transient) Ca^{2+} channel. The fast Ca^{2+} channel activates

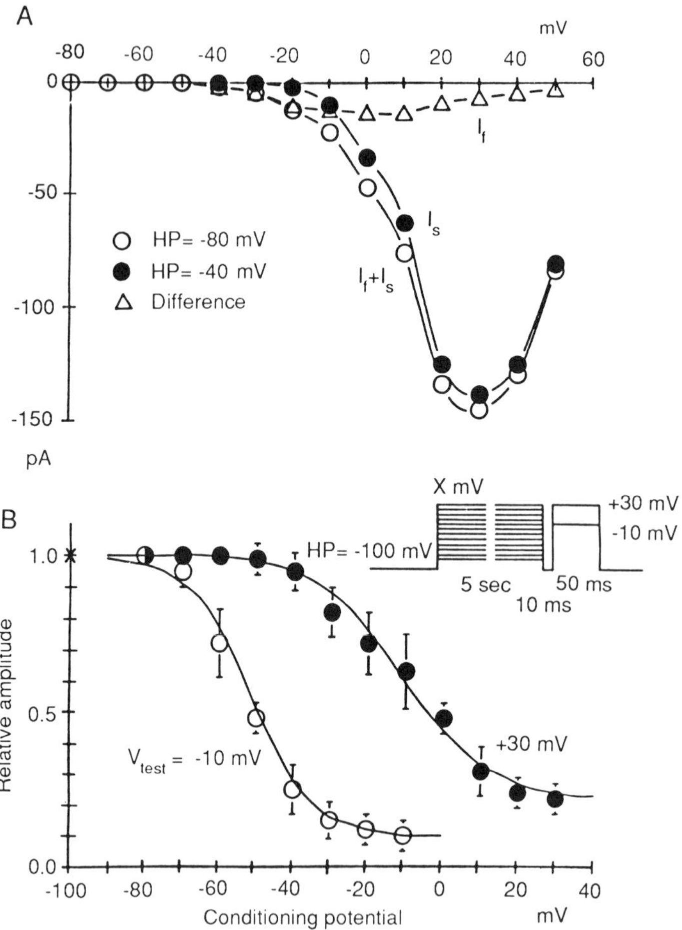

Fig. 11 Two types of Ca^{2+} channel currents recorded from freshly isolated single smooth muscle cells of guinea pig mesenteric arteries. (A) *I*/*V* curves of the peak current amplitudes obtained from HPs of −80 mV ($I_f + I_S$, open circles) and −40 mV (I_S, solid circles), and the difference curve (I_f) between these two currents (open triangles). (B) Steady-state inactivation curves for the fast current (I_f, open circles) and slow current (I_S, solid circles). Conditioning pulses (5 s) to various voltages were applied before application of the test pulse (to −10 mV for the fast current and to +30 mV for the slow current); interval between the conditioning pulse and test pulse was 10 ms. The HP was kept at −100 mV. Data are shown as mean ± SD (n = 3–5). Curves were drawn by fitting the data to a Boltzmann distribution. The pipet contained high Cs^+ and 5 m*M* ATP, and the bath contained isotonic Ba^{2+} solution. Data taken from Ohya and Sperelakis (1989a).

at more negative membrane potentials (low threshold), has an equal preference for Ca^{2+} and Ba^{2+}, and is insensitive to the dihydropyridines.

The properties of the two types of Ca^{2+} channels in freshly isolated cells from guinea pig small mesenteric arteries (resistance vessels) were examined using the whole-cell voltage clamp technique (Ohya and Sperelakis, 1989a; Sperelakis and Ohya, 1991). Fast and slow Ca^{2+} channel currents were recorded, and their characteristics were the same as previously reported (Fig. 11). Injection of ATP (0.3–5 m*M*), using an intracellular perfusion method, modified the slow current, but did not affect the fast current (Fig. 12). The $K_{0.5}$ value was 0.3 m*M* ATP. When the production of ATP was inhibited by adding cyanide and 2-deoxy-D-glucose to the bath (in absence of glucose), the Ca^{2+} slow current was abolished within 10 min, whereas the fast Ca^{2+} current was nearly unaffected. These results indicate that only the slow Ca^{2+} channel is metabolically dependent. The ATP dependence of the activity of the Ca^{2+} slow channels was also shown at the single-channel level (Ohya and Sperelakis, 1989b) (Fig. 13).

AMP-PNP (β,γ-imidoadenosine 5′-triphosphate) could not substitute

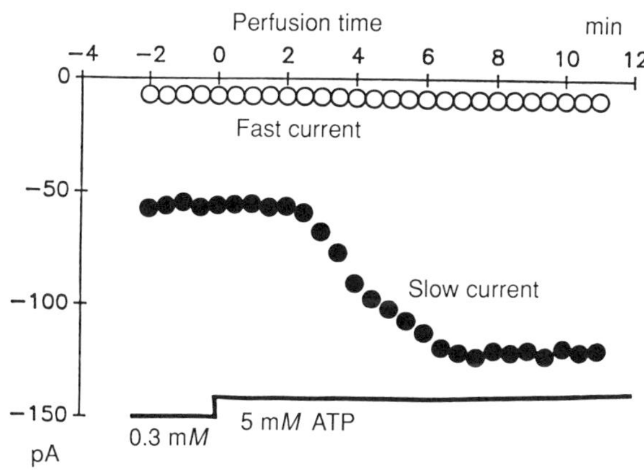

Fig. 12 Effects of intracellularly perfused ATP on the separated fast and slow Ca^{2+} channel currents. The concentration of ATP was increased from 0.3 to 5 m*M* using the intracellular perfusion technique. The time after step switching the ATP level is indicated on the abscissa. The slow current was isolated by using a HP of −40 mV to voltage-inactivate the fast channels; command potentials to +30 mV were applied every 30 s. The fast current was recorded using a HP of −80 mV and a command potential to −20 mV. The peak magnitudes of the fast (open circles) and slow (solid circles) Ca^{2+} channel currents were plotted as a function of time following switching to 5 m*M* ATP. Fast and slow currents obtained from two different cells. Data taken from Ohya and Sperelakis (1989a).

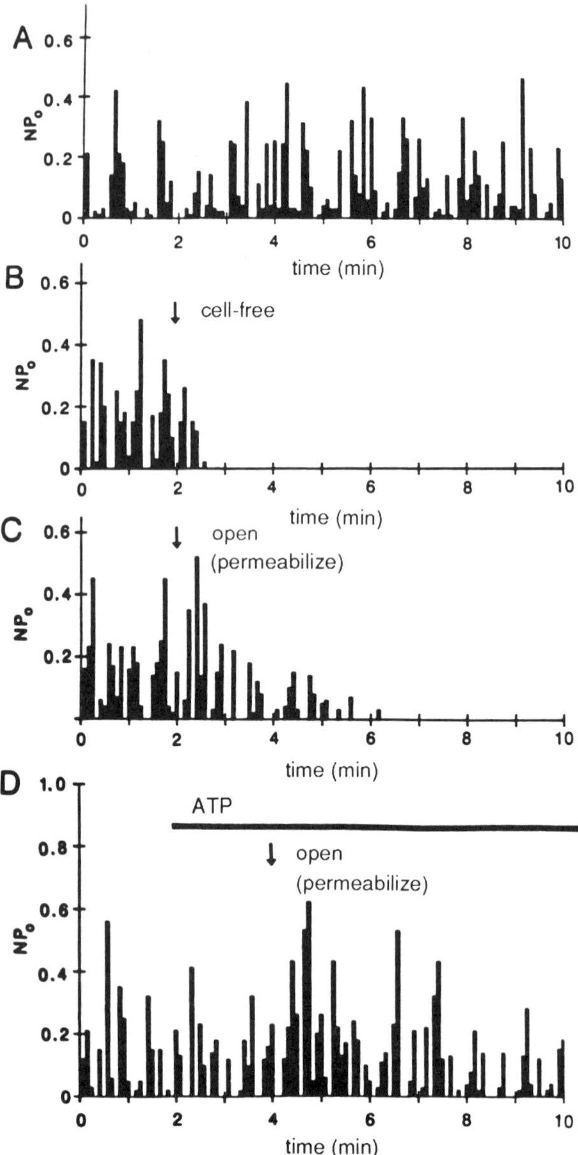

Fig. 13 Histogram of Ca^{2+} slow channel activity to show changes over time in cell-attached patch (A), isolated inside-out patch (cell-free) (B), and cell-attached patch with cell opened distally without ATP (C) and with ATP (D) in the bath. (A) The values of NP_O (ordinates) were calculated for each sweep (from 5 to 100 ms after onset of the depolarizing pulse) and were plotted against time (abscissa). Holding potential of −70 mV and stepped to 0 mV (at 0.2 Hz). (B) Channel activity recorded in the cell-attached configuration initially, and the patch was excised (isolated patch) at the arrow. (C) Cell-attached configuration; at the arrow, one end of the cell was mechanically

for ATP in sustaining the slow Ca^{2+} current in intestinal smooth muscle cells (Ohya *et al.*, 1987) and in guinea pig portal vein VSM cells, but ATP-γ-S did (Ohya and Sperelakis, 1989b). There may be an obligatory binding site for ATP on the inner surface of the channel that must be occupied for channel activity, or ATP may be required for phosphorylation of ion channels or associated regulatory proteins by PK-C (see Fig. 17). However, the relatively high $K_{0.5}$ value of 0.3 mM argues against the phosphorylation alternative, since the K_m value for phosphorylation is usually in the low micromolar range.

C. Modulation of Ca^{2+} Channels by Agonists

1. Regulation by Angiotensin

Agonists, such as norepinephrine (NE) or Ang-II, may induce contraction of VSM via several mechanisms, including (*a*) stimulation of Ca^{2+} influx through receptor-operated channels (ROCs); (*b*) stimulation of Ca^{2+} entry through voltage-dependent Ca^{2+} channels, opened indirectly by the depolarization resulting from an increase in membrane conductance for other ions; and (*c*) release of intracellular Ca^{2+} from store sites (for review, see Bolton, 1979; Johansson and Somlyo, 1980; Kuriyama *et al.*, 1982; Sperelakis and Ohya, 1989). Some agonists (NE, Ang-II, etc.) may modify the voltage-dependent Ca^{2+} channels of VSM cells. The reported results with NE on Ca^{2+} channel current have been contradictory: one group reported enhancement in rabbit ear artery (Aaronson *et al.*, 1986), whereas another group reported inhibition (Droogmans *et al.*, 1987). Another group reported that NE enhanced the fast-type of Ca^{2+} channel, but inhibited the slow-type in cultured rat portal vein cells (Pacaud *et al.*, 1987). Others did not observe a significant change in mesenteric artery and saphenous vein (Bean *et al.*, 1986; Yatani *et al.*, 1987b). Hence, agonists such as NE may affect the Ca^{2+} channel activity in VSM cells via a number of different mechanisms. It is possible than an intracellular factor that modifies the response to agonists is lost during perfusion of the cell for whole-cell voltage clamp.

Ang-II, applied by bolus perfusion to cultured VSM cells from rat aorta, elicited a transient depolarization of up to 30 mV, which sometimes triggered an AP (Zelcer and Sperelakis, 1981) (Fig. 14, top). The Ang-II-induced depolarization, disappeared in Na^+-free solution, suggesting that

---

disrupted (using another electrode) to permeabilize the membrane. (D) 5 mM ATP was applied to the bath solution before the membrane was permeabilized (at the arrow). The pipet contained Ba^{2+} (100 mM) and the bath was K^+-rich. Data taken from Ohya and Sperelakis (1989b).

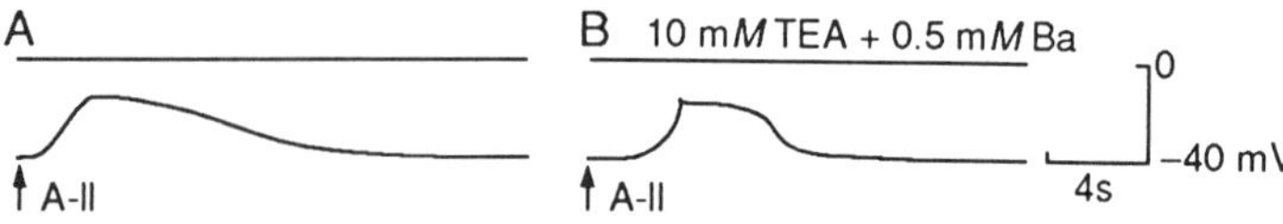

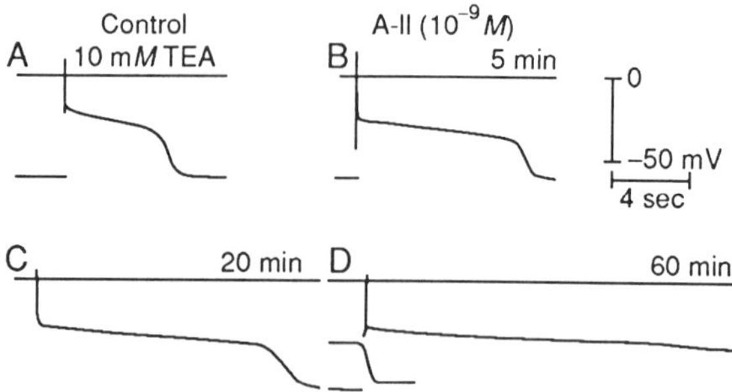

Fig. 14 Effects of angiotensin (A-II) on the membrane potential and TEA-induced action potentials (APs) in cultured smooth muscle cells from rat aorta. (Top) Bolus exposure (SHR preparations). (A) A bolus of A-II (20 ml of 10^{-6} *M*) rapidly depolarized the cell. (B) In the presence of 10 m*M* TEA and 0.5 m*M* Ba^{2+}, an A-II bolus depolarized and induced an AP. (Bottom) Steady-state exposure. (A) Control AP induced by electrical stimulation in presence of 10 m*M* TEA. (B–D) Bath addition of 10^{-9} *M* A-II (continuous exposure) progressively increased the AP duration at 5 min (B), 20 min (C), and 60 min (D). All records were from the same impalement. Data taken from Zelcer and Sperelakis (1981) (top) and Johns and Sperelakis (1990) (bottom).

it was primarily due to an increase in Na^+ conductance, with resulting Na^+ influx. Continuous superfusion of cultured aortic reaggregates with Ang-II (10^{-6} *M*) also produced depolarization, which triggered a long-lasting AP (Johns and Sperelakis, 1990). The membrane resistance was decreased by Ang-II, consistent with an increase in conductance for an inward depolarizing current. Low concentrations of Ang-II (10^{-9} *M*) caused a marked prolongation of the Ca^{2+}-dependent APs without depolarizing the membrane (Fig. 14, bottom).

The effects of Ang-II on Ca^{2+} and K^+ channels in cultured VSM cells from rabbit aorta were examined by whole-cell voltage clamp (Bkaily *et al.*, 1988a). The bath solution contained 50 m*M* Ba^{2+} as charge carrier (with TEA and 4-AP to block the outward current), and the holding potential was −80 mV. Addition of Ang-II caused a substantial increase in the I_{Ba} current at all depolarizing steps, and an Ang-II antagonist blocked this effect. Ang-II (1–5 × 10^{-8} *M*) also suppressed the delayed-rectifier outward K^+ current. These effects of Ang-II on the Ca^{2+} and K^+ channel currents can explain the depolarization and AP prolongation found in aortic cells. These observations suggest that agonists control vascular tone, at least in part, by modifying the Ca^{2+} channel and K^+ channel activities, and thereby changing $[Ca]_i$.

Ang-II also greatly stimulated the Ca^{2+} slow channel current in isolated single cells from guinea pig portal vein (Ohya and Sperelakis, 1991a,b) (Fig. 15). This effect of Ang-II was mediated via a G-protein, based on inhibition of the Ang-II effect by GDP-β-S and stimulation of the basal I_{Ca} by GTP-γ-S. However, this G protein was not sensitive to pertussis toxin or to cholera toxin, hence is not G_i like or G_s-like.

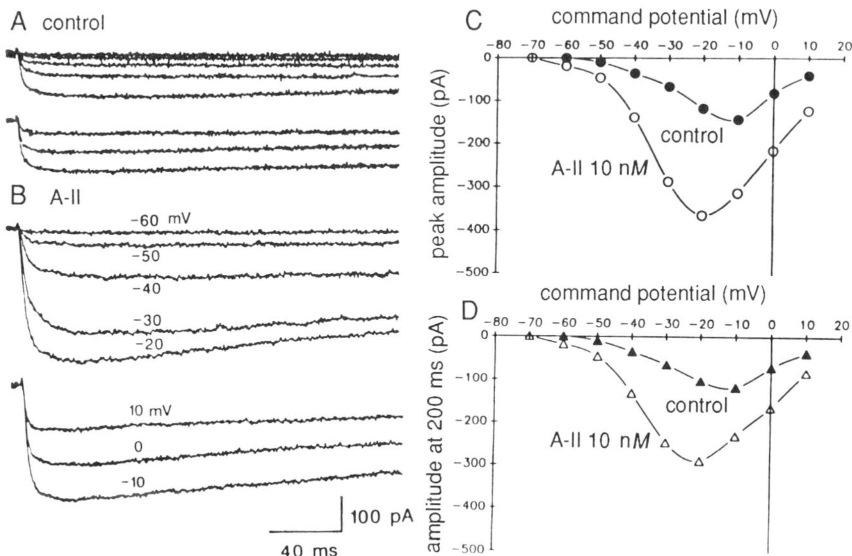

Fig. 15 Effects of angiotensin-II (A-II) on the Ca^{2+} channel current recorded from guinea pig portal vein. (A) Ba^{2+} current recorded before (a) and 3 min after application of 10 n*M* A-II. Step command pulses were applied over the voltage range of −60 to 10 mV from a holding potential (HP) of −90 mV. (B) Peak current amplitude (a) and current amplitude measured at 200 ms (b), measured before (solid symbols) and after (open symbols) 10 n*M* A-II application, were plotted against the command potentials. Data in (A) and (B) were collected from the same cell. Bath solution contained 2 m*M* Ba^{2+} as the charge carrier. Pipet solution contained high Cs with 5 m*M* ATP and 0.1 m*M* GTP. (A), (B), and (C) were taken from three different cells. Reproduced from Ohya and Sperelakis (1991a,b).

2. Regulation by Protein Kinase C

Some neurotransmitters and hormones, including Ang-II, promote the breakdown of membrane phospholipids and the formation of two putative second messengers, IP_3 and DAG (see review by Nishizuka, 1984). One of the major roles of IP_3 is to release Ca^{2+} from intracellular stores (for review, see Berridge and Irvine, 1984), including the SR of VSM cells (Somlyo *et al.*, 1985; Suematsu *et al.*, 1985). It has been suggested that PK-C is involved in agonist-induced Ca^{2+} influx (Rasmussen *et al.*, 1987), which is responsible for the tonic phase of smooth muscle contraction. The phorbol esters, which activate PK-C, produce a slowly developing sustained contracture of VSM that is partially dependent on $[Ca]_o$ (Danthuluri and Deth, 1986; Gleason and Flaim, 1986; Chiu *et al.*, 1987; Itoh and Lederis, 1987).

Superfusion of cultured aortic reaggregates with a phorbol ester, phorbol 12,13-diacetate (PDA, 0.1–5 μM), produced a gradual depolarization and allowed the appearance of an active (plateau-type) AP response to electrical stimulation that could be abolished by verapamil (Ousterhout and Sperelakis, 1992) (Fig. 16). Prolonged exposure (20–30 min) to high concentrations of PDA depressed the APs, perhaps due to depolarization-induced inactivation of the Ca^{2+} channels. The depolarization produced

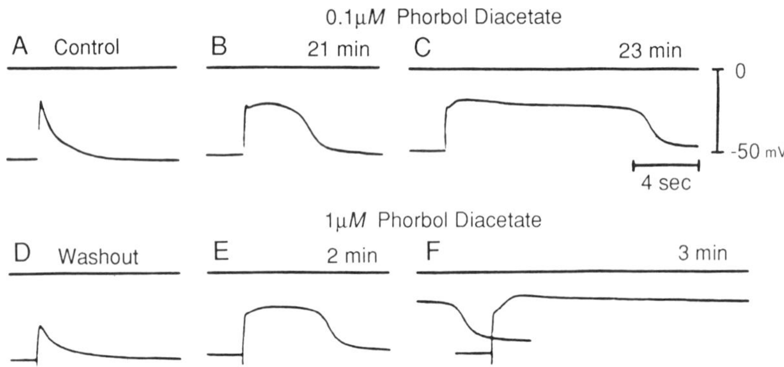

Fig. 16 Induction of APs by phorbol diacetate in cultured rat aortic cells. (A) Control record showing a lack of response to electrical stimulation in normal physiological salt solution. (B) Induction of an AP after exposure to 10^{-7} *M* phorbol diacetate for 21 min. (C) Further prolongation of the plateau component (23 min). (D) Lack of response to stimulation after washout for 2.5 h. (E) Induction of the AP after exposure to 10^{-6} *M* phorbol diacetate for 2 min. (F) Further prolongation of the plateau component (repolarization shown in subsequent sweep superimposed). Data taken from Ousterhout and Sperelakis (1992).

by the phorbol esters could be due to inhibition of K^+ channels, and the induction of APs could result from this plus the activation of a Ca^{2+} conductance, possibly mediated by phosphorylation. In cultured aortic cells (A7r5 cell line), phorbol esters increased the slow (sustained) Ca^{2+} channel current (Fish *et al.*, 1988). In cultured neurons, phorbol esters produced dual effects on the Ca^{2+}-dependent APs, and it was suggested that either Ca^{2+} or K^+ conductances could be depressed, depending on the membrane potential (Werz and MacDonald, 1987).

3. Direct Regulation by G Protein

Direct regulation of ionic channels by GTP-binding protein (G protein) has been identified in several tissues (see reviews by Brown and Birnbaumer, 1988; Rosenthal *et al.*, 1988). In some neurons, G protein (G_o) is thought to mediate the modulation (inhibition) of Ca^{2+} channels produced by receptor activation (Hescheler *et al.*, 1987b). In cardiac cells, G

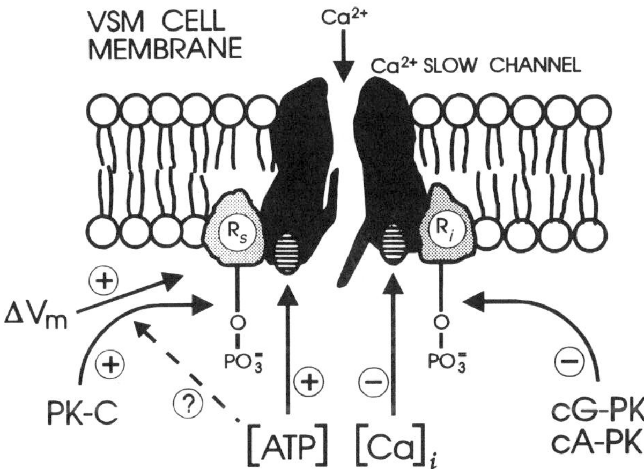

Fig. 17 Diagram depicting some of the factors that regulate the voltage-dependent Ca^{2+} slow channels in VSM cells. As shown, phosphorylation of a site on the channel [or an inhibitory protein (R_i)] by cA-PK or cG-PK inhibits channel activity, whereas phosphorylation of a second site [or a stimulatory protein (R_S)] by PK-C stimulates channel activity. Therefore, agents that elevate cAMP or cGMP inhibit the channels, and thereby decrease Ca^{2+} influx and produce vasodilation. Agonists that increase DAG production stimulate the channels, and thereby increase Ca^{2+} influx and produce vasoconstriction. As shown, ATP is required for activity of the channels ($K_{0.5}$ of ca. 0.3 mM). There may be a binding site for ATP on the channel that must be occupied, or ATP may be required for phosphorylation. Ohya *et al.* (1987) have shown that internal Ca^{2+} inhibits channel activity.

protein (G_s) was reported to enhance directly the activity of Ca^{2+} channels (Yatani *et al.*, 1987a). The effects of GTP-γ-S on Ca^{2+} channels in freshly isolated single VSM cells from guinea pig portal vein were examined in cell-attached patch configuration with 100 m*M* Ba^{2+} and Bay-K-8644 in the pipet and high K^+ solution in the bath (Ohya and Sperelakis, 1988).

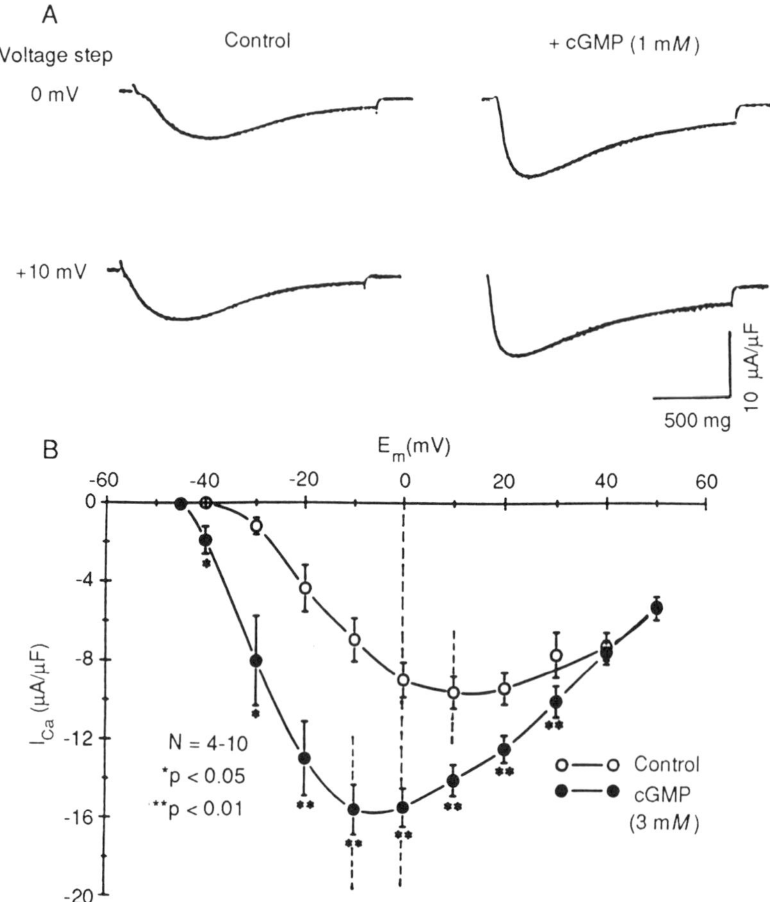

Fig. 18 (A) Slow calcium currents recorded from one bullfrog skeletal muscle fiber. Voltage steps were applied to the potentials indicated. Holding potential was −90 mV. (Left) Control records; (right) records obtained 30 min after internal application of cGMP (1 m*M*) via the end pools. (B) Current/voltage curves for peak I_{Ca} in the absence (control, open circles) and presence of 3 m*M* cGMP (solid circles). Each point is the mean ± SEM of 5–10 voltage steps from 10 fibers (Kokate *et al.*, 1993).

The single-channel conductance was 18–22 pS. After permeabilization of one end of the cell, GTP-γ-S (0.1 m*M*) applied to the bath (for access into the cell interior) enhanced the channel activity. This observation suggests that a G protein may be one of the factors regulating Ca^{2+} channels in VSM cells also. However, it is not known whether G protein acts directly or indirectly on the Ca^{2+} channel.

D. Summary and Conclusions

Vascular tone is regulated by a variety of neurotransmitters, vasoactive hormones and autacoids, and vasoactive drugs. These actions are mediated, at least in part, by actions on the membrane ion channels, exerted either directly or indirectly. Evidence was presented that three different protein kinase systems (protein kinases A, G, and C) act on and modulate the Ca^{2+} slow channels in some VSM cells, and that ATP is required for activity of these channels (Fig. 17). The slow Ca^{2+} current is inhibited by cAMP and cGMP, and stimulated by activation of PK-C.

IV. Skeletal Muscle Fibers

Cyclic AMP, and agents that act to elevate cyclic AMP, stimulate I_{Ca} in frog skeletal muscle fibers (Arreola *et al.*, 1987). In preliminary experiments on single bullfrog semitendinosus fibers examined by the vaseline-gap technique (Hille and Campbell, 1976), cyclic GMP was found to stimulate

Table IV

Effects of Various Agents on Slow Calcium Currents in Bullfrog Skeletal Muscle Fibers

Compound	% Increase in peak $I_{Ca(s)}$	% Decrease in time to peak current	Mean shift in peak $I_{Ca(s)}$ (mV)	*n*
8-Br-cGMP (1 m*M*)	32.4 ± 5.1	20.1 ± 5.4	−11.0 ± 2.2	5
cGMP (1 m*M*)	44.1 ± 9.0	27.3 ± 5.9	−11.3 ± 2.1	8
cGMP (3 m*M*)	72.3 ± 7.4	32.8 ± 3.1	−13.0 ± 1.6	10
8-Br-cAMP (1 m*M*)	18.4 ± 4.5	34.6 ± 4.3	−18.0 ± 2.3	5
cAMP (1 m*M*)	25.0 ± 3.7	21.8 ± 4.7	−11.9 ± 2.6	8
cAMP (3 m*M*)	57.3 ± 12.3	25.2 ± 2.3	−12.1 ± 1.4	7
Isoproterenol[a] (0.01 m*M*)	17.7 ± 2.1	22.0 ± 2.8	−6.3 ± 2.0	4

Note. Values are expressed as the mean ± standard error of the mean. All changes were statistically significant ($p < 0.05$). Data taken from Kokate *et al.* (1993).

[a] Isoproterenol was applied to the external pool. All other agents were applied internally via the end pools.

Table V

Summary of Effect of Cyclic Nucleotides on Ca^{2+} Slow Channels in Cardiac Muscle, Vascular Smooth Muscle, and Skeletal Muscle

	Cardiac muscle	Vascular smooth muscle cells	Skeletal muscle[a]
cAMP	Stimulation	Inhibition	Stimulation
cGMP	Inhibition	Inhibition	Stimulation

[a] From bullfrog.

I_{Ca} in a manner similar to that of cyclic AMP (Kokate *et al.,* 1993) (Fig. 18A). In fact, cyclic GMP was more potent than cyclic AMP in this action (Table IV). The 8-bromo cyclic nucleotides had similar, but less potent, effects (Table IV). As can be seen in Figure 18 and Table IV, the time to peak current was also decreased by the cyclic nucleotides. There was also a small shift to the left (more negative potentials) of the *I*/*V* curve (Fig. 18B; Table IV). In summary, in skeletal muscle, both cyclic nucleotides act in the same direction, namely to stimulate I_{Ca}. Although the physiological significance of this action of cyclic nucleotides on force of contraction, twitch or tetanic, is unclear at present, it is possible that tetanic force would be increased.

V. Summary

The effects of cAMP and cGMP on the slow Ca^{2+} channels in cardiac muscle, VSM, and skeletal muscle fibers are summarized in Table V. As shown, in cardiac muscle, cAMP stimulates and cGMP inhibits. In VSM, both cAMP and cGMP inhibit. In skeletal muscle, both cAMP and cGMP stimulate.

Acknowledgments

These research projects were supported by NIH Grants HL-31942 and HL-22619. We thank Rhonda S. Hentz for typing the manuscript, Glenn Doerman for making some figures, and Anthony Sperelakis for photography.

References

Aaronson, P. I., Benham, C. D., Bolton, T. B., Hess, P., Lang, R. J., and Tsien, F. W. (1986). Two types of single-channel and whole-cell calcium or barium currents in single smooth muscle cells of rabbit ear artery and the effects of noradrenaline. *J. Physiol. (London)* **377,** 36.

Armstrong, D., and Eckert, R. (1987). Voltage-activated calcium channels that must be phosphorylated to respond to membrane depolarization. *Proc. Natl. Acad. Sci. U.S.A.* **84,** 2518–2522.

Arreola, J., Calvo, J., Garcia, M. C., and Sanchez, J. A. (1987). Modulation of calcium channels of twitch skeletal muscle fibers of the frog by adrenaline and cyclic adenosine monophosphate. *J. Physiol. (London)* **393,** 307–330.

Bean, B. (1985). Two kinds of calcium channels in canine atrial cells. Differences in kinetics, selectivity, and pharmacology. *J. Gen. Physiol.* **86,** 1–30.

Bean, B. P., Nowycky, M. C., and Tsien, R. W. (1984). Beta-adrenergic modulation of calcium channels in frog ventricular heart cells. *Nature (London)* **307,** 371–375.

Bean, B. P., Sturek, M., Puga, A., and Hermsmeyer, K. (1986). Calcium channels in muscle cells isolated from rat mesenteric arteries: Modulation by dihydropyridine drugs. *Circ. Res.* **59,** 229–235.

Berridge, M. J., and Irvine, R. F. (1984). Inositol trisphosphate, a novel second messenger in cellular signal transduction. *Nature (London)* **312,** 315–321.

Bkaily, G., and Sperelakis, N. (1984). Injection of protein kinase inhibitor into cultured heart cells blocks calcium slow channels. *Am J. Physiol.* **246,** H630–H634.

Bkaily, G., and Sperelakis, N. (1985). Injection of cyclic GMP into her cells blocks the slow action potentials. *Am. J. Physiol.* **248,** H745–H749.

Bkaily, G., and Sperelakis, N. (1986). Calmodulin is required for a full activation of the calcium slow channels in heart cells. *J. Cyclic Nucleotide Protein Phosphorylation Res.* **11,** 25–34.

Bkaily, G., Sperelakis, N., and Eldefrawi, M. (1984). Effects of the calmodulin inhibitor, trifluoperazine, on membrane potentials and slow action potentials of cultured heart cells. *Eur. J. Pharmacol.* **105,** 23–31.

Bkaily, G., Peyrow, M., Sculptoreanu, A., Jacques, D., Chahine, M., Regoli, D., and Sperelakis, N. (1988a). Angiotensin II increased I_{si} and blocks I_K in single aortic cell of rabbit. *Pfluegers Arch.* **412,** 448–450.

Bkaily, G., Yamamoto, T., Peyrow, M., Sculptoreanu, A., Jacques, D., and Sperelakis, N. (1988b). Macroscopic Ca^{2+} and K^+ currents in single heart and aortic cells. *Mol. Cell. Biochem.* **80,** 59–72.

Bolton, T. B. (1979). Mechanisms of action of transmitters and other substances on smooth muscle. *Physiol. Rev.* **59,** 606–718.

Brown, A. M., and Birnbaumer, L. (1988). Direct G protein gating of ion chanels. *Am. J. Physiol.* **254,** H401–H410.

Brown, J. H., Buxton, I. L., and Brunton, L. L. (1985). Alpha 1-adrenergic and muscarinic cholinergic stimulation of phosphoinositide hydrolysis in adult rat cardiomyocytes. *Circ. Res.* **57,** 532–537.

Bruckner, R., and Scholz, H. (1984). Effects of alpha-adrenoceptor stimulation with phenylephrine in the presence of propranolol on force of contraction, slow inward current and cyclic AMP content in the bovine heart. *Br. J. Pharmacol.* **82,** 223–232.

Cachelin, A. B., dePeyer, J. E., Kokubun, S., and Reuter, H. (1983). Ca^{2+} channel modulation by 8-bromo-cyclic AMP in culture heart cells. *Nature (London)* **304,** 462–464.

Chad, J. E., and Eckert, R. J. (1986). An ezymatic mechanism for calcium current inactivation in dialysed Helix neurones. *J. Physiol. (London)* **378,** 31–51.

Chiu, A. T., Bozarth, J. M., Forsyth, M. S., and Timmermans, P. B. M. W. M. (1987). Ca^{2+} utilization of the constriction of rat aorta to stimulation of protein kinase C by phorbol dibutyrate. *J. Pharmacol. Exp. Ther.* **242,** 934–939.

Cuppoletti, J., Thakkar, J., Sperelakis, N., and Wahler, G. (1988). Cardiac sarcolemmal substrate of the cGMP-dependent protein kinase. *Membr. Biochem.* **7,** 135–142.

Danthuluri, N. R., and Deth, R. D. (1986). Acute desensitization to angiotensin II: Evidence for a requirement of agonist-induced diacylglycerol production during tonic contraction of rat aorta. *Eur. J. Pharmacol.* **125,** 1103–1109.

DiFrancesco, D., and Tromba, C. (1988). Muscarinic control of the hyperpolarization-activated current (i_f) in rabbit sino-atrial node myocytes. *J. Physiol.* (*London*) **405,** 493–510.

Dosemeci, A., Dhalla, R. S., Cohen, N. M., Lederer, W. J., and Rogers, T. B. (1988). Phorbol ester increases calcium current and stimulates the effects of angiotensin II on cultured neonatal rat heart myocytes. *Circ. Res.* **62,** 347–357.

Droogmans, G., Declerck, I., and Casteel, R. (1987). Effect of adrenergic agonists on Ca^{2+}-channel currents in single vascular smooth muscle cells. *Pfluegers Arch.* **409,** 7–12.

Ehara, T., and Ishihara, K. (1990). Anion channels activated by adrenaline in cardiac myocytes. *Nature* (*London*) **347,** 284–286.

Fischmeister, R., and Hartzell, R. C. (1987). Cyclic guanosine 3′,5′-monophosphate regulates the calcium current in single cells from frog ventride. *J. Physiol.* (*London*) **387,** 455–472.

Fish, R. D., Sperti, G., Colucci, W. S., and Clapham, D. E. (1988). Phorbol ester increases the dihydropyridine-sensitive calcium conductance in a vascular smooth muscle cell line. *Circ. Res.* **62,** 1049–1054.

Friedman, M. E., Suarez-Kurtz, G., Karzorowski, G. J., Katz, G. M., and Reuben, J. P. (1986). Two calcium currents in a smooth muscle cell line. *Am. J. Physiol.* **250,** H699–H703.

George, W. J., Polson, J. B., O'Toole, A. G., and Goldberg, N. D. (1970). Elevation of guanosine 3′,5′-cyclic phosphate in rat heart after perfusion with acetylcholine. *Proc. Natl. Acad. Sci. U.S.A.* **66,** 398–403.

Gleason, M. M., and Flaim, S. F. (1986). Phorbol ester contracts rabbit thoracic aorta by increasing intracellular calcium and by activating calcium influx. *Biochem. Biophys. Res. Commun.* **138,** 1362–1369.

Goldberg, N. D., Haddox, M. K., Nicol, S. E., Glass, D. B., Sanford, C. H., Kuehl, F. A., Jr., and Estensen, R. (1975). Biological regulation through opposing influences of cyclic GMP and cyclic AMP: The Yin Yang hypothesis. *Adv. Cyclic Nucleotide Res.* **5,** 307–330.

Hartzell, H. C., and Fischmeister, R. (1986). Opposite effects of cyclic GMP and cyclic AMP on Ca^{2+} current in single heart cells. *Nature* (*London*) **323,** 273–275.

Hescheler, J., Kameyama, M., and Trautwein, W. (1986). On the mechanism of muscarinic inhibition of the cardiac Ca current. *Pfluegers Arch.* **407,** 182–189.

Hescheler, J., Kameyama, M., Trautwein, W., Mieskes, G., and Soling, H. D. (1987a). Regulation of the cardiac calcium channel by protein phosphatases. *Eur. J. Biochem.* **165,** 261–266.

Hescheler, J., Rosenthal, W., Trautwein, W., and Schultz, G. (1987b). The GTP-binding protein, G_o, regulates neuronal calcium channels. *Nature* (*London*) **325,** 445–447.

Hescheler, J., Mieskes, G., Ruegg, J. C., Takai, A., and Trautwein, W. (1988). Effects of a protein phosphatase inhibitor, okadaic acid, on membrane currents of isolated guinea-pig cardiac myocytes. *Pfluegers Arch.* **412,** 248–252.

Hille, B., and Campbell, D. T. (1976). An improved vaseline gap voltage-clamp for skeletal muscle fibers. *J. Gen. Physiol.* **67,** 265–293.

Irisawa, H., and Kokobun, S. (1983). Modulation by intracellular ATP and cyclic AMP of the slow inward current in isolated single ventricular cells of the guinea-pig. *J. Physiol.* (*London*) **338,** 321–327.

Itoh, H., and Lederis, K. (1987). Contraction of rat thoracic aorta strips induced by phorbol 12-myristate 13-acetate. *Am. J. Physiol.* **252,** C244–C247.

Johansson, B., and Somlyo, A. P. (1980). Electrophysiology and excitation-contraction coupling. *In* "Handbook of Physiology" (R. M. Berne, A. P. Somlyo, and H. V. Sparks, eds.), Sect. 2, Vol. 2, pp. 301–323. Am. Physiol. Soc., Bethesda, MD.

Johns, D. W., and Sperelakis, N. (1990). Angiotensin-II stimulation of Ca^{2+}-dependent action potentials in cultured rat aortic smooth muscle cells. *Eur. J. Pharmacol.* **187,** 183–191.

Johnson, J. C., Wittenauer, L. A., and Nathan, R. D. (1983). Calmodulin, Ca^{2+}-antagonists and Ca^{2+}-transporters in nerve and muscle. *J. Neural Transm. Suppl.* **18,** 97–111.

Josephson, I., and Sperelakis, N. (1978). 5′-Guanylimidodiphosphate stimulation of slow Ca^{2+} current in myocardial cells. *J. Mol. Cell. Cardiol.* **10,** 1157–1166.

Kameyama, M., Hofmann, F., and Trautwein, W. (1986). On the mechanism of β-adrenergic regulation of the Ca^{2+} channel in the guinea-pig heart. *Pfluegers Arch.* **405,** 285–293.

Kohlhardt, M., and Haap, K. (1978). 8-Bromo-guanosine-3′,5′-monophosphate mimics the effect of acetylcholine on slow response action potential and contractile force in mammalian atrial myocardium. *J. Mol. Cell. Cardiol.* **10,** 573–578.

Kokate, T. G., Heiny, J. A., and Sperelakis, N. (1993). Stimulation of the slow calcium current in bullfrog skeletal muscle fibers by cAMP and cGMP. *Am. J. Physiol.* **265,** C47–C53.

Kuriyama, H., Ito, Y., Suzuki, H., Kitamura, T., and Itoh, T. (1982). Factors modifying contraction-relaxation cycle in vascular smooth muscle. *Am. J. Physiol.* **243,** H641–H662.

Levi, R. C., Alloatti, G., and Fischmeister, R. (1989). Cyclic GMP regulates the Ca-channel current in guinea pig ventricular myocytes. *Pfluegers Arch.* **413,** 685–687.

Li, T., and Sperelakis, N. (1983). Stimulation of slow action potentials in guinea pig papillary muscle cells by intracellular injection of cAMP, Gpp(NH)p, and cholera toxin. *Circ. Res.* **52,** 111–117.

MacLeod, K. M., and Diamond, J. (1986). Effects of the cyclic GMP lowering agent LY83583 on the interaction of carbachol with forskolin in rabbit isolated cardiac preparations. *J. Pharmacol. Exp. Ther.* **238,** 313–318.

Mehegan, J. P., Muir, W. W., Unverferth, D. V., Fertel, R. H., and McGuirk, S. M. (1985). Electrophysiological effects of cyclic GMP on canine cardiac Purkinje fibers. *J. Cardiovasc. Pharmacol.* **7,** 30–35.

Mery, P. F., Lohmann, S. M., Walter, U., and Fischmeister, R. (1991). Ca^{2+} current is regulated by cyclic GMP-dependent protein kinase in mammalian cardiac myocytes. *Proc. Natl. Acad. Sci. U.S.A.* **88,** 1197–1201.

Nargeot, J., Nerbonne, J. M., Engels, J., and Lester, H. A. (1983). Time course of the increase in the myocardial slow inward current after a photochemically generated concentration jump of intracellular cAMP. *Proc. Natl. Acad. Sci. U.S.A.* **80,** 2395–2399.

Nawrath, H. (1977). Does cyclic GMP mediate the negative inotropic effect of acetylcholine in the heart? *Nature (London)* **267,** 72–74.

Nilius, B., Hess, P., Lansman, J. B., and Tsien, R. W. (1985). A novel type of cardiac calcium channel in ventricular cells. *Nature (London)* **316,** 443–446.

Nishizuka, Y. (1984). The role of protein kinase C in cell surface signal transduction and tumor promotion. *Nature (London)* **308,** 693–698.

Ohya, Y., and Sperelakis, N. (1988). Guanosine triphosphate dependent stimulation of L-type calcium channels of vascular smooth muscle. *Physiologist* **31,** A88.

Ohya, Y., and Sperelakis, N. (1989a). ATP regulation of the slow calcium channels in vascular smooth muscle cells of guinea pig mesenteric artery. *Circ. Res.* **64,** 145–154.

Ohya, Y., and Sperelakis, N. (1989b). Modulation of single slow (L-type) calcium channels by intracellular ATP in vascular smooth muscle cells. *Pfluegers Arch.* **414,** 257–264.

Ohya, Y., and Sperelakis, N. (1991a). Involvement of a GTP-binding protein in stimulating action of angiotension II on calcium channels in vascular smooth muscle cells. *Circ. Res.* **68,** 763–777.

Ohya, Y., and Sperelakis, N. (1991b). Agonist modulation of voltage-dependent calcium channels in vascular smooth muscles. *In* "Electrophysiology and Ion Channels of Vascular Smooth Muscle and Endothelial Cells" (N. Sperelakis and H. Kuriyama, eds.), pp. 39–46. Elsevier, Amsterdam.

Ohya, Y., Kitamura, K., and Kuriyama, H. (1987). Modulation of ionic currents in smooth muscle balls of the rabbit intestine by intracellularly perfused ATP and cyclic AMP. *Pfluegers Arch.* **408,** 465–473.

Ono, K., and Trautwein, W. (1991). Potentiation by cyclic GMP of β-adrenergic effect on Ca^{2+} current in guinea pig ventricular cells. *J. Physiol.* (*London*) **443,** 387–404.

Ono, K., Kiyosue, T., and Arita, M. (1989). Isoproterenol, DBcAMP, and forskolin inhibit cardiac sodium current. *Am. J. Physiol.* **256,** C1131–C1137.

Osterrieder, W., Brum, G., Hescheler, J., Trautwein, W., Flockerzi, V., and Hofmann, F. (1982). Injection of subunits of cyclic AMP-dependent protein kinase into cardiac myocytes modulates Ca^{2+} current. *Nature* (*London*) **298,** 576–578.

Ousterhout, J. M., and Sperelakis, N. (1987). Cyclic nucleotides depress action potentials in cultured aortic smooth muscle cells. *Eur. J. Pharmacol.* **144,** 7–14.

Ousterhout, J. M., and Sperelakis, N. (1992). Dual effects of phorbol esters on the excitability of cultured aortic smooth muscle cells. (unpublished).

Pacaud, P., Lorrand, G., Mironneau, C., and Mironneau, J. (1987). Opposing effects of noradrenaline on the two classes of voltage-dependent calcium channels of single vascular smooth muscle cells in short-term primary cultures. *Pfluegers Arch.* **410,** 557–559.

Rasmussen, H., Takuwa, Y., and Park, S. (1987). Protein kinase C in the regulation of smooth muscle contraction. *FASEB J.* **1,** 177–185.

Reuter, H. (1983). Calcium channel modulation by neurotransmitters, enzymes, and drugs. *Nature* (*London*) **301,** 569–574.

Reuter, H., and Scholz, H. (1977). The regulation of the calcium conductance of cardiac muscle by adrenaline. *J. Physiol.* (*London*) **264,** 49–62.

Reuter, H., Stevens, C.-F., Tsien, R. W., and Yellen, G. (1982). Properties of single calcium channels in cardiac cell culture. *Nature* (*London*) **297,** 501–504.

Rosenthal, W., Hescheler, J., Trautwein, W., and Schultz, G. (1988). Control of voltage-dependent Ca^{2+} channels by G protein-coupled receptor. *FASEB J.* **2,** 2784–2790.

Schneider, J. A., Shigenobu, K., and Sperelakis, N. (1976). Valinomycin inhibition of the inward slow current of cardiac muscle. *Recent Adv. Stud. Card. Struct. Metab.* **9,** 33–52.

Shigenobu, K., and Sperelakis, N. (1972). Ca^{2+} current channels induced by catecholamines in chick embryonic hearts whose fast Na^{+} channels are blocked by tetrodotoxin or elevated K^{+}. *Circ. Res.* **31,** 932–952.

Somlyo, A. V., Bond, M., Somlyo, A. P., and Scarps, A. (1985). Inositol trisphosphate-induced calcium release and contraction in vascular smooth muscle. *Proc. Natl. Acad. Sci. U.S.A.* **82,** 5231–5236.

Spah, F. (1984). Forskolin, a new positive inotropic agent, and its influence on myocardial electrogenic cation movements. *J. Cardiovasc. Pharmacol.* **6,** 99–106.

Sperelakis, N. (1988). Regulation of calcium slow channels of cardiac muscle by cyclic nucleotides and phosphorylation. *J. Mol. Cell. Cardiol.* **20,** 75–105.

Sperelakis, N., and Ohya, Y. (1989). Electrophysiology of vascular smooth muscle. *In* "Physiology and Pathophysiology of the Heart," (N. Sperelakis, ed.), 2nd ed., pp. 773–811. Kluwer Academic Press, Boston.

Sperelakis, N., and Ohya, Y. (1991). Regulation of calcium slow channels in vascular smooth muscle cells. *In* "Electrophysiology and Ion Channels of Vascular Smooth Muscle and

Endothelial Cells" (N. Sperelakis and H. Kuriyama, eds.), pp. 27–38. Elsevier, Amsterdam.

Sperelakis, N., and Schneider, J. A. (1976). A metabolic control mechanism for calcium ion influx that may protect the ventricular myocardial cell. *Am. J. Cardiol.* **37,** 1079–1085.

Sperelakis, N., Inoue, Y., Nozaki, M., and Ishikawa, S. (1991). Neuromuscular transmission at adrenergic nerve terminals with vascular smooth muscle in guinea-pig mesenteric artery. *In* "Electrophysiology and Ion Channels of Vascular Smooth Muscle and Endothelial Cells" (N. Sperelakis and H. Kuriyama, eds.), pp. 3–15. Elsevier, Amsterdam.

Suematsu, E., Hirata, M., Sasaguri, T., Hashimoto, T., and Kuriyama, H. (1985). Roles of Ca^{2+} on the inositol 1,4,5-trisphosphate-induced release of Ca^{2+} from saponin-permeabolized single cells of the porcine coronary artery. *Comp. Biochem. Physiol. A* **82A,** 645–649.

Tareen, F. M., Ono, K., Noma, A., and Ehara, T. (1991). β-adrenergic and muscarinic regulation of the chloride current in guinea-pig ventricular cells. *J. Physiol.* (*London*) **440,** 225–241.

Thakkar, J., Tang, S., Sperelakis, N., and Wahler, G. (1988). Inhibition of cardiac slow action potentials by 8-bromo-cyclic GMP occurs independent of changes in cyclic AMP levels. *Can. J. Physiol. Pharmacol.* **66,** 1092–1095.

Tohse, N., and Sperelakis, N. (1990). Long-lasting openings of single slow (L-type) Ca^{2+} channels in chick embryonic heart cells. *Am. J. Physiol.* **259,** H639–H642.

Tohse, N., and Sperelakis, N. (1991a). Cyclic GMP inhibits the activity of single calcium channels in embryonic chick heart cells. *Circ. Res.* **69,** 325–331.

Tohse, N., Kameyama, M., Sakiguchi, K., Shearman, M. S., and Kanno, M. (1990). Protein kinase C activation enhances the delayed rectifier potassium current in guinea-pig heart cells. *J. Mol. Cell. Cardiol.* **22,** 725–734.

Tohse, N., Conforti, L., and Sperelakis, N. (1991b). Bay K 8644 enhances Ca^{2+} channel activities in embryonic chick heart cells without prolongation of open times. *Eur. J. Pharmacol.* **203,** 307–310.

Tohse, N., Meszaros, J., and Sperelakis, N. (1992a). Developmental changes in long-opening behavior of L-type Ca^{2+} (slow) channels in embryonic chick heart cells. *Circ. Res.* **71,** 376–384.

Tohse, N., Masuda, H., and Sperelakis, N. (1992b). Novel isoform of Ca^{2+} channel in rat fetal cardiomyocytes. *J. Physiol.* (*London*) **451,** 295–306.

Tohse, N., Nakaya, H., Takeda, Y., and Kanno, M. (1992c). Inhibitory effect of human atrial natriuretic peptide on cardiac L-type Ca channels. *Jpn. J. Pharmacol.* **58,** 184P.

Trautwein, W., and Hoffman, F. (1983). Activation of calcium current by injection of cAMP and catalytic subunit of cAMP-dependent protein kinase. *Proc. Int. Union Physiol. Sci.* **15,** 75–83.

Trautwein, W., Taniguchi, J., and Noma, A. (1982). The effect of intracellular cyclic nucleotides and calcium on the action potential and acetylcholine response of isolated cardiac cells. *Pfluegers Arch.* **392,** 307–314.

Tripathi, O., and Sperelakis, N. (1991). Effects of 8-Bromo cyclic GMP on slow channel mediated action potentials of 3-days-old embryonic chick ventricle. *J. Dev. Physiol.* **16,** 309–316.

Tsien, R. W., Giles, W., and Greengard, P. (1972). Cyclic AMP mediates the action of adrenaline on the action potential plateau of cardiac Purkinje fibers. *Nature* (*London*) **240,** 181–183.

Vogel, S., and Sperelakis, N. (1981). Induction of slow action potentials microiontophoresis of cyclic AMP into heart cells. *J. Mol. Cell. Cardiol.* **13,** 51–64.

Vogel, S., Sperelakis, N., Josephson, J., and Brooker, G. (1977). Fluoride stimulation of slow Ca^{2+} current in cardiac muscle. *J. Mol. Cell. Cardiol.* **9,** 461–475.

Wahler, G. M., and Sperelakis, N. (1985). Intracellular injection of cyclic GMP depresses cardiac slow action potentials. *J. Cyclic Nucleotide Protein Phosphorylation Res.* **10,** 83–95.

Wahler, G. M., and Sperelakis, N. (1986). Cholinergic attenuation of the electrophysiological effects of forskolin. *J. Cyclic Nucleotide Protein Phosphorylation Res.* **11,** 1–10.

Wahler, G. M., Rusch, N. J., and Sperelakis, N. (1990). 8-bromo-cyclic GMP inhibits the calcium channel current in embryonic chick ventricular myocytes. *Can. J. Physiol. Pharmacol.* **68,** 531–534.

Watanabe, A. M., Green, F., and Ahmad, Z. (1989). Studies on the cellular mechanisms of action of positive and negative inotropic agents. *Basic Res. Cardiol.* **84,** 19–22.

Werz, M. A., and MacDonald, R. L. (1987). Phorbol esters: Voltage-dependent effects on calcium-dependent action potentials of mouse general and peripheral neurons in cell culture. *J. Neurosci.* **7,** 1639–1647.

Yatani, A., Condina, J., Imoto, Y., Reeves, J. P., Birnbaumer, L., and Brown, A. M. (1987a). A G protein directly regulates mammalian calcium channels. *Science* **238,** 1288–1292.

Yatani, A., Seidel, C. L., Allen, J., and Brown, A. M. (1987b). Whole-cell and single-channel calcium currents of isolated smooth muscle cells from saphenous vein. *Circ. Res.* **60,** 523–533.

Yazawa, K., and Kameyama, M. (1990). Mechanism of receptor-mediated modulation of the delayed outward potassium current in guinea-pig ventricular myocytes. *J. Physiol. (London)* **421,** 135–150.

Zelcer, E., and Sperelakis, N. (1981). Angiotensin induction of active responses in cultured reaggregates of rat aortic smooth muscle cells. *Blood Vessels,* **18,** 263–279.

Effect of Cyclic GMP on Intestinal Transport

Arie B. Vaandrager and Hugo R. De Jonge
Department of Biochemistry
School of Medicine
Erasmus University, Rotterdam
3000 DR Rotterdam, The Netherlands

I. Introduction

Cyclic GMP (cGMP) was established as an important regulator of intestinal ion transport in the late seventies, when it was shown to be the intracellular mediator of salt and water secretion induced by *Escherichia coli* heat-stable enterotoxin (STa) (Field *et al.*, 1978; Hughes *et al.*, 1978). Heat-stable enterotoxins, together with cAMP-raising toxins from *Vibrio cholerae* and *E. coli,* are the mediators of severe secretory diarrhea, causing the death of millions of infants in developing countries (Giannella, 1981; Hirschhorn and Greenough, 1991). In the developed world, STa is associated with travellers diarrhea, "Turista," and diarrhea in domestic animals. It was subsequently discovered that the intestine contained unique isozymes of both guanylyl cyclase (GC-C) and cGMP-dependent protein kinase (PK-G II) (Kuno *et al.*, 1986; Schulz *et al.*, 1990; De Jonge, 1981, 1984b). The recent finding of a distinctive endogenous activator of intestinal guanylyl cyclase, guanylin (Currie *et al.*, 1992), affirms that the intestine possesses a unique cGMP-signaling pathway for the physiological regulation of ion transport. Intracellular cAMP- and Ca-signals were already known to stimulate salt and water secretion in many epithelia including the intestine (Field, 1981). The regulation of fluid secretion by cGMP, however, was thought to be restricted to the intestine. However, the

Advances in Pharmacology, Volume 26

recent detection of some of the components of the "intestinal cGMP pathway" (GC-C and guanylin) in other tissues (Forte *et al.*, 1988; Schulz *et al.*, 1992; Laney *et al.*, 1992) suggests that it is expressed more generally than previously assumed.

II. Synthesis of Cyclic GMP in the Intestine

A. Intestinal Form of Guanylyl Cyclase (Type C)

Intestinal epithelial cells contain a unique form of membrane/cytoskeleton-bound guanylyl cyclase, which can be stimulated by STa and by guanylin (Kuno *et al.*, 1986; Currie *et al.*, 1992). Since STa binding could be separated from the guanylyl cyclase activity by gel filtration, it was believed that the cyclase and the STa receptor were different proteins (Kuno *et al.*, 1986). However, molecular cloning techniques revealed an intestinal guanylyl cyclase possessing a STa binding domain within the same protein (Schulz *et al.*, 1990). This receptor–guanylyl cyclase showed a considerable homology to the natriuretic peptide receptor–guanylyl cyclase GC-A and GC-B, and was therefore named GC-C (Wong and Garbers, 1992; see also Chapters 2 and 5). GC-C expressed in human embryonic kidney cells was shown to be a glycoprotein with a M_r of 160 kDa (fully processed form) or 140 kDa (high mannose form). STa was shown to activate GC-C in the absence of other specific regulatory proteins by inducing a conformational change in a preexisting GC-C oligomer. Activated GC-C was protected against a subsequent inactivation by a direct interaction with ATP (Vaandrager *et al.*, 1993b). In the brush border membrane of intestinal cells, GC-C is present as a proteolytically modified form, consisting of a 45- to 80-kDa N-terminal fragment, which binds STa, and an 85-kDa C-terminus, constituting the cyclase domain. Under native conditions these fragments are still linked, presumably by inter- and intramolecular disulfide bridges, forming a functionally active, STa-stimulatable, guanylyl cyclase complex (De Sauvage *et al.*, 1992; Vaandrager *et al.*, 1993a; Cohen *et al.*, 1993). The following observations make it very likely that GC-C is the major (if not sole) form of guanylyl cyclase in intestinal epithelial cells: (a) similarity between recombinant GC-C and guanylyl cyclase from intestinal brush border membranes in their sensitivity to STa, stabilization by ATP, and gel filtration elution pattern (De Sauvage *et al.*, 1992a; Vaandrager *et al.*, 1993a,b; A. B. Vaandrager, unpublished results); (b) high abundance of GC-C and mRNA for GC-C in intestinal epithelial cells and colocalization of GC-C with STa binding and intestinal guanylyl cyclase activity (see Section II,B); (c) virtual absence of natriuretic peptide-stimulated guanylyl cyclases and soluble guanylyl cyclase in intestinal epithelial cells (De Jonge, 1975; Vaandrager *et al.*, 1992a). The endogenous activator of intestinal guanylyl cyclase is thought to be guanylin, a low-molecular-weight peptide that is synthesized as a large precursor form by paracrinic cell types in the intestine and bears some homology to STa (Currie *et al.*, 1992; Karnaky *et al.*, 1992; Wiegeland *et al.*, 1992; Schulz *et al.*, 1992; De Sauvage *et al.*, 1992b).

B. Localization of Guanylyl Cyclases in the Intestine

STa-stimulatable guanylyl cyclase activity could be demonstrated in all segments of the small and large intestine, but could not be detected in other tissues of the rat or rabbit (Rao *et al.*, 1980; Guerrant *et al.*, 1980). Likewise, GC-C or mRNA for GC-C was predominantly found in epithelial cells of the small intestine, cecum, and proximal colon of the rat (Fig. 1;

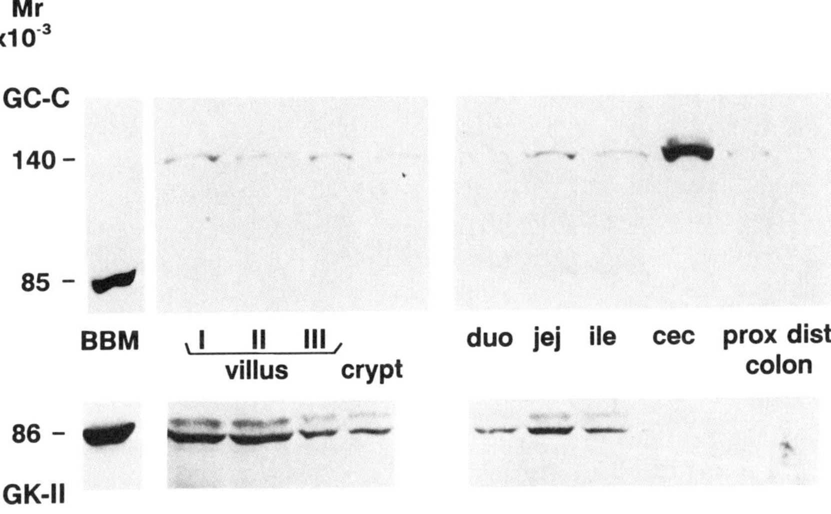

Fig. 1 Localization of guanylyl cyclase-C (GC-C) and cGMP-dependent protein kinase type II (PK-G II) in rat intestine. Epithelial cells were isolated from various segments of intestine or along the villus-to-crypt axis (De Jonge, 1975). The samples were analyzed on immunoblots with antibody against the C-terminal peptide of GC-C (top) or antibody raised against purified pig PK-G II (= GK-II, bottom) (for methods, see Vaandrager *et al.*, 1993). BBM, brush border membranes from jejunal enterocytes; villus I, II, III, cells from upper, mid, and bottom regions of the jejunal villi, respectively; duo, duodenum; ile, ileum; jej, jejunum; cec, cecum. (Top) The 85-kDa band found in BBM is a proteolytic fragment of GC-C, whereas the 140-kDa band observed in whole cells is a not fully processed form of GC-C (Vaandrager *et al.*, 1993). (Bottom) The doublet of PK-G II observed in whole cells, but not in brush border membranes, probably reflects different states of post-translational modification (e.g., myristylation as described for PK-G I).

Laney *et al.*, 1992). Surprisingly the highest levels of GC-C were found in the cecum. High levels of STa-stimulatable guanylyl cyclase/GC-C were also found in the human colon carcinoma cell lines T84 and CaCo-2, which are frequently used as model systems in intestinal ion transport studies (Huott *et al.*, 1988; Mann *et al.*, 1993; Vaandrager *et al.*, 1992a). Besides in humans or domestic mammals, a STa-stimulatable guanylyl cyclase was detected in the intestine of the North American opossum and the chicken (Forte *et al.*, 1988; Katwa and White, 1992). Interestingly the opossum possesses relatively high levels of STa receptors in nonintestinal epithelia (Krause *et al.*, 1990). Recently, low levels of GC-C and guanylin have been detected also in other tissues from the rat (Schulz *et al.*, 1992; Laney *et al.*, 1992).

Remarkably, the nontransformed intestinal cell lines IEC-6 and INT407, which are undifferentiated stem cell-like enterocytes isolated originally from, respectively, rat and human small intestine, have been shown to express natriuretic peptide-stimulatable guanylyl cyclase instead of GC-C (Crane *et al.*, 1990; Vaandrager *et al.*, 1992a). Whole mammalian intestinal mucosa was also found to contain atrial natriuretic peptide-stimulatable and soluble guanylyl cyclase activity localized in nonepithelial cells in the lamina propria (Waldman *et al.*, 1984; Vaandrager *et al.*, 1992a; Wilson *et al.*, 1993b).

In the rat small intestine, the expression of STa receptor, guanylyl cyclase activity, GC-C protein, and GC-C mRNA displays a similar gradient along the villus-to-crypt axis, the upper villus cells showing the highest, and the crypt cells the lowest, but clearly detectable levels of expression (Fig. 1; De Jonge, 1975; Cohen *et al.*, 1992). In rat proximal colon a STa-induced rise in cGMP levels was detected by histological staining in surface, crypt, and mucus-containing goblet cells (Vaandrager *et al.*, 1992a).

Within the villus cell, guanylyl cyclase and STa binding sites are enriched in the brush border membrane (De Jonge, 1975; Walling *et al.*, 1978; Guarino *et al.*, 1987). Little guanylyl cyclase activity and STa binding were found in the basolateral membrane fraction. The relative high abundance of the 140-kDa not fully processed form of GC-C and STa receptors in crude microsomal membranes but not in brush border membranes suggests that a large fraction of GC-C is also present in intracellular membranes (Fig. 1; Guarino *et al.*, 1987; Vaandrager *et al.*, 1993a). The intracellular distribution of guanylyl cyclase is clearly distinct from the localization of adenylyl cyclase, which is exclusively present in the basolateral membrane of the enterocyte (Murer *et al.*, 1976; Walling *et al.*, 1978), or from the localization of a G-protein-coupled phospholipase C (the enzyme that generates the second messengers diacyl glycerol and inositol trisphosphate), which was found in both brush border and basolateral membranes (Vaandrager *et al.,* 1990).

III. Effects of Cyclic GMP on Intestinal Transport in Mammals

A. Effects of Cyclic GMP in Epithelial Cells

In humans and a variety of domestic mammals, the elevation of cGMP levels in intestinal epithelial cells leads to net fluid secretion into the intestinal lumen. Since the epithelial layer is highly permeable for water, the fluid flow will follow the net flow of solutes. In the intestine the main solutes involved in the regulation of fluid absorption and secretion are Na and Cl ions (Powell, 1987). Most studies on the mechanisms of ion transport in intestinal epithelium are performed in a Ussing-type chamber. In this setup, a mucosal preparation, from which the adherent muscle layer is stripped away, is mounted between two half-chambers, which are perfused with an oxygenated Ringer-like medium. In the Ussing chamber both neutral and electrogenic ion fluxes across the tissue can be determined simultaneously by means of radioactive isotope and electrical measurements. Studies in Ussing chambers have demonstrated that cGMP increases salt and thus fluid secretion by two distinct mechanisms: (*a*) inhibition of electroneutral active Na and Cl absorption and (*b*) stimulation of electrogenic Cl^- secretion. These processes are thought to occur independently of each other and are spatially separated (Field, 1981). A model illustrating the effects of cGMP on intestinal ion transport is depicted in Fig. 2.

1. Effect of cGMP on Na and Cl Absorption

Active absorption of ions in the intestine is predominantly performed by the villus cells in the small intestine and by the surface cells in the colon. The driving force for the active absorption is provided by the steep electrochemical gradient for Na across the cell membrane, which is maintained by the extrusion of Na by the Na/K ATPase in the basolateral membrane. Na absorption across the apical membrane is coupled to other solutes, or involves yet unspecified leak pathways and transport through an amiloride-inhibitable Na channel (expressed predominantly in distal colon). From the coupled transporters, the Na/H antiporter (in ileum and proximal colon in combination with a Cl/CO_3 antiporter) and the Na/glucose and Na/aminoacid cotransporters (restricted to the small intestine) make the largest contribution to active Na absorption (Powell, 1987; Field *et al.,*

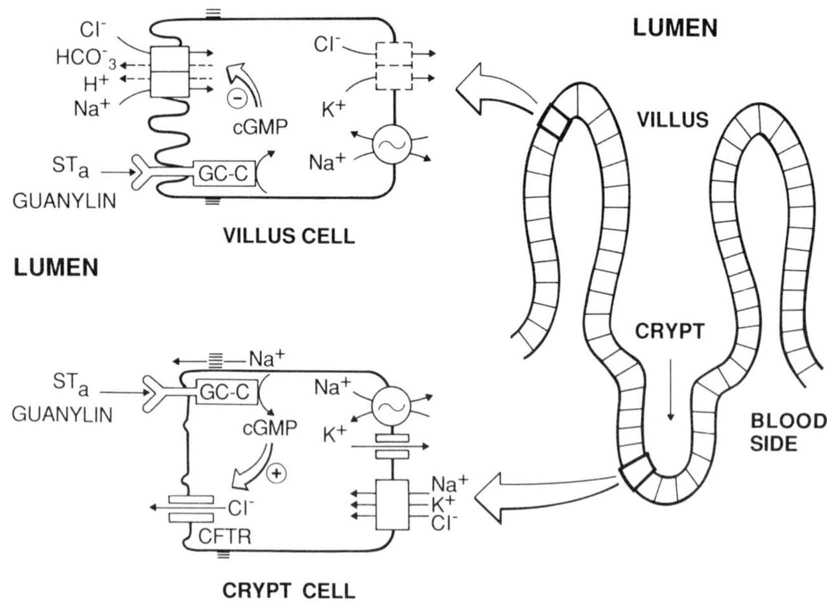

Fig. 2 Model for the stimulation of net salt secretion by cGMP. The model, illustrating the inhibitory effect of cGMP on Na and Cl absorption and the stimulatory effect of cGMP on electrogenic Cl secretion, is discussed in Section III. CFTR, cystic fibrosis transmembrane conductance regulator; ST_a, *Escherichia coli* heat-stable enterotoxin; GC-C, guanylyl cyclase-C. Adapted from Field (1981).

1989). The presence of at least four different isoforms of cloned Na/H exchangers (NHE-1, NHE-2, NHE-3, and NHE-4) has been demonstrated in the intestine. The ubiquitous NHE-1 isoform was found predominantly in the basolateral membrane and is presumed to be involved in intracellular pH regulation (Tse *et al.*, 1991). The NHE-3 isoform is most likely responsible for active Na absorption, since it is expressed specifically in intestine and kidney and the level of its mRNA is increased by glucocorticoids, known stimulators of Na absorption in the ileum (Tse *et al.*, 1992; Yun *et al.*, 1993). The NHE-2 isoform is also a candidate to function as an apical membrane Na/H exchanger in intestinal and renal epithelial cells, but its relative contribution to active Na absorption is presently unclear (Tse *et al.*, 1993).

Chloride is actively absorbed in the ileum and the colon, but not in the jejunum (Field *et al.*, 1989). The absorption of Cl in these tissues is linked to the absorption of Na and (in part) to the extrusion of HCO_3. A Cl/HCO_3 antiporter coupled to the Na/H antiporter by a transmembrane pH gradient is shown to function as an electro-neutral NaCl import system in the ileum and probably plays a major role in active Cl absorption (Powell, 1987). A protein related to the "band 3" anion exchanger in erythrocytes has recently been cloned from rabbit ileum, and probably represents the Cl/HCO_3 exchanger involved in active Cl absorption in the intestine, as it is localized exclusively in the brush border membrane (Chow *et al.*, 1992). However, additional Cl uptake pathways, like a Na–Cl cotransporter may exist (Field *et al.*, 1989). The absence of a Cl/HCO_3 antiporter in brush border membranes of the jejunum may explain the absence of active Cl uptake in this segment (Vaandrager and De Jonge, 1988). The absorbed Cl leaves the epithelial cell at the basolateral membrane by an unknown mechanism, possibly a Cl conductance or a K–Cl cotransporter.

The inhibition of active Na and Cl absorption by cGMP is thought to occur mainly at the level of the uptake systems in the apical membrane. The Na–glucose cotransporter and the amiloride-sensitive Na channel in the distal colon, however, are not inhibited by cGMP (Field *et al.*, 1989). This lack of sensitivity of the glucose transporter to cyclic nucleotide signals is exploited in the oral rehydration therapy for secretory diarrhea based on oral administration of glucose–electrolyte solutions (Hirschhorn and Greenough, 1991). A more likely target of the cGMP-induced inhibition is the Na/H antiporter. This is corroborated by the following observations. (*a*) In rat jejunum STa and 8-Br-cGMP increase the surface pH, which is kept relatively acidic by the action of the Na/H antiporter (McKie *et al.*, 1988; Shimada and Hoshi, 1988). (*b*) In chicken enterocytes cGMP is shown to inhibit the Na/H exchange (Semrad *et al.*, 1990). (*c*) In rabbit ileum the Na/H antiporter could be inhibited by cAMP (Sundaram *et al.*, 1991) or by activation of protein kinase C (PK-C) (Cohen *et al.*, 1991). Activation of PK-C has effects on NaCl absorption similar to those of cGMP and cAMP and is therefore thought to inhibit the same transport systems. The molecular form of the Na/H antiporter susceptible to inhibition by cGMP is not known with certainty but may become identified in future studies using NHE-1–NHE-4-transfected cells or NHE-isoform-specific antibodies.

Furthermore, regulation by cGMP of other transport systems, e.g., the Cl/HCO_3 antiporter or a Na–Cl cotransporter, cannot be ruled out and deserves further exploration.

2. Effect of cGMP on Cl and HCO_3 Secretion

The active secretion of Cl by the intestine mainly arises from the crypts, although small intestinal villus cells may also contribute to electrogenic Cl secretion (Donowitz and Welsh, 1987; Stewart and Turnberg, 1989).

In the secretory cells Cl is accumulated across the basolateral membrane by a bumetanide-inhibitable Na–K–2Cl cotransporter above its electrochemical equilibrium. This is a consequence of the coupling of Cl to the uptake of Na, for which a large driving force is created by the Na/K ATPase, whereas the other cotransported ion, K, is in electrochemical equilibrium by the relatively high K permeability of the membrane. Opening of a Cl channel by cGMP will therefore lead to an efflux of this ion. The cGMP-stimulated channels are located in the apical membrane of the epithelial cells, causing electrogenic secretion of Cl into the intestinal

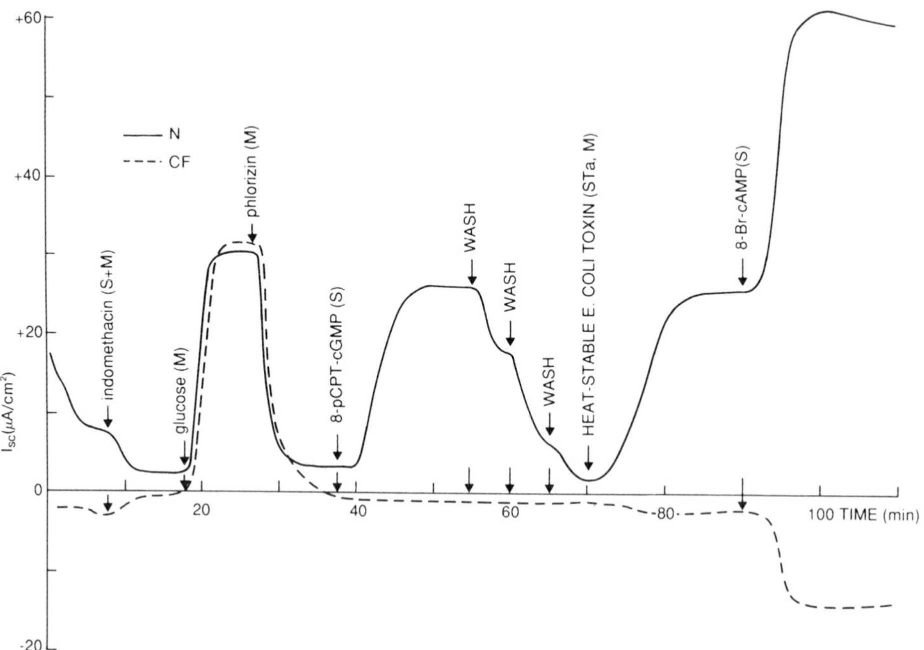

Fig. 3 Effect of cGMP in normal human ileum and in ileum from cystic fibrosis patients. Control ileum (N) or ileum from patients with cystic fibrosis (CF) was obtained during small bowel resections. Mucosal preparations were mounted in an Ussing chamber. The increase in short-circuit current (I_{sc}) after addition of glucose is a measure for the electrogenic Na-coupled glucose absorption and serves as an indicator of the viability of the tissue. The Na–glucose transporter is inhibited by phlorizin. The increase in I_{sc} observed after the addition of the cAMP and cGMP analogues and STa was shown to be caused by electrogenic Cl secretion, since it was not observed in the absence of Cl and it could be inhibited by serosal addition of bumetanide and barium. The addition of indomethacin prevents indirect effects of secretagogues, mediated by prostaglandin formation in the submucosa. M, mucosal addition; S, serosal addition; 8-pCPT-cGMP, 8-parachlorophenylthio-cGMP, a specific activator of cGMP-dependent protein kinases.

lumen. To compensate for the surplus of negative charge in the lumen, Na will follow Cl by a paracellular pathway. The Cl channel, which can be regulated by both cGMP and cAMP has been identified recently as the cystic fibrosis transmembrane conductance regulator (CFTR), the product of the gene mutated in cystic fibrosis patients (see Section VII). That the opening of CFTR Cl channels can fully account for the cyclic nucleotide-elicited Cl secretion in the intestine is clearly demonstrated by the complete absence of electrogenic Cl secretion provoked by cAMP and cGMP analogues and STa in intestinal epithelium from cystic fibrosis patients (De Jonge *et al.*, 1987, 1989, 1992; Berschneider *et al.*, 1988; Taylor *et al.*, 1988; Baxter *et al.*, 1988; Veeze *et al.*, 1991; O'Loughlin *et al.*, 1991; see Fig. 3). So far it is not known whether cGMP additionally stimulates other transporters that could contribute to an increase in Cl secretion, such as the Na–K–2Cl cotransporter or K-channels. A stimulation of these transporters was observed in human colonic cell lines during cAMP-induced Cl secretion (Matthews *et al.*, 1992; Vaandrager *et al.*, 1992b). However, the localization of these transporters in the basolateral membrane in contrast to the preferential localization of the membrane-bound intestinal cGMP-dependent protein kinase in the apical membrane renders a direct activation unlikely.

It seems plausible that *in vivo,* a rise in intracellular cGMP, analogous to cAMP action (Field *et al.*, 1989), may also provoke a net secretion of HCO_3. Studies in airway epithelium suggest that HCO_3, similar to Cl, can be secreted across the apical membrane through the CFTR anion channel, after entering the cell by a Na-dependent import system in the basolateral membrane (Smith and Welsh, 1992).

3. Localization of the Effect of cGMP, Comparison with cAMP

In the rabbit a stimulation of electrogenic Cl secretion by cGMP was found throughout the intestine with the exception of the proximal segment, the duodenum, and the distal colon. Both cGMP-unresponsive segments were capable of active Cl secretion in response to cAMP. The Cl currents elicited by 8-Br-cGMP in the jejunum, the ileum, and the proximal colon were significantly smaller than those induced by 8-Br-cAMP. However, in the cecum, cAMP and cGMP appeared equally effective (Rao *et al.*, 1980). Similar differences between cAMP and cGMP effects on electrogenic Cl secretion were found in rat intestinal segments (H. R. De Jonge, unpublished results). The smaller stimulation of Cl secretion by cGMP compared to cAMP observed in most segments may be explained by a less efficient coupling of the cGMP signal to the CFTR Cl channel (see

Section VII) or to other transport systems involved in transepithelial Cl secretion, i.e., the Na–K–2Cl cotransporter and the basolateral K conductance (see Section III,A,2).

In contrast to their different potency in stimulating Cl secretion, cGMP and cAMP inhibited NaCl absorption in rabbit ileum to a similar extent (Guandalini *et al.*, 1982). Cyclic GMP and cAMP were also equipotent in inhibiting Na/H exchange in rat jejunum, as judged by their similar effect on surface pH (Shimada and Hoshi, 1988). In rat distal colon cGMP inhibited NaCl absorption but did not stimulate Cl secretion (Nobles *et al.*, 1991).

STa also induced an increase in electrogenic Cl secretion in the human colon carcinoma cell line T84 (Huott *et al.*, 1988; Forte *et al.*, 1992; Levine *et al.*, 1991). However, the STa response in T84 cells was not always mimicked by the addition of 8-Br-cGMP (H. R. De Jonge, unpublished results; Forte *et al.*, 1992; Levine *et al.*, 1991). These differences most plausibly result from a different expression level of cGMP-dependent protein kinase in the various T84 cultures (see Section V,C).

B. Effects of Cyclic GMP in Nonepithelial Cell Types

Natriuretic peptides, known ligands of GC-A and GC-B receptor cyclases, were found to provoke electrogenic Cl secretion in colon and jejunum *in vitro* (Moriarty *et al.*, 1990; Sharkey *et al.*, 1991; Vaandrager *et al.*, 1992a). In contrast to STa, the effects of ANP and BNP could be inhibited by blockers of neurotransmission, indicating that the natriuretic peptides do not exert direct effects on epithelial cells. Instead they may stimulate neuroendocrine cells or trigger neuronal circuits in the subepithelial layer, resulting in the release of endogenous secretagogues. It is known that the intestinal mucosa can produce a variety of endogenous regulators of ion transport (Donowitz and Welsh, 1987). Both acetylcholine (in the distal colon) and serotonin (jejunum) have been implicated in the effects of the natriuretic peptides. The effect of ANP on intestinal fluid transport *in vivo* however is less dramatic compared to STa, since no significant increase in fluid secretion could be detected in rat intestine after infusion with ANP (Kaufman and Monckton, 1988). Sodium nitroprusside, a stimulator of soluble guanylyl cyclase, has been shown to induce net Cl secretion and to inhibit NaCl absorption in rat distal colon (Wilson *et al.*, 1993a). Similar to the ANP response, the rise in cGMP levels in response to nitroprusside is restricted to subepithelial cell types and Cl secretion is sensitive to inhibitors of neurotransmission and prostaglandin formation (Wilson *et al.*, 1993b).

IV. Effects of Cyclic GMP on Intestinal Transport in Winter Flounder

There is a general paucity of studies describing the effects of cGMP on ion transport in nonmammalian intestine. One exception, however, is the winter flounder, where cGMP has been shown to inhibit NaCl absorption (Rao *et al.,* 1984). The intestine of this teleost has a relatively simple morphology. It contains uniform columnar epithelial cells and it lacks crypts. The flounder intestine actively absorbs Na and Cl in a way resembling the mechanism by which Cl is secreted by mammalian crypt cells, but then acting in the opposite direction (i.e., secreting Cl from the lumen to the blood side). Na, K, and Cl are taken up by a bumetanide-sensitive Na–K–2Cl cotransporter in the apical membrane. Na is pumped out across the basolateral membrane by the Na/K–ATPase, thus providing the driving force for the NaCl uptake. K is recycled across the apical membrane through K channels, whereas Cl leaves the cell by Cl channels in the basolateral membrane and by an electroneutral pathway (O'Grady *et al.,* 1988). Cyclic GMP reduced electrogenic NaCl absorption in flounder intestine by inhibiting the apical Na–K–2Cl cotransporter (O'Grady *et al.,* 1985). Furthermore, cGMP was shown to inhibit a voltage-activated K channel in dissociated flounder enterocytes, which is suggested to be involved in recycling across the apical membrane (O'Grady *et al.,* 1991). In clear contrast to their action in mammalian intestine, the effects of cGMP and cAMP in flounder intestine are different. Cyclic AMP does not inhibit the cotransporter or the voltage-activated K channel, but is thought to increase the permeability of the luminal membrane for Cl (Rao and Nash, 1988, O' Grady and Wolters, 1990). Another difference with the mammalian system is the finding that atrial natriuretic factor and not STa elicits the cGMP response in flounder intestine (O'Grady *et al.,* 1985; Rao and Nash, 1988).

V. Mechanisms of Cyclic GMP Action in the Intestine

A. General Mechanisms of Cyclic GMP Action

In mammalian tissues, cGMP has been shown to exert its physiological action through a variety of different mechanisms, as summarized in Fig. 4.

1. In most cell types the effects of cGMP are mediated by a specific family of protein kinases, the cGMP-dependent protein kinases (PK-G).

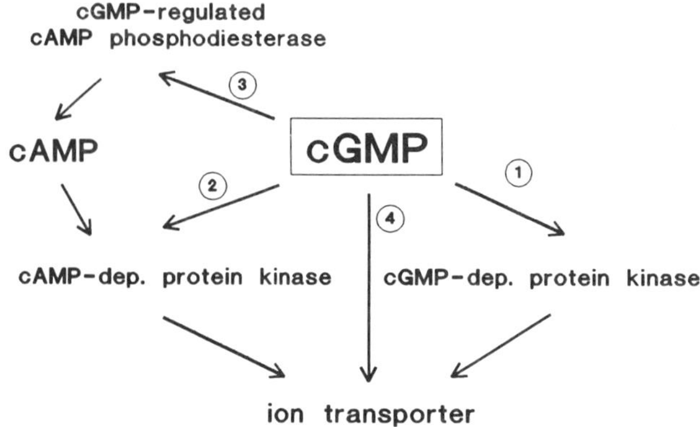

Fig. 4 Potential pathways by which cGMP may modulate ion transport. For explanation see Section V.A.

Intestinal cells contain a unique membrane-bound isozyme of PK-G, designated type II, which is structurally and functionally different from the more ubiquitous soluble isoforms PK-G Iα and Iβ (see Section VI). As will be discussed in section V,B and V,C, PK-G II is believed to play a key role in both the inhibition of Na and Cl absorption and the stimulation of Cl secretion in the intestine by cGMP, with the exception of the distal colon. Readers are also refered to Chapter 7 for additional discussion of PK-G.

2. When elevated to relatively high concentrations ($> 5\ \mu M$), cGMP can cross-activate the cAMP-dependent protein kinase (PK-A) in the intestine (De Jonge, 1984b; Forte *et al.*, 1992). In this tissue, cAMP has effects similar to those of cGMP on NaCl absorption and generally a larger effect on Cl secretion (see Section III,A,4). Both R_I and R_{II} subunits of PK-A have been identified in intestinal crypt and villus cells (De Jonge and Lohmann, 1985). The type II isozyme of PK-A was enriched in the apical membrane of the villus cells and may therefore be involved in the regulation of NaCl absorption, which is localized in the same compartment (De Jonge and Lohmann, 1985). A promiscuous activation of PK-A by cGMP was demonstrated to be involved in the STa-induced Cl secretion in a variant of the human colon carcinoma cell line T84 (Forte et al., 1992).

3. By stimulating or inhibiting distinct species of cAMP-phosphodiesterase, cGMP can either decrease or increase the intracellular levels of cAMP and thus activate or inhibit the cAMP-signaling pathway (Beavo, 1988). Our present knowledge of the distribution of the various types of cyclic nucleotide phosphodiesterases in the intestine is fairly limited. A

cGMP-inhibitied phosphodiesterase may be responsible for the cGMP inhibition of NaCl absorption in rat distal colon (Nobles *et al.*, 1991). Additional discussion of phosphodiesterases can be found in Chapter 6.

4. In some tissues cGMP is known to regulate the activity of ion channels by a direct allosteric interaction with its target. The gating by cGMP of cation channels in visual and olfactory epithelium (Altenhofen *et al.*, 1991) and of Na channels in the renal inner-medullary collecting duct (Light *et al.*, 1990) are examples of this mechanism. Whether the amiloride-sensitive Na channel in the colon is structurally related to the renal Na channel remains to be resolved. However, there is no indication for a direct regulation of Na and Cl import systems or Cl channels by cGMP in the intestine (see Sections V,B and V,C).

B. Mechanisms of Inhibition of Na and Cl Absorption by Cyclic GMP

The predominant presence of PK-G II in the brush border membrane of the villus cell, in close proximity to the Na and Cl import systems, renders it very likely that PK-G II is the first step in the signal route downstream of cGMP. After luminal exposure to STa *in vivo,* the occupancy of the cyclic nucleotide binding domains of PK-G II in rat brush border membranes by cGMP was increased, whereas the occupancy of the regulatory domain of PK-A appeared unchanged (Van Dommelen and De Jonge, 1986). This finding strengthens the concept that cGMP mediates its effect by PK-G II and not PK-A. A direct allosteric interaction of cGMP with Na or Cl transport proteins seems also unlikely, since no modulation of Na or Cl uptake by cGMP could be detected at the level of isolated brush border membrane vesicles (Vaandrager, 1987). In *in vitro* studies with rat brush border membranes two endogenous substrates for cGMP-dependent phosphorylation have been identified. One is PK-G II itself (see Section VI), and the other is a 25-kDa protein, showing some properties of a proteolipid (De Jonge, 1984b; De Jonge and Rao, 1990). Interestingly the 25-kDa protein was also phosphorylated by PK-A, suggesting that it might act as a convergence point for the cAMP and cGMP pathways. The 25-kDa protein is unlikely to represent the Na/H or Cl/HCO_3 exchanger itself, since these transporters have molecular masses of approx. 90 and 160 kDa, respectively (Tse *et al.*, 1992; Chow *et al.*, 1992). It is feasible, however, that this phosphoprotein functions as a regulator of ion transport systems in the microvillar membrane (see Section VII).

In flounder intestine, the cGMP-mediated inhibition of NaCl absorption could be prevented by addition of the isoquinoline sulfonamide derivative

H-8, an inhibitor of cyclic nucleotide-dependent protein kinases, suggesting the involvement of protein phosphorylation in this process (O'Grady *et al.,* 1988). Furthermore, cGMP was found to regulate both *in vitro* and *in vivo* the phosphorylation state of a 50-kDa protein, which was subsequently identified as a cAMP binding protein (De Jonge *et al.*, 1986; Toskulkao and Rao, 1990). However, the relationship between this protein and the cGMP-inhibited Na–K–2Cl cotransporter in the flounder brush border membrane remains to be clarified.

Our previous attempts to demonstrate effects of cGMP or cAMP on electroneutral Na and Cl transport in cyclic nucleotide- and ATP-loaded brush border membrane vesicles isolated from flounder and rat intestine were unsuccessful (Vaandrager, 1987; A. B. Vaandrager and H. R. De Jonge, unpublished results). The lack of effect may be explained by a fast dephosphorylation of the PK-A and PK-G substrates. Such a mechanism was suggested to explain the increased PK-A-mediated inhibition of a renal Na/H exchanger after reconstitution in proteoliposomes (Weinman *et al.,* 1988). Alternatively, it is conceivable that cGMP-dependent phosphorylation does not affect the Na and Cl importers directly or through a transporter-associated protein, but acts through a more indirect mechanism, e.g., by the generation of a third messenger. Ca has been implicated as the intermediate of cGMP- and cAMP-induced inhibition of Na/H exchange in chicken enterocytes (Semrad and Chang, 1987; Semrad *et al.*, 1990). In mammalian small intestine, Ca is also considered an important regulator of Na and Cl absorption, but a Ca-mediated signaling pathway is probably not involved in the inhibition of Na absorption by cGMP or cAMP in rabbit ileum (Donowitz *et al.,* 1989). However, Ca was suggested to play a role in the decrease in Na and Cl absorption induced by STa in mouse intestine (Goyal *et al.*, 1987). STa was also shown to raise intracellular Ca in isolated rat enterocytes, but it is not known whether this is mimicked by cGMP analogues (Knoop *et al.,* 1991). On the other hand, STa or 8-Br-cGMP did not elevate intracellular Ca levels in the human colonic cell lines T84 and HT29-cl.19A (Huott *et al.,* 1988; Van Den Berghe, 1992).

In the distal colon of the rat, but not in the proximal colon, the cGMP-induced inhibition of Na and Cl absorption was shown to be mediated by an elevation in the level of cAMP by a cGMP-induced inhibition of its hydrolysis (Nobles *et al.,* 1991). The difference in cGMP effects in distal colon and other segments of the intestine may be explained by the absence of PK-G II in the distal colon, as no specific cGMP-dependent phosphorylation was observed in this segment in rabbits (Rao, 1985) and no PK-G II could be detected immunologically in rat distal colon (Fig. 1).

C. Mechanisms of Stimulation of Cl Secretion by Cyclic GMP

In the human colonic carcinoma cell line T84, the STa-induced Cl secretion was found to be mediated by a promiscuous activation of PK-A by cGMP (Forte *et al.,* 1992). Whether this mode of action contributes significantly to the cGMP activation of Cl secretion in native intestine remains to be resolved. However, several lines of evidence strongly suggests that PK-G II rather than PK-A is the major mediator of cGMP-provoked Cl secretion in intestinal mucosa. (*a*) PK-G II, but not PK-G I, is present in intestinal crypt cells, whereas no PK-G II could be detected on immunoblots of 8-BrcGMP-unresponsive T84 and HT29cl.19A colon carcinoma cell lines (H. R. De Jonge, unpublished results). (*b*) Purified PK-G II, but not PK-G I, was able to activate the CFTR Cl channel in isolated membrane patches of CFTR-transfected cells (see Section VII). (*c*) In native intestine, in contrast to the T84 cells used by Forte *et al.,* the effect of STa could be mimicked by 8-Br-cGMP or 8-pCPT-cGMP (Fig. 3; Rao *et al.,* 1980). Some of the membrane-permeable cGMP analogues, in particular 8-pCPT-cGMP, are apparently able to discriminate between PK-G and PK-A in intact cells (Butt *et al.,* 1992). (*d*) In a different subclone of the T84 cell line in which 8-Br-cGMP did act as a potent secretagogue, an endogenous cGMP-dependent protein kinase, apparently associated with the apical membrane, was able to activate a CFTR-like Cl channel in excised membrane patches (Lin *et al.,* 1992). (*e*) In rat proximal colon, the cGMP-induced Cl secretion appeared more sensitive to the kinase inhibitors staurosporine and H-8 than the cAMP-induced secretion, whereas a similar dose of these inhibitors blocked the response to cAMP and STa in 8-Br-cGMP-unresponsive T84 cells (H. R. De Jonge, unpublished). (*f*) The PK-A antagonist Rp-8-Br-cAMPS was found capable of inhibiting STa-provoked Cl secretion in PK-G II-deficient T84 cells, but not in rat proximal colon (H. R. De Jonge, unpublished results).

Furthermore, cGMP is unlikely to induce Cl secretion by inhibiting a cAMP-phosphodiesterase, because inhibitors of this family of enzymes (e.g., amrinone, cilostamide) were unable to elicit electrogenic Cl secretion in rat intestinal epithelium (H. R. De Jonge, unpublished results).

VI. Intestinal Form of Cyclic GMP-Dependent Protein Kinase (Type II)

Most mammalian tissues with the clear exception of intestinal epithelial cells (De Jonge, 1981; De Jonge and Lohmann, 1985) express a cytosolic and peripheral membrane species of cGMP-dependent protein kinase (type I) with a native molecular mass of 153–156 kDa and consisting of two identical subunits arranged in a parallel fashion and linked together by disulfide bridges (Edelman *et al.,* 1987). Cloning and purification of the type I enzyme led to the identification of two subspecies, designated Iα and Iβ (Wolfe *et al.,* 1989; Wernet *et al.,* 1989; Sandberg *et al.,* 1989). The Iβ isozyme differs from the Iα isozyme only in the N-terminus (amino acids 1–104), comprising several autophosphorylation sites, a kinase autoinhibitory sequence, the dimerization site, and a "hinge" region (see Fig. 5). Most likely, the two forms have arisen from a single gene by alternate splicing (Wolfe *et al.,* 1989; for recent reivews, see Hofmann *et al.,* 1992; Butt *et al.,* 1993; also Chapter 7).

The discovery of a cGMP-sensitive 86-kDa phosphoprotein in intestinal brush borders (De Jonge, 1976) and its comigration with a monomeric cGMP receptor protein on one- and two-dimensional gels (De Jonge, 1981) were the first indications for the expression of a different isotype of cGMP-dependent protein kinase (type II) in mammalian intestine. Structural and

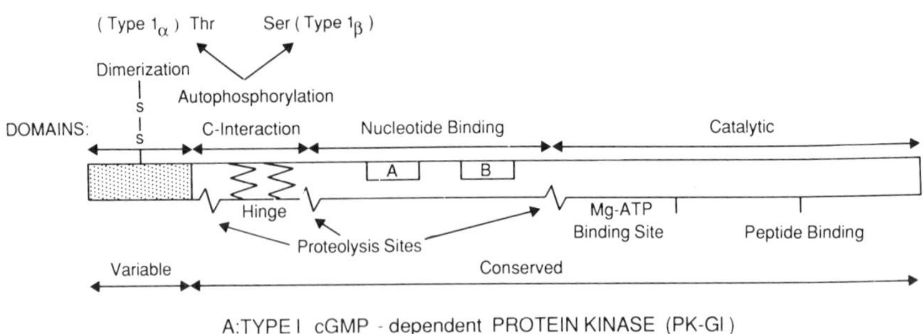

A:TYPE I cGMP - dependent PROTEIN KINASE (PK-GI)

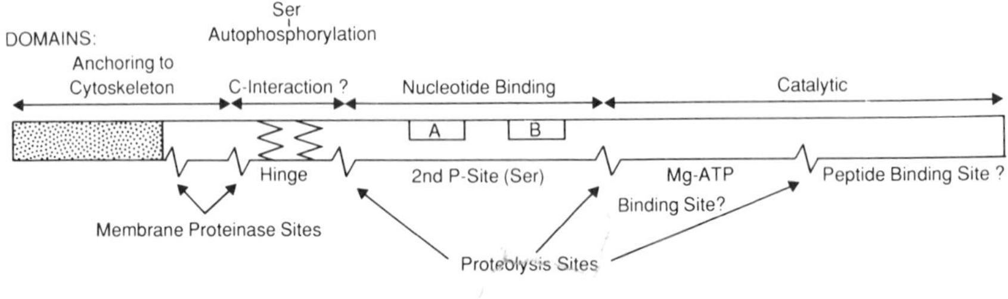

B:TYPE II cGMP - dependent PROTEIN KINASE (PK-GII)

Fig. 5 Comparison of tentative structural models of mammalian cGMP-dependent protein kinase isotypes.

functional analysis of the purified enzyme (reviewed in De Jonge, 1984b; De Jonge and Lohmann, 1985; De Jonge and Rao, 1990) revealed that the type II isozyme resembles PK-G I in the following aspects (Fig. 5): (*a*) the cGMP binding and regulatory domains are covalently coupled, a major structural difference with the class of cAMP-dependent protein kinases; (*b*) it uses arginine-rich histones as *in vitro* substrates; (*c*) it is poorly inhibited by PK-I, the heat-stable inhibitory protein of PK-A; (*d*) it has two distinct cGMP binding sites on the same peptide chain (A and B; Fig. 5); (*e*) it bears an autophosphorylation site in the "hinge" region.

However, PK-G II also shows a number of features that clearly distinguish it from the type I enzyme: (*a*) PK-G II is monomeric, similar to the PK-Gs purified from several unicellular organisms (Hofmann *et al.*, 1992), suggesting that it may have arisen by a very early ancestral gene duplication event; (*b*) the N-terminal region is extended with a 12- to 15-kDa segment apparently serving to anchor the kinase to the microvillar cytoskeleton; this anchoring domain is readily clipped off by exogenous and endogenous proteases; (*c*) contrary to the soluble PK-G I dimer, the intestinal monomer displays hydrophobic properties characteristic for an intrinsic membrane protein and its 71-kDa C-terminal fragment is tightly associated with the brush border membrane. However, the enzyme does not appear to be a transmembrane protein and its functional sites are only accessible from the inner side of the microvillus membrane; (*d*) cGMP was found to stimulate autophosphorylation of PK-G II, but to inhibit autophosphorylation of PK-G I; (*e*) polyclonal antibodies raised against one isoform of PK-G did not cross-react with the other; (*f*) pronounced structural differences between PK-G II and PK-G I were found, as illustrated by dissimilarities in (i) their isolectric points, (ii) phosphopeptide maps, and (iii) their susceptibility to cleavage by a neutral endopeptidase in the intestinal brush border. Recent cloning of PK-G II, in conjunction with protein microsequencing, indeed demonstrates significant differences between PK-G II and PK-G I (results obtained in a collaboration with T. Jarchau, S. M. Lohmann, and U. Walter, Mediz. Univ. Klinik, Würzburg; and J. Vanderkerckhove, University of Ghent).

The intestinal isozyme also displayed important functional dissimilarities with the type I isozymes:

1. Studies of PK-G II autophosphorylation and p-25 phosphorylation in the intestinal brush borders have recently revealed pronounced differences in activation constants for cyclic nucleotide analogues between type II and type I isozymes (H. R. De Jonge; B. Jastorff, University of Bremen; and E. Butt, Mediz. Univ. Klinik, Würzburg, unpublished results). Interestingly, all (Sp)-diastereomers of cGMP and cAMP, which have an axial

exocyclic sulfur atom in the cyclic phosphate moiety, were agonists of PK-G I and PK-G II, and all (Rp)-phosphorothioates of cAMP, which have an equatorial exocyclic sulfur atom, acted as antagonists of both isozymes (cf. Butt *et al.,* 1990). However, all Rp-analogues of cGMP behaved as antagonists of PK-G I but as weak agonists of PK-G II. The different response of PK-G I and PK-G II to cyclic nucleotide analogues confirms the results of structural analysis of the cGMP binding domains and may be advantageously exploited in future studies aimed to assess the role of each isozyme in cGMP regulation of cellular functions.

2. As discussed in detail in Section VII, the activation of the CFTR Cl^- channel and the phosphorylation of PK-G II itself and of its 25-kDa substrate protein in intestinal brush borders appeared isozyme-specific; i.e., the effect of PK-G II could not be mimicked by equal or higher concentrations of purified PK-G I (cf. De Jonge, 1984a,b). This striking difference in substrate-specificity was apparently limited to the natural PK-G II substrates because only minor kinetic differences between both isozymes were found in *in vitro* phosphorylation studies using purified PK-G I substrates including G-substrate, phosphatase inhibitor-1 and DARPP-32 (donated by Dr. A. C. Nairn, Rockefeller University, New York; H. R. De Jonge, unpublished results; cf. Nairn *et al.,* 1985). Sequencing of the phosphorylation domains of PK-G II (Fig. 5) and p-25 (De Jonge and Rao, 1990) is clearly needed to further define the requirements for recognition by PK-G II at the level of their primary structure.

Although it is tempting to postulate a tight physical association of the immobilized PK-G II with other components of the cGMP signaling pathway, i.e., the GC-C, the 25-kDa substrate protein (see Section V,B), or cGMP-regulated transport proteins, direct evidence for such a high-ordered structural arrangement is lacking so far.

VII. The CFTR Cl Channel and Its Regulation by Cyclic GMP

Ussing chamber measurements of transepithelial ion transport in intestinal mucosa from control and cystic fibrosis (CF) patients have suggested a key role for the CFTR Cl channel and the lack of a contribution of other intestinal Cl channels to the cAMP-, cGMP-, and Ca-induced electrogenic Cl secretion in this organ (see Section III,A,2 and Fig. 3).

The identification of the CF gene by positional cloning, the functioning of the CF gene-encoded protein (CFTR) as a Cl channel, and the mechanism of activation of CFTR by cAMP and PK-A have been discussed extensively in recent reviews (Collins, 1992; McIntosh and Cutting, 1992;

Fuller and Benos, 1992). In line with its classification as a member of the superfamily of ATP binding cassette (ABC) transporters, CFTR contains two transmembrane domains, each involving six loops that span the membrane, and two nucleotide binding folds (NBF) (Fig. 6). In addition, it has a unique region, denoted the R (regulatory) domain, which contains many charged residues and multiple potential sites for phosphorylation by PK-A and PK-C. Binding of ATP to the NBFs followed by ATP hydrolysis is an essential step in the mechanism of activation of the CFTR Cl channel following its phosphorylation by PK-A (Anderson and Welsh, 1992). A major missense mutation found in about 70% of the CF alleles results in a phenylalanine deletion at position 508 in NBF 1 (ΔF508). In heterologous CFTR expression systems (Cheng *et al.*, 1990) and in tissues from CF patients homozygous for ΔF508, i.e., sweat duct (Kartner *et al.*, 1992) and airway epithelium (Puchelle *et al.*, 1992), this most frequent mutation causes misprocessing and mislocalization of the CFTR Cl channel rather than a loss of function. Whether a similar maturation and trafficking abnormality underlies the defective Cl secretion in CF intestine is

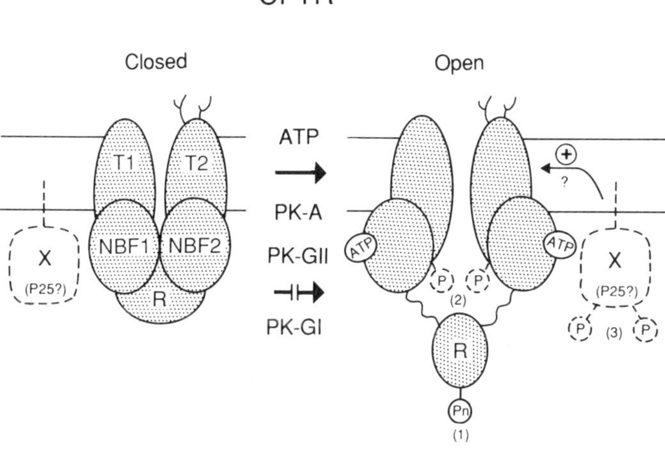

Fig. 6 Hypothetical model illustrating the isotype-specific opening of the CFTR Cl channel by type II cGMP-dependent protein kinase (PK-G II). T1, T2; transmembrane domains; NBF 1, NBF 2; nucleotide binding folds; R, regulatory domain; X, a hypothetical phosphoprotein serving as a substrate for PK-A and PK-G II and promoting CFTR Cl channel activation in its phosphorylated state; p25, a 25-kDa phosphoprotein enriched in intestinal brush borders and serving as a substrate for endogenous PK-G II and PK-A; P, potential phosphorylation sites for protein kinases; PK-G I, the soluble isozyme of cGMP-dependent protein kinase purified from bovine lung; PK-G II: the membranal isozyme of cGMP-dependent protein kinase purified from pig intestinal brush borders.

not known yet but is currently analyzed by immunological techniques (De Jonge *et al.*, 1992). A recent *in situ* hybridization study has demonstrated a high level of CFTR expression in intestinal crypt cells (Trezise and Buchwald, 1991). In HT-29, T84, and CaCo-2 intestinal cell lines the CFTR protein has been localized predominantly in the apical membrane, in accordance with its putative role in intestinal Cl secretion (Denning *et al.*, 1992).

The mechanism by which cAMP and PK-A activate the CFTR Cl channel has been studied in detail. The successful reconstitution of highly purified CFTR in a planar bilayer resulting in the appearance of ATP-dependent and PK-A-activatable Cl channels clearly demonstrates that PK-A regulates CFTR function via direct phosphorylation (Bear *et al.*, 1992). At least four serine residues in the R-domain that are phosphorylated *in vivo* and *in vitro* after activation of PK-A have been identified (Cheng *et al.*, 1991; Picciotto *et al.*, 1992). No one site was essential for channel activity. However, the serine-to-alanine mutation of all four resulted in a significant loss of activatable conductance. Recent studies however show that mutation of all nine consensus phosphorylation sites for PK-A in the R-domain, and the mutation of an additional pre-NBF 1 site, did not completely abolish CFTR channel activation by PK-A, suggesting that its regulation is more complex than previously thought (Chang *et al.*, 1993).

Much less information is yet available about the mechanism by which cGMP is able to activate the CFTR Cl channel. Lin *et al.* (1992) were the first to show that a Cl channel with biophysical properties similar to those of CFTR could be activated in excised membrane patches from 8-Br-cGMP-responsive T84 cells in the presence of cGMP and ATP. Activation could be blocked by H-8, the cyclic nucleotide-dependent protein kinase inhibitor, but not by PK-I, the specific PK-A inhibitor. Their data suggest that an endogenous, membrane-associated form of intestinal PK-G, presumably PK-G II, was able to activate a CFTR-like channel in the absence of cytosolic factors. Compelling evidence for a PK-G II-mediated activation of the CFTR Cl channel was obtained more recently in our laboratory by patch clamp studies on CFTR-transfected IEC-6 cells (a homologous expression system consisting of undifferentiated rat enterocytes lacking endogenous PK-Gs and CFTR; see Bijman *et al.*, 1993) and on CFTR-transfected NIH-3T3 fibroblasts (a heterologous expression system containing hundreds of CFTR Cl channels per excised, inside-out patch; see Anderson *et al.*, 1991). In both expression models, exposure of the patch to cGMP and ATP alone failed to activate the Cl channels. However, the further addition of PK-G II purified from pig intestinal brush borders, or the addition of the catalytic subunit of PK-A resulted in CFTR Cl channel activation (French *et al.*, 1993). Activation of the channel by PK-A but not by PK-G II was blocked by PK-I. Interestingly, and in agreement with recent findings by Berger *et al.* (1993), even excessive amounts of PK-G I were unable to substitute for PK-G II in this channel activation assay. However, rather paradoxically, PK-A and PK-G I were reported to phosphorylate the R-domain of CFTR *in vitro* on the same four sites (Picciotto *et al.*, 1992). This apparent discrepancy between phosphorylation and channel activation experiments can be explained in several ways (see Fig. 6): (*a*) PK-G I, in contrast to PK-G II does not phosphorylate native CFTR *in vivo;* i.e., CFTR phosphorylation by PK-G I is an *in vitro* artifact. (*b*) Phosphorylation of the four serines in the R-domain is insufficient to open the CFTR channel; additional phosphorylation events in R or in other domains of the protein (sites 2, Fig. 6) catalyzed by PK-A and PK-G II but not by PK-G I are required for channel activation. (*c*) PK-G II does not activate the channel through a direct mechanism but phosphorylates a CFTR-associated protein (denoted X in Fig. 6) capable of activating the CFTR Cl channel only in its phosphorylated state. Similar indirect models of ion transport regulation by PK-G substrate proteins associated with the transporter include the stimulation of the Ca–ATPase in sarcoplasmic reticulum by phospholamban (Sarcevic *et al.*, 1989), the stimulation of the smooth muscle Ca-pump through phosphorylation of a 240-kDa membrane protein (Yoshida *et al.*, 1991), and the activation of Ca-dependent K channels by calpromotin, a ubiquitous 23-kDa cytoplasmic protein recently identified as a PK-G I substrate *in vitro* (Moore *et al.*, 1992). The 25-kDa phosphoprotein of intestinal brush borders, acting as a substrate for PK-A and PK-G II but not for PK-G I (see Section VI), might fulfill a similar role in CFTR activation. However, in order to explain the PK-G II activation of CFTR Cl channels in 3T3 fibroblasts, it would be necessary to postulate a more ubiquitous expression of the 25-kDa protein in nonintestinal cell types. The identification of the *in vitro* and *in vivo* phosphorylation sites in CFTR for both PK-G I and PK-G II and a further analysis of p-25 expression and function are clearly needed to discriminate between a direct or indirect model of channel regulation. The study of CFTR regulation by PK-Gs in the lipid bilayer may also help to resolve this issue.

VIII. Function of Cyclic GMP-Induced Electrolyte and Fluid Secretion in the Intestine

In order to prevent the loss of water on one hand, and dehydration of the intestinal contents and consequently obstruction of the intestine on the other, the intestine attempts to maintain a constant fluidity in the lumen. To deal with the potential large variation in the amount of fluid reaching the intestine, this "leaky" epithelium has a high capacity for absorbing and secreting fluid. Both absorption and secretion of salt and water can occur simultaneously and must therefore be carefully balanced (Powell, 1987). The importance of the regulation of ion transport is underscored by the observation that apart from cGMP, also cAMP, Ca/calmodulin, and Ca/PK-C signal transduction pathways play a role as modulators of this process (Donowitz and Welsh, 1987). Furthermore when the electrogenic Cl secretion is impaired, as observed in patients with cystic fibrosis or in transgenic mice with disrupted CF genes, an obstruction of the intestine (i.e., meconium ileus) can be observed (Snouwaert *et al.*, 1992). The pathology of cystic fibrosis also points to another function of fluid secretion, i.e., to act as a defense mechanism against pathogenic agents. Active electrolyte and fluid secretion may keep the epithelial layer clean by a "weep and sweep" mechanism. A number of substances associated with the immune system are indeed capable of inducing fluid secretion (Donowitz and Welsh, 1987).

Since no selective blockers of GC-C or PK-G II, or transgenic mice lacking these enzymes are yet available, one can only speculate about the relative importance of cGMP as a physiological regulator of intestinal transport. The localization of GC-C in the apical membrane implies that the cGMP pathway is activated by luminal signals, whereas the other pathways are preferentially exploited by neuro-hormonal signals acting at the basolateral side (see Section II,B). A luminal receptor might function preferentially as a sensor for monitoring the fluidity of the lumen. One possibility is that an endogenous activator of GC-C, i.e., guanylin, is secreted into the lumen at a relative constant rate. Its steady-state concentration and thus its subsequent effect on GC-C will then depend on the amount of fluid in the lumen; i.e., a high fluidity will result in a small activation of secretion and vice versa. Alternatively, the secretion of guanylin itself may be promoted by as yet unknown luminal or neuro-endocrine stimuli released in response to a low luminal fluidity or produced following the intraluminal accumulation of pathogens or inflammatory cell types. The predominant role of the cGMP system as a regulator of ion secretion in the cecum, in contrast to its minor role in the distal colon (see Section III,A,4), also hints at the importance of cGMP in maintaining intraluminal fluidity. It seems plausible that in the cecum, where the intestinal contents are fermented, maintaining a high fluidity is more important than in the distal colon, where the relatively dry feces is formed.

The concept that the cGMP signaling system is a major determinant of fluid homeostasis in the intestine may also lead to the suggestion that a deficiency or hyperactivity of intestine-specific components of this system (i.e., guanylin, GC-C, or PK-G II) could play a role in the etiology of some cases of idiopathic constipation or diarrhea, respectively. The recent development of immunological methods to determine the levels of each component in intestinal samples, and the availability of electrophysiological techniques to monitor cGMP-provoked Cl secretion in mucosal biopsies (cf. Veeze *et al.,* 1991), will help to identify the role of the cGMP system in the pathophysiology of constipation and diarrheal disease.

Finally, in discussing the cGMP pathway in the intestine in this chapter much emphasis is laid on its role as a regulator of transepithelial ion and fluid transport. However, despite a current lack of experimental evidence, the possibility should be kept open that other functions of the intestine are cGMP-regulated as well. Such functions may either be epithelium-specific (e.g., mucin synthesis and secretion in intestinal goblet cells; paracrinic functions of enterochromaffin cells; crypt cell proliferation and differentiation) or resemble those known to be cGMP-modulated in nonepithelial tissues, e.g., gap-junction permeability (Takens-Kwak and Jongsma, 1992), volume regulation through the Na–K–2Cl cotransporter (Clemo *et al.,* 1992), phospholipase C activity (Takai *et al.,* 1981; Waldmann and Walter, 1989; Ruth *et al.,* 1993), Ca transport (Rashatwar *et al.,* 1987; Yoshida *et al.,* 1991; Sarcevic *et al.,* 1989; Merry *et al.,* 1991), and cytoskeleton-associated functions (Wyatt *et al.,* 1991; Reinhard *et al.,* 1992). It will be a major challenge of future research to sort out these possibilities in order to obtain a more comprehensive picture of cGMP action in intestinal epithelium.

References

Altenhofen, W., Ludwig, J., Eismann, E., Bönigk, W., and Kaupp, U. B. (1991). Control of ligand specificity in cyclic nucleotide-gated channels from rod photoreceptors and olfactory epithelium. *Proc. Natl. Acad. Sci. U.S.A.* **88,** 9868–9872.

Anderson, M. P., and Welsh, M. J. (1992). Regulation by ATP and ADP of CFTR Cl channels that contain mutant nucleotide-binding domains. *Science* **257,** 1701–1704.

Anderson, M. P., Berger, H. A., Rich, D. P., Gregory, R. J., Smith, A. E., and Welsh, M. J. (1991). Nucleoside triphosphates are required to open the CFTR chloride channel. *Cell (Cambridge, Mass.)* **67,** 775–784.

Baxter, P. S., Goldhill, J., Hardcastle, J., and Hardcastle, P. T. (1988). Accounting for cystic fibrosis. *Nature (London)* **355,** 211.

Bear, C. E., Li, C. H., Kartner, N., Bridges, R. J., Jensen, T. J., Ramjeesingh, M., and Riordan, J. R. (1992). Purification and functional reconstitution of the cystic fibrosis transmembrane conductance regulator (CFTR). *Cell (Cambridge, Mass.)* **68,** 809–818.

Beavo, J. A. (1988). Multiple isozymes of cyclic nucleotide phosphodiesterases. *Adv. Second Messengers Protein Phosphorylation Res.* **17,** 1–37.

Berger, H. A., Travis, S. M., and Welsh, M. J. (1993). Regulation of the cystic fibrosis transmembrane conductance regulator Cl channel by specific protein kinases and protein phosphatases. *J. Biol. Chem.* **268,** 2037–2047.

Berschneider, H. M., Knowles, M. R., Azizkhan, R. G., Boucher, R. C., Tobey, N. A.,

Orlando, R. C., and Powell, D. W. (1988). Intestinal electrolyte transport in cystic fibrosis. *FASEB J.* **2,** 2625–2629.

Bijman, J., Dalemans, W., Edixhoven, M. J., Keulemans, J., Hoogeveen, A. H., Scholte, B. J., and De Jonge, H. R. (1992). Protein kinase A (PK-A) and PK-G regulation of CFTR expressed in epithelial and non-epithelial cell lines. *Pediatr. Pulmonol., Suppl.* **8,** 255 (abstr.).

Bijman, J., Dalemans, W., Kansen, M., Keulemans, J., Verbeek, E., Hoogeveen, A., De Jonge, H. R., Wilke, M., Dreyer, D., Lecocq, J.-P., Pavirani, A., and Scholte, B. (1993). Low conductance chloride channels in IEC-6 and CF nasal cells expressing CFTR. *Am. J. Physiol.* **264,** L229–L235.

Butt, E., Van Bemmelen, M., Fischer, L., Walter, U., and Jastorff, B. (1990). Inhibition of cGMP-dependent protein kinase by (R_p)-guanosine 3′,5′-monophosphorothioates. *FEBS Lett.* **263,** 47–50.

Butt, E., Nolte, C., Schulz, S., Beltman, J., Beavo, J. A., Jastorff, B., and Walter, U. (1992). Analysis of the functional role of cGMP-dependent protein kinase in intact human platelets using a specific activator 8-para-chlorophenylthio-cGMP. *Biochem. Pharmacol.* **43,** 2591–2600.

Butt, E., Geiger, J., Jarchau, T., Lohmann, S. M., and Walter, U. (1993). The cGMP-dependent protein kinase—Gene, protein, and function. *Neurochem. Res.* **18,** 27–42.

Chang, X.-B., Tabcharani, J. A., Hou, Y.-X., Jensen, T. J., Kartner, N., Alon, N., Hanrahan, J. W., and Riordan, J. R. (1993). Protein kinase A (PKA) still activates CFTR chloride channel after mutagenesis of all 10 PKA consensus phosphorylation sites. *J. Biol. Chem.* **268,** 11304–11311.

Cheng, S. H., Gregory, R. J., Marshall, J., Paul, S., Souza, D. W., White, G. A., O'Riordan, C. R., and Smith, A. E. (1990). Defective intracellular transport and processing of CFTR is the molecular basis of most cystic fibrosis. *Cell (Cambridge, Mass.)* **63,** 827–834.

Cheng, S. H., Rich, D. P., Marshall, J., Gregory, R. J., Welsh, M. J., and Smith, A. E. (1991). Phosphorylation of the R-domain by cAMP-dependent protein kinase regulates the CFTR chloride channel. *Cell (Cambridge, Mass.)* **66,** 1027–1036.

Chow, A., Dobbins, J. W., Aronson, P. S., and Igarashi, P. (1992). cDNA cloning and localization of a band 3-related protein from ileum. *Am. J. Physiol.* **263,** G345–G352.

Clemo, H. F., Feker, J. J., and Baumgarten, J. J. (1992). Modulation of rabbit ventricular cell volume and Na-K-2Cl cotransport by cGMP and atrial natriuretic factor. *J. Gen. Physiol.* **100,** 89–114.

Cohen, M. B., Mann, E. A., Lau, C., Henning, S. J., and Giannella, R. A. (1992). A gradient in expression of the *Escherichia Coli* heat-stable enterotoxin receptor exists along the villus-to-crypt axis of rat small intestine. *Biochem. Biophys. Res. Commun.* **186,** 483–490.

Cohen, M. B., Jensen, N. J., Hawkins, J. A., Mann, E. A., Thompson, M. R., Lentze, M. J., and Giannella, R. A. (1993). Receptors for *Escherichia coli* heat-stable enterotoxin in human intestine and in a human intestinal cell line (Caco-2). *J. Cell. Physiol.* **156,** 138–144.

Cohen, M. E., Wesolek, J., McCullen, J., Rys-Sikora, K., Pandol, S., Rood, R. P., Sharp, G. W. G., and Donowitz, M. (1991). Carbachol- and elevated Ca^{2+}-induced translocation of functionally active protein kinase C to the brush border of rabbit ileal Na^+ absorbing cells. *J. Clin. Invest.* **88,** 855–863.

Collins, F. S. (1992). Cystic fibrosis: Molecular biology and therapeutic implications. *Science* **256,** 774–779.

Crane, M. R., O'Hanley, P., and Waldman, S. A. (1990). Rat intestinal cell atrial natriuretic peptide receptor coupled to guanylate cyclase. *Gastroenterology* **99,** 125–131.

Currie, M. G., Fok, K. F., Kato, J., Moore, R. J., Hamra, F. K., Duffin, K. L., and Smith,

C. E. (1992). Guanylin: An endogenous activator of intestinal guanylate cyclase. *Proc. Natl. Acad. Sci. U.S.A.* **89,** 947–951.

De Jonge, H. R. (1975). The localization of guanylate cyclase in rat small intestinal epithelium. *FEBS Lett.* **53,** 237–242.

De Jonge, H. R. (1976). Cyclic nucleotide-dependent phosphorylation of intestinal epithelial proteins. *Nature (London)* **262,** 590–593.

De Jonge, H. R. (1981). Cyclic GMP-dependent protein kinase in intestinal brush borders. *Adv. Cyclic Nucleotide Res.* **14,** 315–333.

De Jonge, H. R. (1984a). The mechanism of action of Escherichia coli heat-stable toxin. *Biochem. Soc. Trans.* **12,** 180–184.

De Jonge, H. R. (1984b). Cyclic nucleotide-dependent protein phosphorylation in intestinal epithelium. *Kroc Found. Ser.* **17,** 263–286.

De Jonge, H. R., and Lohmann, S. M. (1985). Mechanisms by which cyclic nucleotides and other intracellular mediators regulate secretion. *Ciba Found. Symp.* **112,** 116–138.

De Jonge, H. R., and Rao, M. C. (1990). Cyclic nucleotide-dependent kinases. *In* "Text Book of Secretory Diarrhea" (E. Lebenthal and M. Duffey, eds.), pp. 191–207. Raven Press, New York.

De Jonge, H. R., Vaandrager, A. B., O'Grady, S. M., and Field, M. (1986). A 50 kDa protein in flounder intestine brush borders is phosphorylated by cGMP and Ca-CaM kinases and is specifically dephosphorylated by a cAMP-activated phosphatase. *Fed. Proc. Fed. Am. Soc. Exp. Biol.* **45,** 4281 (abstr.).

De Jonge, H. R., Bijman, J., and Sinaasappel, M. (1987). Relation of regulatory enzyme levels to chloride transport in intestinal epithelial cells. *Pediatr. Pulmonol. Suppl.* **1,** 54–57.

De Jonge, H. R., Van den Berghe, N., Tilly, B. C., Kansen, M., and Bijman, J. (1989). (Dys)regulation of epithelial chloride channels. *Biochem. Soc. Trans.* **17,** 816–818.

De Jonge, H. R., Bot, A. G. M., Bijman, J., Veeze, H. J., and Sinaasappel, M. (1992). CFTR dysfunction in CF intestine. *Pediatr. Pulmonol. Suppl.* **8,** 272 (abstr.).

Denning, G. M., Ostegaard, L. S., Cheng, S. H., Smith, A. E., and Welsh, M. J. (1992). Localization of cystic fibrosis transmembrane conductance regulator in chloride secretory epithelia. *J. Clin. Invest.* **89,** 339–349.

De Sauvage, F. J., Horuk, R., Bennett, G., Quan, C., Burnier, J. P., and Goeddel, D. V. (1992a). Characterization of the recombinant human receptor for *Escherichia coli* heat-stable enterotoxin. *J. Biol. Chem.* **267,** 6479–6482.

De Sauvage, F. J., Keshav, S., Kuang, W-J., Gillett, N., Henzel, W., and Goeddel, D. V. (1992b). Precursor structure, expression, and tissue distribution of human guanylin. *Proc. Natl. Acad. Sci. USA* **89,** 9089–9093.

Donowitz, M., and Welsh, M. J. (1987). Regulation of mammalian small intestinal electrolyte secretion. *In* "Physiology of the Gastrointestinal Tract" (L. R. Johnson, ed.), 2nd ed., pp. 1351–1388. Raven Press, New York.

Donowitz, M., Cohen, M. E., Gould, M., and Sharp, G. W. G. (1989). Elevated intracellular Ca^{2+} acts through protein kinase C to regulate rabbit ileal NaCl absorption. *J. Clin. Invest.* **83,** 1953–1962.

Edelman, A. M., Blumenthal, D. K., and Krebs, E. G. (1987). Protein serine/threonine kinases. *Annu. Rev. Biochem.* **56,** 567–613.

Field, M. (1981). Secretion of electrolytes and water by mammalian small intestine. *In* "Physiology of the Gastrointestinal Tract" (L. R. Johnson, ed.), 1st ed., pp. 963–982. Raven Press, New York.

Field, M., Graf, L. H., Jr., Laird, W. J., and Smith, P. L. (1978). Heat-stable enterotoxin of *Escherichia coli:* In vitro effects on guanylate cyclase activity, cyclic GMP concentration and ion transport. *Proc. Natl. Acad. Sci. U.S.A.* **75,** 2800–2804.

Field, M., Rao, M. C., and Chang, E. B. (1989). Intestinal electrolyte transport and diarrheal disease (parts I and II). *N. Engl. J. Med.* **321,** 800–806, 879–883.

Forte, L. R., Krause, W. J., and Freeman, R. H. (1988). Receptors and cGMP signalling mechanism for *E. coli* enterotoxin in opossum kidney. *Am. J. Physiol.* **255,** F1040–F1046.

Forte, L. R., Thorne, P. K., Eber, S. L., Krause, W. J., Freeman, R. H., Francis, S. H., and Corbin, J. D. (1992). Stimulation of intestinal Cl transport by heat-stable enterotoxin: Activation of cAMP-dependent protein kinase by cGMP. *Am. J. Physiol.* **263,** C607–C615.

French, P. J., Scholte, B. J., De Jonge, H. R., and Bijman, J. (1993). Protein kinase G regulation of CFTR. *Pediatr. Pulmonol. Suppl.* **9,** 226–227 (abstract).

Fuller, C. M., and Benos, D. J. (1992). CFTR ! *Am. J. Physiol.* **263,** C267–C286.

Giannella, R. A. (1981). Pathogenesis of acute bacterial diarrheal disorders. *Annu. Rev. Med.* **32,** 341–357.

Goyal, J., Ganguly, N. K., Mahajan, R. C., Garg, U. C., and Walia, B. N. S. (1987). Studies on the mechanism of *Escherichia coli* heat-stable enterotoxin-induced diarrhoea in mice. *Biochim. Biophys. Acta* **925,** 341–346.

Guandalini, S., Rao, M. C., Smith, P. L., and Field, M. (1982). cGMP modulation of ileal ion transport: In vitro effects of *Escherichia coli* heat-stable enterotoxin. *Am. J. Physiol.* **243,** G36–G41.

Guarino, A., Cohen, M. B., Overmann, G., Thompson, M. R., and Giannella, R. A. (1987). Binding of *E. coli* heat-stable enterotoxin to rat intestinal brush borders and to basolateral membranes. *Dig. Dis. Sci.* **32,** 1017–1026.

Guerrant, R. L., Hughes, J. M., Chang, B., Robertson, D. C., and Murad, F. (1980). Activation of intestinal guanylate cyclase by heat-stable enterotoxin of *Escherichia coli:* Studies of tissue specificity, potential receptors, and intermediates. *J. Infect. Dis.* **142,** 220–228.

Hirschhorn, N., and Greenough, W. B. (1991). Progress in oral rehydration therapy. *Sci. Am.* **264**(5), 50–56.

Hofmann, F., Dostmann, W., Keilbach, A. Landgraf, W., and Ruth, P. (1992). Structure and physiological role of cGMP-dependent protein kinase. *Biochim. Biophys. Acta* **1135,** 51–60.

Hughes, J. M., Murad, F., Chang, B., and Guerrant, R. L. (1978). Role of cyclic GMP in the action of heat-stable enterotoxin of *Escherichia coli. Nature* (*London*) **271,** 755–756.

Huott, P. A., Liu, W., McRoberts, J. A., Giannella, R. A., and Dharmsathaphorn, K. (1988). Mechanism of action of Escherichia coli heat stable enterotoxin in a human colonic cell line. *J. Clin. Invest.* **82,** 514–523.

Karnaky, K., Jr., King, J., and Currie, M. (1992). Immunocytochemical localization of guanylin in the epithelium of rat intestine. *Pediatr. Pulmonol., Suppl.* **8,** 273 (abstr.).

Kartner, N., Augustinas, O., Jensen, T. J., Naismith, A. L., and Riordan, J. R. (1992). Mislocation of ΔF508 CFTR in cystic fibrosis sweat gland. *Nat. Gene.* **1,** 321–327.

Katwa, L. C., and White, A. A. (1992). Presence of functional receptors for the *Escherichia coli* heat-stable enterotoxin in the gastrointestinal tract of the chicken. *Infect. Immun.* **60,** 3546–3551.

Kaufman, S., and Monckton, E. (1988). Influence of right atrial stretch and atrial natriuretic factor on rat intestinal fluid content. *J. Physiol.* (*London*) **402,** 1–8.

Knoop, F. C., Owens, M., Marcus, J. N., and Murphy, B. (1991). Elevation of calcium in rat enterocytes by *Escherichia coli* heat-stable (STa) enterotoxin. *Curr. Microbiol.* **23,** 291–296.

Krause, W. J., Freeman, R. H., and Forte, L. R. (1990). Autoradiographic demonstration of specific binding sites for *E. coli* enterotoxin in various epithelia of the North American opossum. *Cell Tissue Res.* **260,** 387–394.

Kuno, T., Kamisaki, Y., Waldman, S. A., Gariepy, J., Schoolnik, G., and Murad, F. (1986). Characterization of the receptor for heat-stable enterotoxin from *Escherichia coli* in rat intestine. *J. Biol. Chem.* **261,** 1470–1476.

Laney, D. W., Jr., Mann, E. A., Dellon, S., Perkins, D. R., Giannella, R. A., and Cohen, M. B. (1992). Novel sites for expression of an *Escherichia coli* heat-stable enterotoxin receptor in the developing rat. *Am. J. Physiol.* **263,** G816–G821.

Levine, S. A., Donowitz, M., Watson, A. J. M., Sharp, G. W. G., Crane, J. K., and Weikel, C. S. (1991). Characterization of the synergistic interaction of *Escherichia coli* heat-stable enterotoxin and carbachol. *Am. J. Physiol.* **261,** G592–G601.

Light, D. B., Corbin, J. D., and Stanton, B. A. (1990). Dual ion-channel regulation by cyclic GMP and cyclic GMP-dependent protein kinase. *Nature* (*London*) **344,** 336–339.

Lin, M., Nairn, A. C., and Guggino, S. (1992). cGMP-dependent protein kinase regulation of a chloride channel in T84 cells. *Am. J. Physiol.* **262,** C1304–C1312.

Mann, E. A., Cohen, M. B., and Giannella, R. A. (1993). Comparison of receptors for *Escherichia coli* heat-stable enterotoxin: novel receptor present in IEC-6 cells. *Am. J. Physiol.* **264,** G172–G178.

Matthews, J. B., Awtrey, C. S., and Madara, J. L. (1992). Microfilament-dependent activation of Na/K/2Cl cotransport by cAMP in intestinal epithelial monolayers. *J. Clin. Invest.* **90,** 1608–1613.

McIntosh, I., and Cutting, G. R. (1992). Cystic fibrosis transmembrane conductance regulator and the etiology and pathogenesis of cystic fibrosis. *FASEB J.* **6,** 2775–2782.

McKie, A. T., Kusel, M., McEwan, G. T. A., and Lucas, M. L. (1988). The effect of heat-stable *Escherichia coli* enterotoxin, theophylline, and forskolin on cyclic nucleotide levels and mucosal surface (acid microclimate) pH in rat proximal jejunum in vivo. *Biochim. Biophys. Acta* **971,** 325–331.

Merry, P. F., Lohmann, S. M., Walter, U., and Fischmeister, R. (1991). Ca current is regulated by cyclic GMP-dependent protein kinase in mammalian cardiac myocytes. *Proc. Natl. Acad. Sci. U.S.A.* **88,** 1197–1201.

Moore, R. B., Hulgan, T. M., Lincoln, T. M., Jenkins, L. D., and Shriver, S. K. (1992). Calpromotin, an activator of Ca^{2+}-dependent K^+ transport, is a substrate for cGMP-dependent protein kinase. *Proc. Inter. Conf. Second Messenger Phosphoprotein Res., 8th,* Glasgow, *1992,* Abstr. No. D145T.

Moriarty, K. J., Higgs, N. B., Lees, M., Tonge, A., Wardle, T. D., and Warhurst, G. (1990). Influence of atrial natriuretic peptide on mammalian large intestine. *Gastroenterology* **98,** 647–653.

Murer, H., Ammann, E., Biber, J., and Hopfer, U. (1976). The surface membrane of the small intestinal epithelial cell. I. Localization of adenyl cyclase. *Biochim. Biophys. Acta* **433,** 509–519.

Nairn, A. C., Hemmings, H. C., Jr., and Greengard, P. (1985). Protein kinases in the brain. *Annu. Rev. Biochem.* **54,** 931–976.

Nobles, M., Diener, M., and Rummel, W. (1991). Segment-specific effects of the heat-stable enterotoxin of *E. coli* on electrolyte transport in the rat colon. *Eur. J. Pharmacol.* **202,** 201–211.

O'Grady, S. M., and Wolters, P. J. (1990). Evidence for chloride secretion in the intestine of the winter flounder. *Am. J. Physiol.* **258,** C243–C247.

O'Grady, S. M., Field, M., Nash, N. T., and Rao, M. C. (1985). Atrial natriuretic factor inhibits Na-K-Cl cotransport in teleost intestine. *Am. J. Physiol.* **249,** C531–C534.

O'Grady, S. M., De Jonge, H. R., Vaandrager, A. B., and Field, M. (1988). Cyclic nucleotide-dependent protein kinase inhibition by H-8, effects on ion transport. *Am. J. Physiol.* **254,** C115–C121.

O'Grady, S. M., Cooper, K. E., and Rae, J. L. (1991). Cyclic GMP regulation of a voltage-activated K channel in dissociated enterocytes. *J. Membr. Biol.* **124,** 159–167.

O'Loughlin, E. V., Hunt, D. M., Gaskin, K. J., Stiel, D., Bruzuszcak, I. M., Martin, H. C. O., Bambach, C., and Smith, R. (1991). Abnormal epithelial transport in cystic fibrosis jejunum. *Am. J. Physiol.* **260,** G758–G763.

Picciotto, M. R., Cohn, J. A., Bertuzzi, G., Greengard, P., and Nairn, A. C. (1992). Phosphorylation of the cystic fibrosis transmembrane conductance regulator. *J. Biol. Chem.* **267,** 12742–12752.

Powell, D. W. (1987). Intestinal water and electrolyte transport. *In* "Physiology of the Gastrointestinal Tract" (L. R. Johnson, ed.), 2nd ed., pp. 1267–1305. Raven Press, New York.

Puchelle, E., Gaillard, D., Ploton, D., Hinnrasky, J., Fuchey, C., Boutterin, M.-C., Jacquot, J., Dreyer, D., Pavirani, A., and Dalemans, W. (1992). Differential localization of the cystic fibrosis transmembrane conductance regulator in normal and cystic fibrosis airway epithelium. *Am. J. Respir. Cell Mol. Biol.* **7,** 485–491.

Rao, M. C. (1985). Toxins that activate guanylate cyclase: heat-stable enterotoxins. *Ciba Found. Symp.* **112,** 74–87.

Rao, M. C., and Nash, N. T. (1988). 8-BrcAMP does not affect Na-K-2Cl cotransport in winter flounder intestine. *Am. J. Physiol.* **255,** C246–C251.

Rao, M. C., Guandalini, S., Smith, P. L., and Field, M. (1980). Mode of action of heat-stable *E. coli* enterotoxin: Tissue and subcellular specificities and role of cyclic GMP. *Biochim. Biophys. Acta* **632,** 35–46.

Rao, M. C., Nash, N. T., and Field, M. (1984). Differing effects of cGMP and cAMP on ion transport across flounder intestine. *Am. J. Physiol.* **246,** C167–C171.

Rashatwar, S. S., Cornwell, T. L., and Lincoln, T. M. (1987). Effect of 8-bromo-cGMP on Ca levels in vascular smooth muscle cells: Possible regulation of Ca-ATPase by cGMP-dependent protein kinase. *Proc. Natl. Acad. Sci. U.S.A.* **84,** 5685–5689.

Reinhard, M., Halbrügge, M., Scheer, U., Wiegeland, C., Jockusch, B. M., and Walter, U. (1992). The 46/50 kDa phosphoprotein VASP purified from human platelets is a novel protein associated with actin filaments and focal contacts. *EMBO J.* **11,** 2063–2070.

Ruth, P., Wang, G-X., Boekhoff, I., May, B., Pfeifer, A., Penner, R., Korth, M., Breer, H., and Hofmann, F. (1993). Transfected cGMP-dependent protein kinase suppresses calcium transients by inhibition of inositol 1,4,5-trisphosphate production. *Proc. Natl. Acad. Sci. USA* **90,** 2623–2627.

Sandberg, M., Natarajan, V., Ronender, S., Kalderon, D., Walter, U., Lohmann, S. M., and Jahnsen, T. (1989). Molecular cloning and predicted full length amino acid sequence of the Iβ isozyme of cGMP-dependent protein kinase from human placenta. *FEBS Lett.* **255,** 321–326.

Sarcevic, B., Brookes, V., Martin, T. J., Kemp, B. E., and Robinson, P. J. (1989). Atrial natriuretic peptide-dependent phosphorylation of smooth muscle particulate fraction proteins is modulated by cGMP-dependent protein kinase. *J. Biol. Chem.* **264,** 20648–20654.

Schulz, S., Green, C. K., Yuen, P. S. T., and Garbers, D. L. (1990). Guanyl cyclase is a heat-stable enterotoxin receptor. *Cell (Cambridge, Mass.)* **63,** 941–948.

Schulz, S., Chrisman, T. D., and Garbers, D. L. (1992). Cloning and expression of guanylin: Its existence in various mammalian tissues. *J. Biol. Chem.* **267,** 16019–16021.

Semrad, C. E., and Chang, E. B. (1987). Calcium-mediated cyclic AMP inhibition of Na-H exchange in small intestine. *Am. J. Physiol.* **252,** C315–C322.

Semrad, C. E., Cragoe, E. J., Jr., and Chang, E. B. (1990). Inhibition of Na/H exchange in avian intestine by atrial natriuretic factor. *J. Clin. Invest.* **86,** 585–591.

Sharkey, K. A., Gall, D. G., and MacNaughton, W. K. (1991). Distribution and function of brain natriuretic peptide in the stomach and small intestine of the rat. *Regul. Pept.* **34,** 61–70.

Shimada, T., and Hoshi, T. (1988). Na-dependent elevation of the acidic cell surface pH (microclimate pH) of rat jejunal villus cells induced by cyclic nucleotides and phorbol ester: Possible mediators of the regulation of the Na/H antiporter. *Biochim. Biophys. Acta* **937,** 328–334.

Smith, J. J., and Welsh, M. J. (1992). cAMP stimulates bicarbonate secretion across normal, but not cystic fibrosis airway epithelia. *J. Clin. Invest.* **89,** 1148–1153.

Snouwaert, J. N., Brigman, K. K., Latour, A. M., Malouf, N. N., Boucher, R. C., Smithies, O., and Koller, B. H. (1992). An animal model for cystic fibrosis made by gene targeting. *Science* **257,** 1083–1088.

Stewart, C. P., and Turnberg, L. A. (1989). A microelectrode study of responses to secretagogues by epithelial cells on villus and crypt of rat small intestine. *Am. J. Physiol.* **257,** G334–G343.

Sundaram, U., Knickelbein, R. G., and Dobbins, J. W. (1991). Mechanism of intestinal secretion: Effect of cyclic AMP on rabbit ileal crypt and villus cells. *Proc. Natl. Acad. Sci. USA* **88,** 6249–6253.

Takai, Y., Kaibuchi, K., Matsubara, T., and Nishizuka, Y. (1981). Inhibitory action of guanosine 3′,5′-monophosphate on thrombin-induced phosphatidyl inositol turnover and protein phosphorylation in human platelets. *Biochem. Biophys. Res. Commun.* **101,** 61–67.

Takens-Kwak, B. R., and Jongsma, H. J. (1992). Cardiac gap junctions: Three distinct single channel conductances and their modulation by phosphorylating treatments. *Pflüegers Arch.* **422,** 198–200.

Taylor, C. J., Baxter, P. S., Hardcastle, J., and Hardcastle, P. T. (1988). Failure to induce secretion in jejunal biopsies from children with cystic fibrosis. *Gut* **29,** 957–962.

Toskulkao, C., and Rao, M. C. (1990). Identification of a 50 kDa Ca-, cAMP-, and cGMP-dependent epithelial phosphoprotein as a cAMP regulatory protein. *Am. J. Physiol.* **258,** C889–C901.

Trezise, A. E. O., and Buchwald, M. (1991). *In vivo* cell-specific expression of the cystic fibrosis transmembrane conductance regulator. *Nature* (*London*) **353,** 434–437.

Tse, C. M., Ma, A. I., Yang, V. W., Watson, A. J. M., Levine, S., Montrose, M. H., Potter, J., Sardet, C., Pouyssegur, J., and Donowitz, M. (1991). Molecular cloning and expression of a cDNA encoding the rabbit ileal villus basolateral membrane Na/H exchanger. *EMBO J.* **10,** 1957–1967.

Tse, C. M., Brant, S. R., Walker, S., Pouyssegur, J., and Donowitz, M. (1992). Cloning and sequencing of a rabbit cDNA encoding an intestinal and kidney-specific Na/H exchanger isoform (NHE-3). *J. Biol. Chem.* **267,** 9340–9346.

Tse, C. M., Levine, S. A., Yun, C. H. C., Montrose, M. H., Little, P. J., Pouyssegur, J., and Donowitz, M. (1993). Cloning and expression of a rabbit cDNA encoding a serum-activated ethylisopropylamiloride-resistant epithelial Na/H exchanger isoform (NHE-2). *J. Biol. Chem.* **268,** 11917–11924.

Vaandrager, A. B. (1987). Ion transport regulation in intestinal brush border membranes. Ph.D. Thesis, Erasmus University, Rotterdam.

Vaandrager, A. B., and De Jonge, H. R. (1988). A sensitive technique for the determination of anion exchange activities in brush-border membrane vesicles. Evidence for two exchangers with different affinities for HCO_3 and SITS in rat intestinal epithelium. *Biochim. Biophys. Acta* **939,** 305–314.

Vaandrager, A. B., Ploemacher, M. C., and De Jonge, H. R. (1990). Phosphoinositide metabolism in intestinal brush borders: Stimulation of IP_3 formation by guanine nucleotides and Ca^{2+}. *Am. J. Physiol.* **259,** G410–G419.

Vaandrager, A. B., Bot, A. G. M., De Vente, J., and De Jonge, H. R. (1992a). Atriopeptins and *Escherichia coli* enterotoxin STa have different sites of action in mammalian intestine. *Gastroenterology* **102,** 1161–1169.

Vaandrager, A. B., Van Den Berghe, N., Bot, A. G. M., and De Jonge, H. R. (1992b). Phorbol esters stimulate and inhibit Cl^- secretion by different mechanisms in a colonic cell line. *Am. J. Physiol.* **262,** G249–G256.

Vaandrager, A. B., Schulz, S., De Jonge, H. R., and Garbers, D. L. (1993a). Guanylyl cyclase-C is an N-linked glycoprotein receptor that accounts for multiple heat-stable enterotoxin binding proteins in the intestine. *J. Biol. Chem.* **268,** 2174–2179.

Vaandrager, A. B., Van der Wiel, E., and De Jonge, H. R. (1993b). Heat-stable enterotoxin activation of immunopurified guanylyl cyclase C: Modulation by adenine nucleotides. *J. Biol. Chem.* **268,** 19598–19603.

Van Den Berghe, N. (1992). Signal transduction pathways involved in intestinal salt and water secretion. Ph.D. Thesis, Erasmus University, Rotterdam.

Van Dommelen, F. S., and De Jonge, H. R. (1986). Local changes in fractional saturation of cGMP and cAMP-receptors in intestinal microvilli in response to cholera toxin and heat-stable *Escherichia coli* toxin. *Biochim. Biophys. Acta* **886,** 135–142.

Veeze, H. J., Sinaasappel, M., Bijman, J., Bouquet, J., and De Jonge, H. R. (1991). Ion transport abnormalities in rectal suction biopsies from children with cystic fibrosis. *Gastroenterology* **101,** 398–403.

Waldman, S. A., Rapoport, R. M., and Murad, F. (1984). Atrial natriuretic factor selectively activates particulate guanylate cyclase and elevates cyclic GMP in rat tissues. *J. Biol. Chem.* **259,** 14332–14334.

Waldmann, R., and Walter, U. (1989). Cyclic nucleotide-elevating vasodilators inhibit platelet aggregation at an early step of the activation cascade. *Eur. J. Pharmacol.* **159,** 317–320.

Walling, M. W., Mircheff, A. K., Van Os, C. H., and Wright, E. M. (1978). Subcellular distribution of nucleotide cyclases in rat intestinal epithelium. *Am. J. Physiol.* **235,** E539–E545.

Weinman, E. J., Dubinsky, W. P., and Shenolikar, S. (1988). Reconstitution of cAMP-dependent protein kinase regulated renal Na-H exchanger. *J. Membr. Biol.* **101,** 11–18.

Wernet, W., Flockerzi, V., and Hofmann, F. (1989). The cDNA of the two isoforms of bovine cGMP-dependent protein kinase. *FEBS Lett.* **251,** 191–196.

Wiegeland, R. C., Kato, J., and Currie, M. G. (1992). Rat guanylin cDNA: characterization of the precursor of an endogenous activator of intestinal guanylate cyclase. *Biochem. Biophys. Res. Commun.* **185,** 812–817.

Wilson, K. T., Xie, Y., Musch, M. W., and Chang, E. B. (1993a). Sodium nitroprusside stimulates anion secretion and inhibits chloride absorption in rat colon. *J. Pharmacol. Exp. Ther.* **266,** 224–230.

Wilson, K. T., Vaandrager, A. B., De Vente, J., Musch, M. W., De Jonge, H. R., and Chang, E. B. (1993b). Second messenger pathways involved in sodium nitroprusside-stimulated colonic electrolyte transport: Localization of cyclic nucleotide and PGE_2 production in the subepithelium. *Gastroenterology,* **104,** A290.

Wolfe, L., Corbin, J. D., and Francis, S. H. (1989). Characterization of a novel isozyme of cGMP-dependent protein kinase from bovine aorta. *J. Biol. Chem.* **264** 7734–7741.

Wong, S. K. F., and Garbers, D. L. (1992). Receptor guanylyl cyclases. *J. Clin. Invest.* **90,** 299–305.

Wyatt, T. A., Lincoln, T. M., and Pryzwansky, K. B. (1991). Vimentin is transiently co-localized with and phosphorylated by cyclic GMP-dependent protein kinase in formyl-peptide-stimulated neutrophils. *J. Biol. Chem.* **266,** 21274–21280.

Yoshida, Y., Sun, H. T., Cai, J. Q., and Imai, S. (1991). Cyclic GMP-dependent protein kinase stimulates the plasma membrane Ca pump ATPase of vascular smooth muscle via phosphorylation of a 240 kDa protein. *J. Biol. Chem.* **266,** 19819–19825.

Yun, C. H. C., Gurubhagavatula, S., Levine, S. A., Montgomery, J. L. M., Brant, S. R., Cohen, M. E., Cragoe, E. J., Jr., Pouyssegur, J., Tse, C. M., and Donowitz, M. (1993). Glucocorticoid stimulation of ileal Na absorptive cell brush border Na/H exchange and association with an increase in message for NHE-3, an epithelial Na/H exchanger isoform. *J. Biol. Chem.* **268,** 206–211.

Cyclic GMP in Lower Forms[1]

Joachim E. Schultz and Susanne Klumpp
Abteilung Biochemie
Pharmazeutisches Institut der Universität
72076 Tübingen, Germany

I. Introduction

Essentially all chapters of this volume document the enormous progress that has been achieved during the last decade in our understanding of the regulation and possible function of cGMP and guanylyl cyclases in mammalian systems. In contrast, a single chapter suffices to summarize what is known about the cGMP system in lower organisms. This obvious imbalance more or less correctly reflects the current status in our knowledge of this second messenger system in higher and lower forms of life. In a speculative way one may take this imbalance as an indication for a less prominent biological role of cGMP in lower organisms, at least a less obvious one. It represents a challenging problem to explore in more detail when and where in the evolution the cGMP system developed as a distinct regulatory system. One would also like to know the kind of functions that are affected and regulated by those primordial cGMP systems. Finally, it would be interesting to compare enzymes of cGMP metabolism such as guanylyl cyclase and cGMP phosphodiesterase from lower organisms with those isolated and cloned from mammals.

In the context of this chapter lower forms of life will mainly, but not exclusively, be defined as single-celled organisms. We will briefly review the rather scanty literature on cGMP in bacteria, which stems in its majority from the seventies, and then describe two systems that have been studied for quite some time, the slime mold *Dictyostelium discoideum* and the protozoans *Paramecium* and *Tetrahymena*. Finally, a few studies on other lower forms are evaluated.

[1] This chapter is dedicated to Professor Dr. H. J. Roth on the occasion of his 65th birthday.

Advances in Pharmacology, Volume 26

II. Bacteria

The early recognition of the presence and importance of cAMP in virtually all cells led to the assumption that a similar universality will be applicable to the guanine congener of cAMP (Goldberg *et al.*, 1973). Using analytical methods for cGMP available in the early seventies, cGMP was found to be present at 5 to 40 n*M* concentrations of *Escherichia coli, Bacillus licheniformis,* and *B. megaterium* (Bernlohr *et al.*, 1974; Gonzales and Peterkofsky, 1975; Setlow and Setlow, 1978). Taking into account the volume of a bacterial cell of approximately 1 fL (Gonzales and Peterkofsky, 1975), this cGMP concentration translates into 2–15 molecules per cell. In later investigations more sophisticated analytical techniques were used to quantify cGMP in order to exclude as much as possible assay artifacts due to unspecific cGMP immunoreactivity (Setlow and Setlow, 1978; Shibuya *et al.*, 1977). In these studies minute quantities of cGMP were identified within cells (1–4 pmol cGMP/mg of protein, 1 mg of protein corresponding to about 10^{13} cells) and, in some instances, in the extracellular medium (200 n*M* at the end of the exponential growth phase) (Shibuya *et al.*, 1977; Cook *et al.*, 1980). Usually, the ratios of cAMP to cGMP levels were 100 : 1 to 1000 : 1. Some of the early contradictory data concerning fluctuations of cGMP levels during the bacterial cell cycle may be understood by those extremely low concentrations of cGMP and the methodological problems involved in their reliable quantitation. In 1977, Shibuya *et al.* convincingly demonstrated that in *E. coli* the adenylyl cyclase is most likely also responsible for biosynthesis of the small quantities of cGMP. These authors used various adenylyl cyclase deletion and overexpression mutants of *E. coli* and showed that the concentration of cGMP was directly correlated to the amount of functional adenylyl cyclase. Therefore, the conclusion seems warranted that the miniscule quantities of cGMP in *E. coli* (and perhaps in other eubacteria) reflect a small side activity of the bacterial adenylyl cyclase, which uses GTP as a substrate. Understandably then, a relevant physiological function for cGMP in eubacteria is unknown. This statement is valid although the purification of a guanylyl cyclase from *E. coli* to apparent homogeneity has been reported. The pure 30-kDa enzyme had a specific activity of 2.6 nmol/mg min^{-1}. It did not convert ATP to cAMP (Macchia *et al.*, 1975). The true function of this protein in cellular metabolism has not yet been revealed. The same applies to a partially purified guanylyl cyclases from *B. licheniformis* and *Caulobacter crescentus* (Clark and Bernlohr, 1972; Sun *et al.*, 1974) and for the reported guanylyl cyclase associated with hemagglutinating virus of Japan (Sendai virus) (Kimura *et al.*, 1981).

To establish a possible physiological role for cGMP in bacterial metabolism, the effect of the membrane-permeable dibutyryl derivative of cGMP or even cGMP itself on selected cellular events was studied (Schmidt and Samuelson, 1972; Lim *et al.*, 1979; Majumdar and Bose, 1985). Although exogenous cGMP appeared to exert strikingly specific effects compared with those of cAMP, the concentrations of cGMP used in these experiments (0.1 to 5 m*M*) make any suggestions about its physiological role highly speculative.

Using dibutyryl cGMP and the phosphodiesterase inhibitor IBMX (3-isobutyl-1-methylxanthine) Schimz and Hildebrand (1987; Schimz *et al.*, 1989) have attempted to link transduction of a light signal in the archaebacterium *Halobacterium halobium* to a cGMP system. The working hypothesis that a G-protein controlled cGMP phosphodiesterase might be involved in sensory transduction in *Halobacterium* is somewhat reminiscent to established light processing reactions in vision. Data indicate that low levels of cGMP might activate, and high levels might inhibit, the flagellar motor switch (Schimz and Hildebrand, 1987; Schimz *et al.*, 1989). Clearly, more work is required until this hypothesis can be considered. What is needed foremost is unequivocal evidence that specific cGMP metabolizing enzymes, i.e., guanylyl cyclase and cGMP phosphodiesterase, are present in *Halobacterium*.

III. Slime Molds and Fungi

A. *Dictyostelium discoideum*

1. Receptor-Mediated Intracellular cGMP Formation

The slime mold *D. discoideum* is widely used as a model to study selected steps in differentiation. During the vegetative, amoeboid phase cells live independently from each other. On starvation these single cells transgress into an aggregation-competent stage and set up an amazing oscillatory system for intercellular communication. cAMP and cGMP are key players in this process, which ends in spore formation. Several substances have been identified to function as extracellular first messenger between cells, e.g., cAMP, folic acid, and the peptide glorin. Among the intracellular second messengers, which are involved in the aggregation process, cGMP plays an important role (Mato *et al.*, 1977a,b; Wurster *et al.*, 1977). Intracellular cGMP formation in *D. discoideum* is triggered by extracellular agents such as cAMP or folate via separate receptor pathways (Mato *et al.*, 1977a,b; Wurster *et al.*, 1977). Approximately 10^{-8} *M* cAMP or 7×10^{-6} *M* folate are necessary to elicit a half-maximal increase in cGMP (Fig. 1A) (Mato *et al.*, 1977a,b; Wurster *et al.*, 1977, 1979; van Haastert, 1983). The extracellular receptors for cAMP or folate, which are linked to activation of either intracellular adenylyl cyclase or guanylyl cyclase,

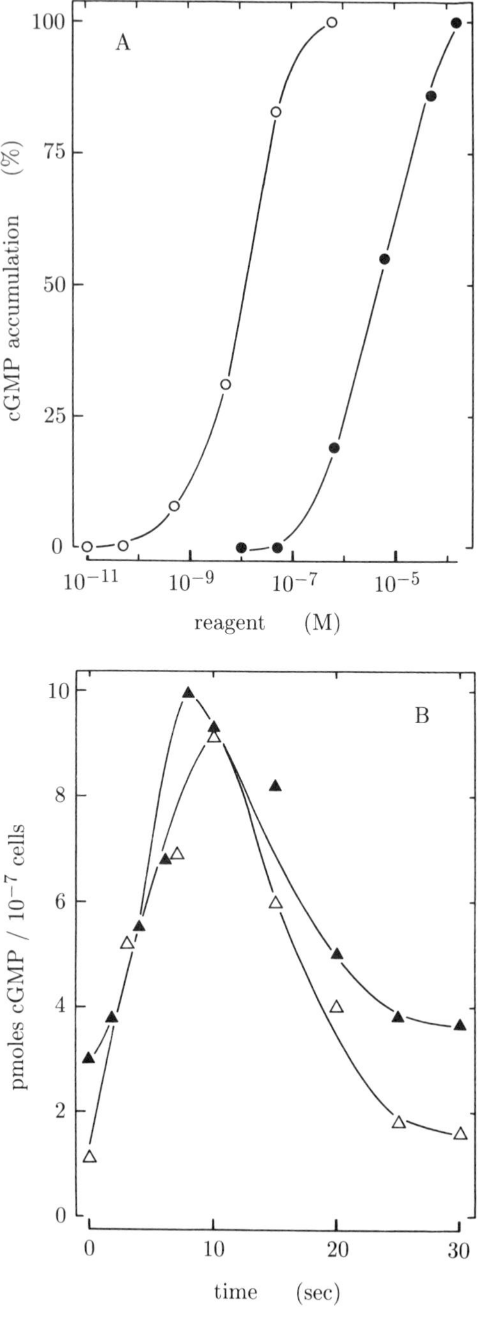

Fig. 1 cGMP accumulation in vegetative cells of *Dictyostelium discoideum*. (A) Concentration dependence of cAMP-mediated (○) and folic acid-mediated (●) cGMP formation. (B) Time course after stimulation with 5×10^{-8} *M* cAMP (△) and 5×10^{-5} *M* folic acid (△) (data replotted from Mato *et al.*, 1977a,b).

are distinct. For example, in starving, preaggregating cells folate will only stimulate cGMP formation; it has no effect on cAMP levels. The cAMP response to folate will, however, develop in these cells when pulses of folate are given at regular intervals over a period of 5 h (Wurster *et al.*, 1979).

Stimulated formation of intracellular cGMP shows a burst-like time course. For example, after addition of cAMP as stimulant, levels of cGMP increase about sevenfold within 10 s and decline to prestimulation levels after a further 20 s time period; i.e., the cGMP response is completed within only 30 s (Fig. 1B). This shortness of the cGMP response is not due to rapid removal of extracellular cAMP by degradation since the time course is unaltered when the nonhydrolyzable cAMP analogue adenosine-3′,5′-monophosphorothioate is used (van Haastert and van der Heijden, 1983; de Witt *et al.,* 1987). Similar data were obtained with folate as a stimulant. The transient nature of the cGMP increase is caused by a rapid fall in the rate of cGMP biosynthesis, which is due to receptor desensitization (van Haastert, 1983), and by the swift hydrolysis of intracellular cGMP by the cells' phosphodiesterases. Desensitization of cAMP receptors coupled to cGMP formation is almost complete within 30 s after addition (homologous desensitization). As cAMP is degraded by the extracellular phosphodiesterase, resensitization occurs within a few minutes. These processes have also been termed adaptation and deadaptation (van Haastert *et al.,* 1983, 1986). There is no cross-desensitization (heterologous desensitization) of cAMP and folate receptors coupled to cGMP formation. This demonstrates that both agents act via different surface receptors. Nevertheless, if maximal stimuli of cAMP and folate are applied simultaneously, they are not additive. This indicates that both stimulatory pathways converge on the same intracellular guanylyl cyclase (van Haastert, 1983).

2. Guanylyl Cyclase and Phosphodiesterase

Enzymic activity of guanylyl cyclase has been measured in cell-free homogenates from *D. discoideum*. Depending on the method used for preparation of cell homogenates, the ratios of soluble to membrane-bound enzyme activity varied considerably (Mato *et al.,* 1978a; Janssens *et al.,* 1987, 1989). It seems that guanylyl cyclase is only superficially attached to the membrane and that the amount which remains associated with the membrane after homogenization depends on the developmental stage of the cells (vegetative or aggregating) and the procedures used to prepare the homogenate (Mato *et al.,* 1978a; Janssens *et al.,* 1987). Guanylyl cyclase activity from *Dictyostelium* is reported to be very unstable (Ward and Brenner, 1977; Janssens and de Jong, 1988; Mato *et al.*, 1978a). The enzyme uses Mn^{2+} and Mg^{2+} as metal cofactor with equal efficiency (Janssens and de Jong, 1988; Mato *et al.*, 1978b). Specific activities of up to 100 pmol/mg min^{-1} have been reported (Janssens and de Jong, 1988; Janssens *et al.*, 1989). The activity measured with Mn-GTP is stimulated by micromolar concentrations of ATP and AMPPNP (Mato, 1979; Mato and Malchow, 1978; Padh and Brenner, 1984). On the other hand, guanylyl cyclase assayed with Mg-GTP as a substrate is activated by 1 μM GTPγS or 100μM guanosine 5′-(β, γ imido)triphosphate. GTP, guanosine mono- and diphosphates, or adenine nucleotides were inactive. These data provoked speculations about an involvement of G-proteins in guanylyl cyclase regulation (Janssens *et al.*, 1989). Ca^{2+}, even at submicromolar concentrations, appeared to inhibit strongly the Mg-dependent guanylyl cyclase of *D. discoideum* (Janssens *et al.*, 1989). The effect of Ca^{2+} ions was opposite that reported using permeabilized cells (Small *et al.*, 1987; Europe-Finner and Newell, 1985). Taken together it seems obvious that an effort must be made to purify the guanylyl cyclase from *Dictyostelium* in order to (*a*) obtain more reliable data about its regulation, (*b*) see how the coupling to an extracellular cAMP receptor to an intracellular guanylyl cyclase is accomplished, and (*c*) allow a meaningful comparison of the enzyme from the slime mold with isozymes obtained from metazoans.

Dictyostelium discoideum contains an intracellular phosphodiesterase that is highly specific for cGMP. This enzyme with a tentative molecular mass of 70 kDa is distinct from extracellular and intracellular cAMP phosphodiesterases (Dicou and Brachet, 1980; Bulgakov and van Haastert, 1983). On cell breakage, most cGMP phosphodiesterase activity is soluble. Enzyme characterization has not yet been carried out with a purified protein. The enzyme seems to have two binding sites for its substrate, which display disparate preferences for different cGMP derivatives. One site supposedly is used for allosteric activation of the phosphodiesterase and the other site represents the catalytic center. Activation by cGMP is envisaged to stabilize the enzyme in a state that has a higher affinity for cGMP at the catalytic center, i.e., decreasing K_m (Kesbeke *et al.*, 1985). cGMP activation of phosphodiesterase is half-maximal at 0.1 μM, whereas the K_m of the activated enzyme for cGMP is around 3 μM. It remains to be proven whether (*a*) the complex regulation and kinetics will also be apparent when a purified enzyme is available for testing and (*b*) whether this is of physiological importance. A mutant of *D. discoideum* that lacks intracellular cAMP phosphodiesterase exists (Darmon *et al.*, 1978). This should be very helpful for purification of the cGMP phosphodiesterase. In another mutant, cGMP phosphodiesterase activity is lacking (van Haastert *et al.*, 1982a; Ross and Newell, 1981; Coukell and Cameron,

1986; Menz *et al.*, 1991). This cell line, which shows an extended time course of stimulated cGMP formation, may prove valuable for studies concerned with the function of cGMP in *Dictyostelium*.

How is the cGMP signal intracellularly translated into a physiological reaction? For a long time, it was impossible to prove the occurrence of a cGMP-regulated protein kinase in *Dictyostelium*. Instead, the existence of intracellular soluble cGMP receptor proteins, which showed high affinity and specificity for this nucleotide, was demonstrated ($K_D = 1.4 \times 10^{-9}$ *M*, 3000 binding sites/cell; Mato *et al.*, 1978b; van Haastert *et al.*, 1982b). The physiological role of this cGMP-binding protein remains enigmatic. In 1990, Wanner and Wurster were able to obtain unequivocal evidence for the presence of a cGMP-activated protein kinase in *D. discoideum*. These authors isolated via cGMP-affinity chromatography a protein kinase of 72 kDa, which was, however, completely unresponsive to stimulation by cyclic nucleotides. An antiserum against this protein was raised and the immunoglobulin fraction was used to prepare an immunoaffinity matrix. From the crude supernatant of a cell homogenate an unstable protein kinase could thus be extracted,which was preferentially activated by cGMP. Therefore, it is likely that cGMP in *Dictyostelium* will regulate cellular processes via a cGMP-activated protein kinase. Whether this kinase is related to the binding proteins mentioned above has not been investigated. A future question will be the identification of specific protein substrates for the cGMP-activated protein kinase. Preliminary evidence exists which indicates that cGMP is involved in the regulation of the association of myosin heavy chains with the cytoskeleton via phosphorylation/dephosphorylation reactions (Liu and Newell, 1988, 1991). This would implicate cGMP in the regulation of mechanical processes, which are required during differentiation, such as development of pseudopods, directional chemotactic movement, and stalk formation.

B. Others

In addition to *D. discoideum* and some related *Dictyostelium* strains the presence of cGMP has also been reported in the slime molds *Physarum polycephalum* (Lovely and Threlfall, 1976) and *Polyspondylium violaceum* (Wurster *et al.*, 1978). In the latter cells glorin (*N*-propionyl-γ-L-glutamyl-L-ornithine-δ-lactam ethylester) stimulates intracellular cGMP formation via specific surface receptors much like cAMP does in *D. discoideum* (Wurster *et al.*, 1978; de Witt *et al.*, 1988).

In fungi, the presence of cGMP is not well established. Eckstein (1988) reported on the presence of cGMP in *Saccharomyces cerevisiae* (6–20 pmol/10^9 cells, depending on growth stage). Two phycomycetes, *Mucor racemosus* (Orlowski and Sypherd, 1976) and *Blastocladiella* (Silverman and Epstein, 1975), have also been reported to contain cGMP. The presumed presence and variations of cGMP in *Neurospora crassa* has been investigated by Rosenberg and Pall (1978, 1979) and by Hasunuma *et al.* (1987). The data do not permit reasonable speculations about a possible physiological role for cGMP in these organisms.

IV. Cyclic GMP in the Protozoans *Tetrahymena* and *Paramecium*

A. Regulation of Intracellular Cyclic GMP Formation

The somewhat related protozoans *Tetrahymena* and *Paramecium* are model systems for a variety of purposes. As far as regulation and function of the cGMP system are concerned, different approaches have been taken.

In *Tetrahymena,* levels of cGMP *in vivo* are about fourfold lower than those of cAMP. In heat-shock synchronized cells cGMP concentrations were found to fluctuate during the cell cycle between 1 and 6 pmol/10^6 cells (Graves *et al.*, 1976; Gray *et al.*, 1977; Zimmerman *et al.*, 1981). Zimmerman *et al.* (1981) used *Tetrahymena* in a pharmacological study and showed that 9 μM Δ^9-tetrahydrocannabinol significantly reduced levels of cGMP (and cAMP) during the cell cycle. Taken together, the data are not sufficient to seriously project a role for cGMP in the cell cycle.

In *Paramecium,* regulation of cGMP formation *in vivo* is being studied in the context of electrophysiological processes that govern the swimming behavior of this unicell (Majima *et al.*, 1986; Schultz *et al.*, 1986; Schultz and Schade, 1989a,b). In response to defined external stimuli, cells either hyperpolarize or generate regenerative and graded Ca^{2+}/K^+ action potentials that resemble the Na^+/K^+ action potential in higher systems such as nerve and muscle cells (Saimi and Kung, 1987). For example, Ba^{2+} ions provoke a violent electrophysiological response in *Paramecium:* trains of action potentials are fired (Oertel *et al.*, 1977). Ba^{2+} addition also triggers a rapid increase of cGMP levels, the extent of which strictly depends on the concentration of extracellular Ca^{2+}. Ba^{2+} ions enter the cells together with Ca^{2+} as charge carrier and inhibit the rapid inactivation of the Ca^{2+} channels. Thus, more Ca^{2+} can permeate into the cell. Because of the dependence of intracellular cGMP formation on external Ca^{2+} it can be concluded that actually that Ca^{2a+}, which enters the cell as inward current, is causally related to enhanced cGMP formation. This conclusion is supported by studies of cGMP metabolism in *Paramecium* mutants. A group of mutants with a defect in the opening of Ca^{2+} channels (pawns) cannot respond electrophysiologically and behaviorally to a Ba^{2+} stimulation (Oertel *et al.*, 1977). Likewise, cGMP is not increased in the mutants under these conditions (Schultz *et al.*, 1986). Veratridine, known to act as a Na^+ channel opener in metazoans, is capable of opening the defective Ca^{2+} channel in the pawn mutants, resulting in enhanced cGMP formation and backward swimming (Schultz and Schade, 1989a). In another group of mutants, dancer, normal inactivation of the voltage-operated Ca^{2+} channel is impaired. This leads to a prolonged Ca^{2+} influx during stimulation (Hinrichsen *et al.*, 1984). The enhanced Ca^{2+} influx into dancer mutant cells is accompanied by an enhanced cGMP formation and a correspondingly prolonged backward swimming response (Schultz *et al.*, 1986). The data obtained with these electrophysiologically characterized Ca^{2+} channel mutants of *Paramecium* unequivocally establish a cause relation effect between the influx of Ca^{2+} through the voltage-dependent Ca^{2+} channels, which are localized in the ciliary membrane (Dunlap, 1977), and intracellular formation of cGMP.

The turnover rates of cGMP in *Paramecium* are extremely high. When cGMP formation is elicited by a depolarizing Ba^{2+} stimulus in the presence of 5 m*M* IBMX, a competitive phosphodiesterase inhibitor, maximal cGMP levels of up to 500 pmol/mg are observed within 30 s compared to about 40 pmol/mg in the absence of the inhibitor (Fig. 2; Schultz and Schade, 1989b). The rate of rise is 25 pmol/s per mg. Addition of the inhibitor alone has only minor effects. This demonstrates that, on stimulation with Ba^{2+}, the rate of cGMP turnover is very high indeed. The peak level of about 40 pmol/mg, which is usually found 10 s after a standard stimulation (1 m*M* Ba^{2+} in the presence of 50 μM Ca^{2+} and in the absence of a phosphodiesterase inhibitor), turned over at least once every 2 s (Schultz and Schade, 1989b). This is reminiscent of the high turnover of cGMP in the retina, which was determined by incorporation of ^{18}O into GMP on hydrolysis of cGMP by ^{18}O-labeled water (Goldberg *et al.*, 1980). Turnover measurements of cGMP with the latter methodology in *Paramecium* fully substantiated the conclusion based on the experiments with the phosphodiesterase inhibitor IBMX (N. D. Goldberg, T. F. Walseth, and J. E. Schultz, unpublished).

Elevated cGMP levels return to prestimulation values within 5 min despite the continued presence of 1 m*M* Ba^{2+} as a stimulant (Fig. 2). Reinforcement of the stimulus by a further fivefold increase in Ba^{2+} is without effect. The lack of a cGMP response coincides with the absence of a physiological backward swimming response. This indicates that the voltage-operated Ca^{2+} channels have been inactivated during the initial stimulation and remain in this state during the continued presence of the stimulant. After stimulus removal Ca^{2+} channels recover time- and

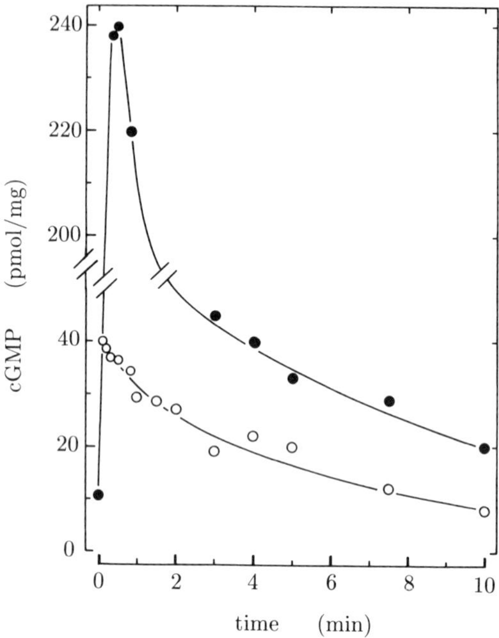

Fig. 2 Time course of Ba^{2+}-stimulated cGMP formation in *Paramecium*. Potentiation by the phosphodiesterase inhibitor isobutylmethylxanthine: (○) 5 m*M* Ba^{2+}; (●) 5 m*M* Ba^{2+} + 5 m*M* isobutylmethylxanthine.

temperature-dependently from an inactivated (≡ inoperable) to a closed (≡ operable) state. This is evident by the gradual recurrence of cGMP formation and of the behavioral response on renewed stimulation (Schultz and Schade, 1989b). These data once more demonstrate the causal link between the depolarizing Ca^{2+} influx and the enhancement of intracellular cGMP formation in *Paramecium*. It is tempting to construe a functional role for cGMP in inactivation of Ca^{2+} channels. However, several observations are not compatible with such a hypothesis, e.g., the time course of cGMP formation (range of seconds) and channel inactivation (range of minutes). A function for cGMP in ciliary motility of *Paramecium* was proposed by Majima *et al.* (1986) and by Bonini and Nelson (1988). Although this is a possibility, final conclusions concerning the role of cGMP in *Paramecium* must await further studies.

In concluding this section it should be noted that *Tetrahymena* has been reported to display properties of electrical excitability comparable to those of *Paramecium*, e.g., generation of a Ca^{2+}-dependent regenerative action potential (Onimaru *et al.*, 1980; Takahashi *et al.*, 1980). It may be reasonably anticipated that results of regulation of cGMP biosynthesis in *Tetrahymena in vivo* will display similarities to that of *Paramecium*. This expectation is supported by the many known properties that are shared by the guanylyl cyclases from both protozoans.

B. Guanylyl Cyclases

In *Tetrahymena* as well as in *Paramecium*, guanylyl cyclase activity is localized in membranes of cilia and cell bodies (Schultz *et al.*, 1983; Muto *et al.*, 1985). In *Tetrahymena*, Muto *et al.* (1985) reported a growth-dependent increase in the ratio of guanylyl cylase activity in cell body and ciliary fractions. In *Paramecium*, about 15% of total guanylyl cyclase activity is localized in the cilia; the remainder is found in the pellet of a centrifuged homogenate from deciliated cells. Fractionation of ciliary components by a discontinuous sucrose gradient showed that guanylyl cylase activity in both protozoans is associated with the membrane and not with axonemal components (Schultz *et al.*, 1983; Muto *et al.*, 1985). The specific activities in the ciliary membranes, using MgGTP as a substrate, is up to 1 nmol/mg min^{-1}.

In many metazoan cells, a requirement for Ca^{2+} in the regulation of cGMP levels *in vivo* has been demonstrated. However, in most cell-free preparations of guanylyl cyclase the concentrations of Ca^{2+} necessary to obtain inhibitory or stimulatory effects are in the millimolar range (Bartfai *et al.*, 1978; Chrisman *et al.*, 1975). A notable exception is the guanylyl cyclase from rod outer segments, a modified sensory 9 + 0 microtubular cilium, where guanylyl cyclase is inhibited by submicromolar Ca^{2+} (Koch and Stryer, 1988; Lambrecht and Koch, 1991). When the membranes of *Tetrahymena* are prepared in the presence of the Ca^{2+} chelator EGTA, the resulting guanylyl cyclase activity is 10 pmol/mg min^{-1} (Kakiuchi *et al.*, 1981). For restoration of full activity, addition of both Ca^{2+} and calmodulin is required. This effect is most specific for calmodulins from *Tetrahymena* and the related protozoan *Paramecium*. Calmodulins from other sources, e.g., bovine brain, sea anemone, *Dictyostelium*, spinach, or scallop, were inactive (Kakiuchi *et al.*, 1981; Kudo *et al.*, 1981, 1982). In fact, bovine brain calmodulin at 10 to 50 μg/assay inhibited the guanylyl cyclase activation by 1 μg of *Tetrahymena* calmodulin (Kudo *et al.*, 1983). It must be noted that the amount of protozoan calmodulin needed to activate *Tetrahymena* guanylyl cyclase is between 100- and 1000-fold higher than that needed for activation of several other calmodulin-dependent enzymes (Kakiuchi *et al.*, 1981; Kudo *et al.*, 1982). Another peculiarity is the rather high concentration of Ca^{2+} (80 μM), which is required for optimal activation (Kudo *et al.*, 1981). In comparison, calmodulin-dependent phosphodiesterase from bovine brain is maximally activated at 3 μM Ca^{2+} (Kakiuchi *et al.*, 1981).

The guanylyl cyclase from *Paramecium* is similar to the one from *Tetrahymena*. Removal of Ca^{2+} from the assay mixture reduces activity to approximately 20%. Ca^{2+} dose-dependently restores enzyme activity and half-maximal activation is obtained at a 10 μM concentration. Ca^{2+} concentrations exceeding 100 μM are inhibitory; Ca^{2+} increases V_{max} and K_m remains unaltered (Klumpp and Schultz, 1982). Ca^{2+} activation is very specific; Sr^{2+} is only half as active and Ba^{2+} is completely inactive. Furthermore, even a 100-fold excess of Ba^{2+} over Ca^{2+} in the incubations has no effect on the 50% effective dose for Ca^{2+} activation (Schultz *et al.*, 1986). Unlike *Tetrahymena*, preparation of ciliary membranes from *Paramecium* in the presence of 50 μM EGTA does not result in an enzyme preparation that is inactive, due to depletion of a Ca^{2+}-dependent regulator protein, e.g., calmodulin. Usage of 2 mM EGTA permanently inactivates guanylyl cyclase from *Paramecium*. Addition of calmodulin from *Tetrahymena* or *Paramecium* to ciliary membranes enhances guanylyl cyclase activity by about 20%. This small, but very reproducible, effect is absolutely restricted to calmodulins from the protozoans and is, in principle, comparable to the much larger effect on guanylyl cyclase from *Tetrahymena* (Schultz and Klumpp, 1984). Inspection of the amino acid sequences of *Paramecium* and *Tetrahymena* calmodulins and comparisons with calmodulins from other sources do not yield a ready clue for the unusual specificities of the protozoan calmodulins in activation of the guanylyl cyclase (Schaefer *et al.*, 1987; Yazawa *et al.*, 1981).

An interesting observation on the side is the fact that a brief incubation of ciliary membrane from *Paramecium* with micromolar concentrations of La^{3+} generates an inactive guanylyl cyclase that can be reactivated by calmodulins from many sources, not just protozoans, and by calmodulin fragments prepared from bovine brain calmodulin (Klumpp and Schultz, 1982; Klumpp *et al.*, 1987). Most surprisingly, reactivation is also accomplished by mixtures of the chelators EDTA or EGTA with either Ca^{2+} or Sr^{2+}. Antibodies generated against calmodulin from *Paramecium* or bovine brain inhibited membrane-bound and solubilized guanylyl cyclase from both protozoans by about 70% (Schultz *et al.*, 1983).

In summary, many experimental results indicate that calmodulin may be the physiologically relevant protein that confers Ca^{2+} senstivity to the protozoan guanylyl cyclases, which so far are the only ones activated by physiologically meaningful Ca^{2+} concentrations *in vitro*. However, given the existing oddities mentioned above, and in view of the existence of Ca^{2+} regulator proteins similar in structure, size, and function to calmodulin, e.g., recoverin and visinin (Lambrecht and Koch, 1991; Dizhoor *et al.*, 1991; Yamagata *et al.*, 1990), this should be reinvestigated in detail with a purified catalyst.

Twenty-two percent of the membrane-bound guanylyl cyclase from *Tetrahymena* can be solubilized by 0.5% digitonin and 20% glycerol. Attempts toward purification of the solubilized enzyme have been reported (Nagao and Nozawa, 1987). The *Paramecium* guanylyl cyclase seems to be glued into the membrane since it can be partially solubilized only by use of detergent and concomitant sonication (Schultz and Klumpp, 1991). Purification efforts are seriously hampered by the lack of a sufficient amount of starting material. So far, the following purification steps of the solubilized enzyme have been established: (*a*) DEAE-Trisacryl, (*b*) phenyl-sepharose, (*c*) hydoxylapatite, (*d*) S-300 gel filtration, (*e*) GTP-affinity chromatography, and (*f*) polylysine–agarose.

C. Phosphodiesterase and Cyclic GMP-Dependent Protein Kinase

Both *Tetrahymena* and *Paramecium* contain highly specific cGMP phosphodiesterases localized in cell bodies as well as in cilia (Kudo *et al.*, 1980, 1986; Schultz *et al.*, 1990). Detailed characterization and comparison of the protozoan enzymes with their metazoan counterparts have not yet been carried out. Further, cGMP-dependent protein kinases have been identified in cilia and cell bodies of both protozoans (Murofushi, 1974; Schultz and Jantzen, 1980; Lewis and Nelson, 1980; Eistetter *et al.*, 1983). The enzyme from *Paramecium* has now been purified and characterized (Miglietta and Nelson, 1988). Specific substrate proteins for cGMP-dependent protein kinases have not yet been identified in *Tetrahymena* or in *Paramecium*.

V. Cyclic GMP in Other Lower Forms

Surprisingly few studies have been carried out concerning cGMP in other lower organisms. In the context of growth and mating, cGMP levels were determined in the green alga *Chlamydomonas* (Sharaf and Rooney, 1982, 1985; Pasquale and Goodenough, 1987). No dramatic patterns of change during these processes were observed.

cGMP has been suggested as an intracellular mediator for actions of the eclosion hormone in the tabacco hornworm *Manduca sexta* and, similarly, in the giant silk moth *Antheraea polyphemus* (Morton and Truman, 1988; Schwartz and Truman, 1984). cGMP and guanylyl cyclase have also been identified in the pheromone-sensitive antennae of the silkmoths

Antheraea and *Bombyx mori* (Ziegelberger *et al.*, 1990). Female sex hormones applied to freshly isolated, living antennae of male silkmoths led to an increase on cGMP. However, based on these data, a physiological role for cGMP in these insects is not definable. Finally, Murtaugh *et al.* (1985) reported in a very peculiar extracellular localization of cGMP in the male accessory reproductive gland of the house cricket *Acheta domesticus* and its fate in mating. High concentrations of cGMP are stored extracellularly in a specialized structure, the handle-capillary tube, and are excreted with the insemination fluid at a concentration of 20 μM. The function of the secreted cGMP is unknown at present.

References

Bartfai, T., Breakefield, X. O., and Greengard, P. (1978). Regulation of synthesis of guanosine 3′:5′-cyclic monophosphate in neuroblastoma cells. *Biochem. J.* **176,** 119–127.

Bernlohr, R. W., Haddox, M. K., and Goldberg, N. D. (1974). Cyclic guanosine 3′:5′-monophosphate in *Escherichia coli* and *Bacillus licheniformis*. *J. Biol. Chem.* **249,** 4329–4331.

Bonini, N. M., and Nelson, D. L. (1988). Differential regulation of *Paramecium* ciliary motility by cAMP and cGMP. *J. Cell Biol.* **106,** 1615–1623.

Bulgakov, R., and van Haastert, P. J. M. (1983). Isolation and partial characterization of a cyclic GMP-dependent cyclic GMP-specific phosphodiesterase from *Dictostelium discoideum*. *Biochim. Biophys. Acta* **756,** 56–66.

Chrisman, T. D., Garbers, D. L., Parks, M. A., and Hardman, J. G. (1975). Characterization of particulate and soluble guanylate cyclases from rat lung. *J. Biol. Chem.* **250,** 374–381.

Clark, V. L., and Bernlohr, R. W. (1972). Guanyl cyclase of *Bacillus licheniformis*. *Biochem. Biophys. Res. Commun.* **46,** 1570–1575.

Cook, W. R., Kalb, V. F., Peace, A. A., and Bernlohr, R. W. (1980). Is cyclic guanosine 3′,5′-monophosphate a cell cycle regulator. *J. Bacteriol.* **141,** 1450–1453.

Coukell, M. B., and Cameron, A. M. (1986). Characterization of revertants of stmF mutants of *Dictyostelium discoideum:* Evidence that stmF is the structural gene of the cGMP-specific phosphodiesterase. *Dev. Genet.* **6,** 163–177.

Darmon, M., Barra, J., and Brachet, P. (1978). The role of phosphodiesterase in aggregation of *Dictyostelium discoideum*. *J. Cell Sci.* **31,** 233–243.

de Witt, R. J. W., Bulgakov, R., Bominaar, T. A., and de Witt, T. F. R. (1987). Differential effects of stimulus termination on excitation and desensitization of folic acid receptors and guanylate cyclase in *Dictyostelium discoideum*. *Biochim. Biophys. Acta* **930,** 1–9.

de Witt, R. J. W., van Bemmelen, M. X., Penning, L. C., Pinas, J. E., Calandra, T. D., and Bonner, J. T. (1988). Studies of cell-surface glorin receptors, glorin degradation, and glorin-induced cellular responses during development of *Polysphondylium violaceum*. *Exp. Cell Res.* **179,** 332–343.

Dicou, E., and Brachet, P. (1980). A separate phosphodiesterase for the hydrolysis of cyclic guanosine 3′,5′-monophosphate in growing *Dictyostelium discoideum* amoebae. *Eur. J. Biochem.* **109,** 507–514.

Dizhoor, A. M., Ray, S., Kumar, S., Niemi, G., Spencer, M., Brolley, D., Walsh, K. A., Philipov, P. P., Hurley, J. B., and Stryer, L. (1991). Recoverin: A calcium sensitive activator of retinal rod guanylate cyclase. *Science* **251,** 915–918.

Dunlap, K. (1977). Localization of calcium channels in *Paramecium caudatum*. *J. Physiol. (London)* **271,** 119–133.

Eckstein, H. (1988). 3′:5′-cyclic GMP in the yeast *Saccharomyces cerevisiae* at different metabolic conditions. *FEBS Lett.* **232,** 121–124.

Eistetter, H., Seckler, B., Bryniok, D., and Schultz, J. E. (1983). Phosphorylation of endogenous proteins of cilia from *Paramecium tetraurelia* in vitro. *Eur. J. Cell Biol.* **31,** 220–226.

Europe-Finner, G. N., and Newell, P. C. (1985). Inositol 1,4,5-trisphosphate induces cyclic GMP formation in *Dictyostelium discoideum. Biochem. Biophys. Res. Commun.* **130,** 1115–1122.

Goldberg, N. D., O'Dea, R. F., and Haddox, M. K. (1973). Cyclic GMP. *Adv. Cyclic Nucleotide Res.* **3,** 155–214.

Goldberg, N. D., Walseth, T. F., Stephenson, J. H., Krick, T. P., and Graff, G. (1980). ^{18}O-labeling of guanosine monophosphate upon hydrolysis of cyclic guanosine 3′:5′-monophosphate by phosphodiesterase. *J. Biol. Chem.* **255,** 10344–10347.

Gonzalez, J. E., and Peterkofsky, A. (1975). Diverse directional changes of cGMP relative to cAMP in *E. coli. Biochem. Biophys. Res. Commun.* **67,** 190–198.

Graves, M. G., Dickinson, J. R., and Swoboda, B. E. P. (1976). Cyclic GMP and cyclic GMP phosphodiesterase in the cell cycle of *Tetrahymena pyriformis. FEBS Lett.* **69,** 165–166.

Gray, N. C. C., Dickinson, J. R., and Swoboda, B. E. P. (1977). Cyclic GMP metabolism in *Tetrahymena pyriformis* synchronized by a single hypoxic shock. *FEBS Lett.* **81,** 311–314.

Hasunuma, K., Funadera, K., Shinobara, Y., Furukawa, K., and Watanabe, M. (1987). Circadian oscillation and light-induced changes in the concentration of cyclic nucleotides in *Neurospora. Curr. Genet.* **12,** 127–133.

Hinrichsen, R. D., Saimi, Y., and Kung, C. (1984). Mutants with altered Ca^{2+}-channel properties in *Paramecium tetraurelia:* Isolation, characterization and genetic analysis. *Genetics* **108,** 545–558.

Janssens, P. M. W., and de Jong, C. C. C. (1988). A magnesium-dependent guanylate cyclase in cell-free preparations of *Dictyostelium discoideum. Biochem. Biophys. Res. Commun.* **150,** 405–411.

Janssens, P. M. W., van Essen H. W., Guijt, J. J. M., de Waal, A., and van Driel, R. (1987). Cell fractionation, detergent sensitivity and solubilization of *Dictyostelium* adenylate cyclase and guanylate cyclase. *Mol. Cell. Biochem.* **76,** 55–65.

Janssens, P. M. W., de Jong, C. C. C., Vink, A. A., and van Haastert, P. J. M. (1989). Regulatory properties of magnesium-dependent guanylate cyclase in *Dictyostelium discoideum* membranes. *J. Biol. Chem.* **264,** 4329–4335.

Kakiuchi, S., Sobue, K., Yamazaki, R., Nagao, S., Umeki, S., Nozawa, Y., Yazawa, M., and Yagi, K. (1981). Ca^{2+}-dependent modulator proteins from *Tetrahymena pyriformis,* sea anemone, and scallop and guanylate cyclase activation. *J. Biol. Chem.* **256,** 19–22.

Kesbeke, F., Baraniak, J., Bulgakov, R., Jastorff, B., Morr, M., Petridis, G., Stec, W. J., Seela, F., and van Haastert, P. J. M. (1985). Cyclic nucleotide specificity of the activator and catalytic sites of a cGMP-stimulated cGMP phosphodiesterase from *Dictyostelium discoideum. Eur. J. Biochem.* **151,** 179–186.

Kimura, H., Uchida, T., Futami, Y., Yoshida, K., Shinomiya, T., Tarni, S., and Okada, Y. (1981). Evidence for guanylate cyclase activity associated with hemagglutinating virus of Japan (Sendai Virus). *J. Biol. Chem.* **256,** 2508–2513.

Klumpp, S., and Schultz, J. E. (1982). Characterization of a Ca^{2+}-dependent guanylate cyclase in the excitable ciliary membrane from *Paramecium. Eur. J. Biochem.* **124,** 317–324.

Klumpp, S., Guerini, D., Krebs, J., and Schultz, J. E. (1987). Effect of tryptic calmodulin fragments on guanylate cyclase activity from *Paramecium tetraurelia. Biochem. Biophys. Res. Commun.* **142,** 857–864.

Koch, K.-W., and Stryer, L. (1988). Highly cooperative feedback control of retinal rod guanylate cyclase by calcium ions. *Nature (London)* **334,** 64–66.

Kudo, S., Nakazawa, K., and Nozawa, Y. (1980). Studies on cyclic nucleotide metabolism in *Tetrahymena pyriformis:* Partial characterization of cyclic AMP- and cyclic GMP-dependent phosphodiesterases. *J. Protozool.* **27,** 342–345.

Kudo, S., Ohnishi, K., Muto, Y., Watanabe, Y., and Nozawa, Y. (1981). *Paramecium* calmodulin can stimulate membrane-bound guanylate cyclase in *Tetrahymena. Biochem. Int.* **3,** 255–263.

Kudo, S., Muto, Y., Nagao, S., Naka, M., Hidaka, H., Sano, M., and Nozawa, Y. (1982). Specificity of *Tetrahymena* calmodulin in activation of calmodulin-regulated enzymes. *FEBS Lett.* **149,** 271–276.

Kudo, S., Muto, Y., Nagao, S., and Nozawa, Y. (1983). Inhibitory effect of bovine brain calmodulin on calmodulin-dependent stimulation of plasma membrane-bound guanylate cyclase in *Tetrahymena pyriformis. Biochem. Int.* **7,** 361–367.

Kudo, S., Nagao, S., Muto, Y., Takahashi, M., and Nozawa, Y. (1986). Characterization of cyclic AMP and cyclic GMP phosphodiesterases in *Tetrahymena* cilia. *Comp. Biochem. Physiol. B* **83B,** 99–102.

Lambrecht, H.-G., and Koch, K.-W. (1991). A 26 kD calcium binding protein from bovine rod outer segments as modulator of photoreceptor guanylate cyclase. *EMBO J.* **10,** 793–798.

Lewis, R. M., and Nelson, D. L. (1980). Biochemical studies of the excitable membrane of *Paramecium:* Protein kinase activities of cilia and ciliary membrane. *Biochim. Biophys. Acta* **615,** 341–353.

Lim, S. T., Hennecke, H., and Scott, D. B. (1979). Effect of cyclic guanosine 3′,5′-monophosphate on nitrogen fixation in *Rhizobium japonicum. J. Bacteriol.* **139,** 256–263.

Liu, G., and Newell, P. C. (1988). Evidence that cGMP regulates myosin interaction with the cytoskeleton during chemotaxis of *Dictyostelium. J. Cell Sci.* **90,** 123–129.

Liu, G., and Newell, P. C. (1991). Evidence that cyclic GMP may regulate the association of myosin II heavy chain with the cytoskeleton by inhibiting its phosphorylation. *J. Cell Sci.* **98,** 483–490.

Lovely, J. R., and Threlfall, R. J. (1976). Fluctuations in cyclic adenosine 3′:5′-monophosphate and cyclic guanosine 3′:5′-monophosphate during the mitotic cycle of the acellular slime mould *Physarum polycephalum. Biochem. Biophys. Res. Commun.* **71,** 789–795.

Macchia, V., Varrone, S., Weissbach, H., Miller, D. L., and Pastan, I. (1975). Guanylate cyclase in *Escherichia coli. J. Biol. Chem.* **250,** 6214–6217.

Majima, T., Hamasaki, T., and Arai, T. (1986). Increase in cellular cyclic GMP level by potassium stimulation and its relation to ciliary orientation in *Paramecium. Experientia* **42,** 62–64.

Majumdar, S., and Bose, S. K. (1985). Derepression of sporulation and synthesis of myobacillin and dipicolinic acid by guanosine 3′:5′-cyclic monophosphate under conditions of glucose repression in *Bacillus subtilis. J. Gen. Microbiol.* **131,** 2783–2788.

Mato, J. M. (1979). Activation of *Dictyostelium discoideum* guanylate cyclase by ATP. *Biochem. Biophys. Res. Commun.* **88,** 569–574.

Mato, J. M., and Malchow, D. (1978). Guanylate cyclase activation in response to chemotactic stimulation in *Dictyostelium discoideum. FEBS Lett.* **90,** 119–122.

Mato, J. M., Krens, F. A., van Haastert, P. J. M., and Konijn, T. M. (1977a). 3′:5′-Cyclic AMP-dependent 3′:5′-cyclic GMP accumulation in *Dictyostelium discoideum. Proc. Natl. Acad. Sci. U.S.A.* **74,** 2348–2351.

Mato, J. M., van Haastert, P. J. M., Krens, F. A., Rhijnsburger, E. H., Dobbe, F. C. P. M., and Konijn, T. M. (1977b). Cyclic AMP and folic acid mediated cyclic GMP accumulation in *Dictyostelium discoideum. FEBS Lett.* **79,** 331–336.

Mato, J. M., Roos, W., and Wurster, B. (1978a). Guanylate cyclase activity in *Dictyostelium discoideum* and its increase during cell development. *Differentiation (Berlin)* **10,** 129–132.

Mato, J. M., Woelders, H., van Haastert, P. J. M., and Konijn, T. M. (1978b). Cyclic GMP binding activity in *Dictyostelium discoideum. FEBS Lett.* **90,** 261–264.

Menz, S., Bumann, J., Jaworski, E., and Malchow, D. (1991). Mutant analysis suggests that cyclic GMP mediates the cyclic AMP-induced Ca^{2+} uptake in *Dictyostelium. J. Cell Sci.* **99,** 187–191.

Miglietta, L. A. P., and Nelson, D. L. (1988). A novel cGMP-dependent protein kinase from *Paramecium. J. Biol. Chem.* **263,** 16096–16105.

Morton, D. B., and Truman, J. W. (1988). The EGPs: The eclosion hormone and cyclic GMP-regulated phosphoproteins. I. Appearance and partial characterization in the CNS of *Manduca sexta. J. Neurosci.* **8,** 1326–1337.

Murofushi, H. (1974). Partial purification and characterization of adenosine 3′,5′-monophosphate-dependent and guanosine 3′,5′-monophosphate-dependent protein kinases. *Biochim. Biophys. Acta* **370,** 130–139.

Murtaugh, M. P., Kapoor, C. L., and Denlinger, D. L. (1985). Extracellular localization of cyclic GMP in the house cricket male accessory reproductive gland and its fate in mating. *J. Exp. Zool.* **233,** 413–423.

Muto, Y., Kudo, S., Nagao, S., and Nozawa, Y. (1985). Growth-dependent changes of guanylate and adenylate cyclase activities in cilia and cell bodies of *Tetrahymena pyriformis. Exp. Cell Res.* **159,** 267–271.

Nagao, S., and Nozawa, Y. (1987). Properties of digitonin-solubilized calmodulin-dependent guanylate cyclase from the plasma membranes of *Tetrahymena pyriformis* NT-1 cells. *Arch. Biochem. Biophys.* **252,** 179–187.

Oertel, D., Schein, S. J., and Kung, C. (1977). Separation of membrane currents using a *Paramecium* mutant. *Nature (London)* **268,** 120–124.

Onimaru, H., Ohki, K., Nozawa, Y., and Naitoh, Y. (1980). Electrical properties of *Tetrahymena,* a suitable tool for studies on membrane excitation. *Proc. Jpn. Acad., Ser. B* **56,** 538–543.

Orlowski, M., and Sypherd, P. S. (1976). Cyclic guanosine 3′,5′-monophosphate in the dimorphic fungus *Mucor racemosus. J. Bacteriol.* **125,** 1226–1228.

Padh, H., and Brenner, M. (1984). Studies of the guanylate cyclase of the social amoeba *Dictyostelium discoideum. Arch. Biochem. Biophys.* **229,** 73–80.

Pasquale, S. M., and Goodenough, U. W. (1987). Cyclic AMP functions as a primary sexual signal in gametes of *Chlamydomonas reinhardtii. J. Cell Biol.* **105,** 2279–2292.

Rosenberg, G., and Pall, M. L. (1978). Cyclic AMP and cyclic GMP in germinating conidia of *Neurospora crassa. Arch. Microbiol.* **118,** 87–90.

Rosenberg, G., and Pall, M. L. (1979). Properties of two cyclic nucleotide-deficient mutants of *Neurospora crassa. J. Bacteriol.* **137,** 1140–1144.

Ross, F. M., and Newell, P. C. (1981). Streamers: chemotactic mutants of *Dictyostelium discoideum* with altered cyclic GMP metabolism. *J. Gen. Microbiol.* **127,** 339–350.

Saimi, Y., and Kung, C. (1987). Behavioral genetics of *Paramecium. Annu. Rev. Genet.* **21,** 47–65.

Schaefer, W. H., Lukas, T. J., Blair, I. A., Schultz, J. E., and Watterson, D. M. (1987). Amino acid sequence of a novel calmodulin from *Paramecium tetraurelia* that contains dimethyllysine in the first domain. *J. Biol. Chem.* **262,** 1025–1029.

Schimz A., and Hildebrand, E. (1987). Effects of cGMP, calcium and reversible methylation on sensory signal processing in *Halobacteria. Biochim. Biophys. Acta* **923,** 222–232.

Schimz, A., Hinsch, K.-D., and Hildebrand, E. (1989). Enzymatic and immunological detection of a G-protein in *Halobacterium halobium. FEBS Lett.* **249,** 59–61.

Schmidt, J. M., and Samuelson, G. M. (1972). Effects of cyclic nucleotides and nucleoside triphosphates on stalk formation in *Caulobacter crescentus*. *J. Bacteriol.* **112,** 593–601.

Schultz, J. E., and Jantzen, H.-M. (1980). Cyclic nucliotide-dependent protein kinases from cilia of *Paramecium tetraurelia. FEBS Lett.* **116,** 75–78.

Schultz, J. E., and Klumpp, S. (1984). Calcium/calmodulin-regulated guanylate cyclases in the ciliary membranes from *Paramecium* and *Tetrahymena. Adv. Cyclic Nucleotide Protein Phosphorylation Res.* **17,** 275–283.

Schultz, J. E., and Klumpp, S. (1991). Calcium-regulated guanylyl cyclases from *Paramecium* and *Tetrahymena. In* "Methods in Enzymology" (R. Johnson and J. Corbin, eds.), Vol. 195, pp. 466–474. Academic Press, San Diego, CA.

Schultz, J. E., and Schade, U. (1989a). Veratridine induces a Ca^{2+} influx, cyclic GMP formation, and backward swimming in *Paramecium tetraurelia* wildtype cells and Ca^{2+} current-deficient pawn mutant cells. *J. Membr. Biol.* **109,** 251–258.

Schultz, J. E., and Schade, U. (1989b). Calcium channel activation and inactivation in *Paramecium* biochemically measured by cyclic GMP production. *J. Membr. Biol.* **109,** 259–267.

Schultz, J. E., Schönefeld, U., and Klumpp, S. (1983). Calcium/calmodulin-regulated guanylate cyclase and calcium-permeability in the ciliary membrane from *Tetrahymena. Eur. J. Biochem.* **137,** 89–94.

Schultz, J. E., Pohl, T., and Klumpp, S. (1986). Voltage-gated Ca^{2+} entry into *Paramecium* linked to intraciliary increase in cyclic GMP. *Nature (London)* **322,** 271–273.

Schultz, J. E., Klumpp, S., and Hinrichsen, R. D. (1990). Calcium and membrane excitation in *Paramecium.* In "Calcium as an Intracellular Messenger in Eukaryotic Microbes" (D. H. O'Day, ed.), pp. 124–150. Am. Soc. Microbiol., Washington, DC.

Schwartz, L. M., and Truman, J. W. (1984). Cyclic GMP may serve as a second messenger in peptide-induced muscle degeneration in an insect. *Proc. Natl. Acad. Sci. U.S.A.* **81,** 6718–6722.

Setlow, B., and Setlow, P. (1978). Levels of cyclic GMP in dormant, germinated, and outgrowing spores and growing and sporulating cells of *Bacillus megaterium. J. Bacteriol.* **136,** 433–436.

Sharaf, M. A., and Rooney, D. W. (1982). Changes in cyclic nucleotide levels correlated with growth, division, and morphology in *Chlamydomonas* chemostat culture. *Biochem. Biophys. Res. Commun.* **105,** 1461–1465.

Sharaf, M. A., and Rooney, D. W. (1985). Rhythmic cyclic AMP changes in *Chlamydomonas* cells synchronized by temperature and light cycles in chemostat culture. *Cell Biol. Int. Rep.* **9,** 561–567.

Shibuya, M., Takebe, Y., and Kaziro, Y. (1977). A possible involvement of cya gene in the synthesis of cyclic guanosine 3′:5′-monophosphate in *E. coli. Cell (Cambridge, Mass.)* **12,** 521–528.

Silverman, P. M., and Epstein, P. M. (1975). Cyclic nucleotide metabolism coupled to cytodifferentiation of *Blastocladiella emersonii. Proc. Natl. Acad. Sci. U.S.A.* **72,** 442–446.

Small, N. V., Europe-Finner, G. N., and Newell, P. C. (1987). Adaptation to chemotactic cyclic AMP signals in *Dictostelium* involves the G-protein. *J. Cell Sci.* **88,** 537–545.

Sun, I. C., Shapiro, L., and Rosen, O. M. (1974). Purification and characterization of guanylate cyclase from *Caulobacter crescentus. Biochem. Biophys. Res. Commun.* **61,** 193–203.

Takahashi, M., Onimaru, H., and Naitoh, Y. (1980). A mutant of *Tetrahymena* with nonexcitable membrane. *Proc. Jpn. Acad., Ser. B* **56,** 585–590.

van Haastert, P. J. M. (1983). Relationship between adaption of the folic acid and the cAMP mediated cGMP response in *Dictyostelium*. *Biochem. Biophys. Res. Commun.* **115,** 130–136.

van Haastert, P. J. M., and van der Heijden, P. R. (1983). Excitation, adaptation, and deadaptation of the cAMP-mediated cGMP response in *Dictyostelium discoideum*. *J. Cell Biol.* **96,** 347–353.

van Haastert, P. J. M., van Lookeren Campagne, M. M., and Ross, F. M. (1982a). Altered cGMP-phosphodiesterase activity in chemotactic mutants of *Dictyostelium discoideum*. *FEBS Lett.* **147,** 149–152.

van Haastert, P. J. M., van Walsum, H., and Pasveer, F. J. (1982b). Nonequilibrium kinetics of a cyclic GMP-binding protein in *Dictyostelium discoideum*. *J. Cell Biol.* **94,** 271–278.

van Haastert, P. J. M., van Lookeren Campagne, M. M., and Kesbeke, F. (1983). Multiple degradation pathways of chemoattractant mediated cyclic GMP accumulation in *Dictyostelium*. *Biochim. Biophys. Acta* **756,** 67–71.

van Haastert, P. J. M., de Witt, R. J. W., Janssens, P. M. W., Kesbeke, F., and de Groede, J. (1986). G-protein-mediated interconversions of cell-surface cAMP receptors and their involvement in excitation and desensitization of guanylate cyclase in *Dictyostelium discoideum*. *J. Biol. Chem.* **261,** 6904–6911.

Wanner, R., and Wurster, B. (1990). Cyclic GMP-activated protein kinase from *Dictyostelium discoideum*. *Biochim. Biophys. Acta* **1053,** 179–184.

Ward, A., and Brenner, M. (1977). Guanylate cyclase from *Dictyostelium discoideum*. *Life Sci.* **21,** 997–1008.

Wurster, B., Schubiger, K., Wick, U., and Gerisch, G. (1977). Cyclic GMP in *Dictyostelium discoideum*. *FEBS Lett.* **76,** 141–144.

Wurster, B., Bozzaro, S., and Gerisch, G. (1978). Cyclic GMP regulation and responses of *Polysphondylium violaceum* to chemoattractants. *Cell Biol. Int. Rep.* **2,** 61–69.

Wurster, B., Schubiger, K., and Brachet, P. (1979). Cyclic GMP and cyclic AMP changes in response to folic acid pulses during cell development of *Dictyostelium discoideum*. *Cell Differ.* **8,** 235–242.

Yamagata K., Goto, K., Kuo, C. H., Kondo, H., and Miki, N. (1990). Visinin: A novel calcium binding protein expressed in retinal cone cells. *Neuron* **4,** 469–476.

Yazawa, M., Yagi, K., Toda, H., Kondo, K., Narita, K., Yamazaki, R., Sobue, K., Kakiuchi, S., Nagao, S., and Nozawa, Y. (1981). The amino acid sequence of the *Tetrahymena* calmodulin which specifically interacts with guanylate cyclase. *Biochem. Biophys. Res. Commun.* **99,** 1051–1057.

Ziegelberger, G., van den Berg, M. J., Kaissling, K.-E., Klumpp, S., and Schultz, J. E. (1990). Cyclic GMP levels and guanylate cyclase activity in pheromone-sensitive antennae of the silkmoths *Antheraea polyphemus* and *Bombyx mori*. *J. Neurosci.* **10,** 1217–1225.

Zimmerman, S., Zimmerman, A. M., and Laurence, H. (1981). Effect of Δ^9-tetrahydrocannabinol on cyclic nucleotides in synchronously dividing *Tetrahymena*. *Can. J. Biochem.* **59,** 489–493.

Clinical Relationships of Cyclic GMP

Jean R. Cusson,* Johanne Tremblay,*
Pierre Larochelle,* Ernesto L. Schiffrin,†
Jolanta Gutkowska,*,† and Pavel Hamet*

**Centre de Recherche*
Hôtel-Dieu de Montréal
Montréal, Québec, Canada H2W 1T8
†Institut de Recherches Cliniques de Montréal
Montréal, Québec, Canada H2W 1R7

I. Introduction

Cyclic GMP (cGMP) production through the stimulation of particulate guanylate cyclase is induced by the action of atrial natriuretic peptide (ANP) on target cells. In humans, the use of radioimmunoassay has enabled the determination of plasma and urinary cGMP levels as well as the development of a potential tool to detect and follow physiological and pharmacological changes evoked by ANP. The purpose of this chapter is to review current knowledge on the clinical use of cGMP measurements as markers of endogenous and exogenous ANP effects.

II. Cyclic GMP Measurements in Healthy Humans

In keeping with the significant relationship between atrial stretch and ANP secretion, it has now been shown in numerous human studies that plasma ANP levels increase following physiological maneuvers causing blood volume expansion. This was demonstrated, for example, when blood volume was increased endogenously via head-down tilt (Larose *et al.*, 1985)

Advances in Pharmacology, Volume 26

and head-out water immersion (HOI) (Epstein *et al.*, 1987). It was also observed after exogenous blood volume expansion was induced by intravenous infusion of isotonic saline (Sagnella *et al.*, 1985) or changes in sodium intake (Shenker *et al.*, 1985). In addition, physiological maneuvers that increase blood volume not only cause a rise in plasma ANP levels, but also increase plasma and urinary cGMP excretion.

A. Measurement Conditions

Before describing investigations establishing cGMP measurement conditions, some basic data must be summarized. Plasma cGMP levels in healthy volunteers on their "usual" salt intake regimen, as determined by radioimmunoassay, are in the range 2–7 nmol/liter. For instance, in 77 healthy male and female volunteers ages 22 to 86 years, plasma cGMP levels in the supine position averaged 4.7 ± 1.8 (standard deviation) nmol/liter (Genest *et al.*, 1989). In other studies from our laboratory, plasma cGMP levels in seated or supine healthy male subjects on a 150–200 mmol/day sodium intake regimen averaged about 3 to 5 nmol/liter, and urinary cGMP excretion rates were about 0.3–0.9 nmol/min (Cusson *et al.*, 1987, 1988, 1989, 1990).

Contrary to plasma ANP levels, data on 77 healthy volunteers established that plasma cGMP levels did not correlate with increasing age, as shown in Table I. On the other hand, posture is an important factor to consider when measuring plasma or urinary cGMP concentrations. In-

Table I

Age-Related Variations in Plasma cGMP and C-Terminal ANP Levels (Means ± SEM)

Groups	Age subsets	N	SBP[a] (mm Hg)	DBP[a] (mm Hg)	C-ANP (pmol/liter)	cGMP (nmol/liter)
Healthy subjects	16–30	20	113 ± 2	70 ± 1	7 ± 1	4.9 ± 0.3
	31–45	19	107 ± 3	70 ± 2	9 ± 1	5.2 ± 0.4
	46–60	12	116 ± 4	71 ± 2	13 ± 2*	4.1 ± 0.5
	≥61	26	126 ± 3	72 ± 2	16 ± 1*	4.7 ± 0.4
Untreated hypertensives	16–30	11	155 ± 5	99 ± 3	9 ± 1	4.2 ± 0.9
	31–45	31	154 ± 4	102 ± 2	9 ± 1	4.6 ± 0.5
	46–60	45	162 ± 3	100 ± 1	11 ± 1	5.3 ± 0.4
	≥61	10	166 ± 9	95 ± 2	13 ± 2	7.1 ± 1.1**

Note. SBP and DBP, systolic and diastolic blood pressure.

[a] SBP and DBP were significantly higher in patients ($p < 0.001$).

* $p < 0.05$ vs healthy younger subjects.

** $p < 0.05$ vs healthy subjects of the same age subset.

deed, assuming the supine from the upright posture significantly increases blood flow to the intrathoracic compartment, with enlargement of apparent heart volume (Gauer and Henry, 1976). We have had opportunities to study the effects of postural changes on normal healthy volunteers in two investigations. Initially, we compared upright and supine postures, each maintained for 2 h, in six healthy male volunteers ages 19 to 32 years. Plasma ANP (Gutkowska *et al.*, 1986) and plasma and urinary cGMP (Richman *et al.*, 1980) were measured by radioimmunoassays. In the supine posture, a modest rise in plasma ANP was associated with a mild increase in plasma cGMP and a doubling of its urinary excretion (Table II). As expected, peripheral renin activity (PRA) and plasma aldosterone (PA) decreased by about 50% with urinary volume and sodium excretion.

In a second study, we compared the effect of changing from the upright to the seated position, to that of changing from the seated to the supine position in five healthy male and five female volunteers ages 20 to 44 years. Each posture was maintained for 2 h. Our specific objective was to evaluate whether part of the increase in ANP secretion seen in a supine

Table II

Posture Induced Changes in Cyclic GMP and Other Parameters in Healthy Volunteers: Upright vs Supine Positions

	Healthy controls (N = 6) (means ± SEM)		
Parameter	Upright	Supine	Δ % (SUP–UP)
SBP (mm Hg)	116 ± 4	113 ± 3	−3 ± 4
DBP (mm Hg)	74 ± 3	74 ± 3	−4 ± 3
HR (beats/min)	81 ± 3	59 ± 4*	−31 ± 5
Hematocrit	0.422 ± 0.008	.405 ± 0.008*	−4 ± 1
Serum alb. (g/liter)	47 ± 1	43 ± 1*	−7 ± 2
PRA (ng/ml/hr)	1.6 ± 0.5	0.4 ± 0.1*	−71 ± 8
P. aldo. (pmol/liter)	644 ± 56	302 ± 28*	−51 ± 6
P. ANP (pmol/liter)	3.5 ± 0.6	5.9 ± 1.1*	74 ± 27
P. cGMP (nmol/liter)	4.5 ± 0.5	5.9 ± 0.6*	36 ± 12
UV (ml/min)	0.5 ± 0.1	1.5 ± 0.5*	167 ± 57
UNaV(μmol/min)	48 ± 8	102 ± 18*	128 ± 47
UKV (μmol/min)	43 ± 34	88 ± 17*	94 ± 29
UcGMP (μM/min)	0.44 ± 0.13	0.78 ± 14*	192 ± 113
Cr. clear. (ml/min)	133 ± 9	129 ± 15	−2 ± 5

Note. HR, heart rate; PRA, peripheral renin activity; P. aldo, plasma aldosterone; UV, UNaV, UKV, urinary volume, sodium and potassium excretion; Cr. clear., creatinine clearance.

* $p < 0.01$.

Table IIIA

Posture-Induced Changes in Cyclic GMP and Other Parameters in Healthy Volunteers: Upright to Sitting

	Healthy control (N = 10)		
Parameters	Upright	Sitting	Δ %
PRA (ng/ml/hr)	1.2 ± 0.3	0.7 ± 0.2**	−44 ± 10
P. aldo. (pmol/liter)	866 ± 231	497 ± 99**	−36 ± 8
P. ANP (pmol/liter)	2.9 ± 0.4	2.5 ± 0.3*	−11 ± 5
P. cGMP (pmol/liter)	2.4 ± 0.3	1.5 ± 0.2*	−30 ± 10
UV (ml/min)	1.2 ± 0.3	2.1 ± 0.4**	163 ± 92
UNaV (μmol/min)	98 ± 26	147 ± 22**	183 ± 99
UKV (μmol/min)	74 ± 15	89 ± 14	5 ± 10
UcGMP (μM/min)	0.65 ± 0.09	0.64 ± 0.10	5 ± 10

* $p < 0.05$.
** $p < 0.01$ vs upright.

posture would occur while sitting, even if the inferior limbs are still below the thoracic cage. Although diuresis and natriuresis were heightened by more than 100%, and PRA and plasma aldosterone levels were decreased by about 40% after sitting, there were no increases in either plasma ANP or plasma and urinary cGMP levels (Table IIIA). In the supine position, PRA and aldosterone were further reduced, diuresis and natriuresis in-

Table IIIB

Posture-Induced Changes in Cyclic GMP and Other Parameters in Healthy Volunteers: Sitting to Supine

	Healthy controls (N = 10)		
Parameter	Sitting	Supine	Δ %
PRA (ng/ml/hr)	0.7 ± 0.2	0.4 ± 0.1**	−46 ± 9
P. aldo. (pmol/liter)	497 ± 117	302 ± 272**	−36 ± 5
P. ANP (pmol/liter)	2.5 ± 0.3	3.8 ± 0.4*	60 ± 17
P. cGMP (pmol/liter)	1.5 ± 0.2	3.2 ± 0.5**	173 ± 80
UV (ml/min)	2.1 ± 0.4	3.8 ± 0.6**	93 ± 26
UNaV (μmol/min)	147 ± 22	186 ± 28**	40 ± 17
UKV (μmol/min)	89 ± 14	89 ± 19	10 ± 20
UcGMP (μM/min)	0.64 ± 0.10	1.07 ± 0.02*	65 ± 26

* $p < 0.05$.
** $p < 0.01$ vs sitting.

creased, and then only plasma ANP and cGMP levels rose, respectively, by 60 ± 17% and 173 ± 80%. Urinary cGMP excretion was augmented by 65 ± 26% (Table IIIB).

Another possible confounding factor in the investigation of changes in ANP and cGMP production is the effect of time. Indeed, similarly to other hormones, there is evidence of a circadian cycle for plasma ANP levels (Richards *et al.,* 1987). On the other hand, it has been shown that some modifications in plasma cGMP levels that might be attributed to diurnal variations are in fact due to postural changes (Bell *et al.,* 1990). Our own experience has revealed that in supine healthy volunteers, both plasma ANP and cGMP levels are stable within a 6-h period (see Table IV).

With these data in mind, we will now review published and unpublished results from our laboratory with regard to the effects of endogenous and exogenous blood volume expansion on plasma cGMP levels.

B. Exogenous Blood Volume Expansion

Intravenous volume infusion is another means of expanding blood volume "physiologically" and investigating the ANP system. Data on cGMP from several such studies are available. Most investigators used a rapid (30–60 min) intravenous infusion of isotonic saline (Lewis *et al.,* 1988; Wehling *et al.,* 1989; Sagnella *et al.,* 1990; Furtwängler *et al.,* 1990). As found with physiological maneuvers employing endogenous means, plasma and

Table IV

Changes in Cyclic GMP and Other Parameters in Healthy Volunteers Maintained in the Supine Position for 6 Hours

	Supine healthy controls (N = 6) (means ± SEM)		
Parameter	Time = 0 h	Time = 4 h	Time = 6 h
SBP (mm Hg)	116 ± 4	110 ± 3	112 ± 6
DBP (mm Hg)	71 ± 3	69 ± 3	69 ± 2
HR (beats/min)	66 ± 3	63 ± 4	68 ± 4
Hematocrit	0.43 ± 0.01	0.42 ± 0.01*	0.41 ± 0.01*
Serum alb. (g/liter)	48 ± 1	43 ± 1**	43 ± 1**
PRA (ng/ml/hr)	0.94 ± 0.22	0.45 ± 0.13**	0.45 ± 0.14**
P. aldo. (pmol/liter)	475 ± 44	220 ± 33**	281 ± 36**
P. ANP (pmol/liter)	4.6 ± 0.7	5.6 ± 0.5	6.0 ± 0.5
P. cGMP (nmol/liter)	3.8 ± 0.5	3.7 ± 0.3	3.3 ± 0.5

* $p < 0.05$.
** $p < 0.01$ vs baseline.

urinary GMP levels rose together with plasma ANP, but the relative cGMP elevation was smaller than that of ANP.

Generally, rises of 30–60% and about 100% occur with plasma and urinary cGMP, respectively, whereas plasma ANP concentrations increase by about 200–300%. An example of such data is given in Table V. Here, blood volume expansion was produced by 4-h infusions of 2 liters of isotonic saline, 2 liters of 5% dextrose in water (an hypotonic solution) and 1 liter of a commercially available colloid solution containing 40 mg of glucose, 47 g of albumin, and 140 mmol of sodium per liter (Plasmanate, Cutter, Rexdale, Ontario, Canada). Our specific goal was to produce increasing degrees of blood volume expansion.

It can be seen that there was an escalating degree of blood volume expansion from the hypotonic to the colloid infusion, based on changes in hematocrit. Following blood volume expansions of 6, 8, and 16%, plasma cGMP levels climbed progressively by 23, 39, and 52%, respectively. There was also a gradual "dose-dependent" increase in plasma ANP levels, which was greater in magnitude than that of cGMP.

C. Endogenous Blood Volume Expansion

Following HOI, where plasma ANP rose by 250%, plasma and urinary cGMP have been reported to increase by 50 and 200%, respectively (Gerbes *et al.*, 1988). In our laboratory (Larochelle *et al.*, 1992), after a 2-h HOI maneuver in 13 healthy volunteers in whom plasma ANP levels went up by about 400%, plasma and urinary cGMP both increased by about 50%.

Table V

Changes in Cyclic GMP and Other Parameters in Healthy Volunteers: Effects of Volume Infusions

Parameter	Dextrose 5% in water ×2 liters	Isotonic saline ×2 liters	Plasmanate ×1 liter
SBP (mm Hg)	106 ± 5 (−12%)	123 ± 4 (−3%)	109 ± 4 (−7%)
DBP (mm Hg)	68 ± 3 (−4%)	77 ± 3 (+7%)	69 ± 2 (+3%)
HR (beats/min)	60 ± 4 (−15%)	63 ± 3 (−9%)	66 ± 3 (−5%)
Hematocrit	0.410 ± 0.012 (−6%)	0.368 ± 0.008 (−8%)	0.370 ± 0.006 (−16%)
S. glucose (mg%)	125 ± 8 (+47%)	83 ± 2 (−7%)	80 ± 3 (+9%)
Serum alb. (g/liter)	44 ± 1 (−12%)	38 ± 1 (−18%)	47 ± 1 (−1%)
PRA (ng/ml/h)	0.6 ± 0.2 (−22%)	0.2 ± 0.1 (−85%)	0.2 ± 0.1 (−70%)
P. aldo. (pmol/liter)	470 ± 190 (−3%)	309 ± 22 (−48%)	314 ± 27 (−17%)
P. ANP (pmol/liter)	6.7 ± 2.7 (+123%)	5.6 ± 0.8 (+142%)	8.6 ± 2.1 (+429%)
P. cGMP (nmol/liter)	6.8 ± 1.0 (+23%)	6.5 ± 0.6 (+39%)	7.7 ± 0.8 (+52%)

Note. Data are end-of-infusion values and % change from baseline (in parentheses).

D. Effect of Pressor Doses of Phenylephrine

Phenylephrine, an α-agonist, was administered to six healthy volunteers at constant rates of 0.8 and 1.6 μg/kg/min over 4 h, to increase diastolic blood pressure by about 15 and 25 mm Hg, respectively (Closas *et al.*, 1988). These infusions, aimed at assessing the effect of an acute elevation of blood pressure on ANP secretion, were compared to vehicle infusion. Plasma ANP and cGMP levels increased after the lower dose infusion, by about 200 and 50%, respectively. Following the higher dose, plasma ANP and cGMP concentrations jumped by about 300 and 200%, respectively.

E. Effect of Intravenous ANP Administration: The Nonlinear ANP–Cyclic GMP Relationship

Exogenous ANP can also stimulate cGMP production in humans. This was initially shown by Gerzer *et al.*, (1985) following single bolus injections of ANP. Using increasing bolus doses of ANP, we later demonstrated the dose-dependency of cGMP production with exogenous ANP. However, this relationship was not a simple one, as seen in Table VI.

In fact, from these data, the peak plasma ANP/peak plasma cGMP ratios for the 12.5, 25, 50, and 100 μg doses were approximately 6, 20, 19, and 68. However, and perhaps due to the irreversible stimulation of particulate guanylate cyclase by ANP (Tremblay *et al.*, 1986), there was a persistent elevation of plasma cGMP levels following single bolus doses of ANP, and as the ANP dose was augmented, the area under the plasma cGMP concentration versus time curve increased in a linear fashion ($r = 0.83$, $p < 0.01$) (Cusson *et al.*, 1988).

In other words, the relationship between the agonist dose and the plasma level of the second messenger was not a linear one. This relationship was

Table VI

Peak Plasma cGMP Levels following Intravenous ANP Infusion in Healthy Volunteers: Relationship with Peak Plasma ANP Levels Achieved with Bolus Doses

Dose	Peak [ANP] (pmol/liter)	Peak [cGMP] (nmol/liter)
12.5 μg bolus (N = 1)	180	32
25 μg bolus (N = 1)	340	17
50 μg bolus (N = 3)	650	35 ± 6
100 μg bolus (N = 3)	2500	37 ± 4

Note. From Cusson *et al.* (1988).

later described mathematically, using the "E-max" model (du Souich *et al.*, 1989). This model accounts for (*a*) a cGMP baseline greater than zero, (*b*) a maximal achievable plasma cGMP level, and (*c*) some relationship between plasma cGMP and ANP levels, specifically, an effective concentration producing a half-maximal response (EC_{50}). The relationship can also be adapted to a theoretical "effect compartment" by modeling the nonlinear curve into a linear one. Then, plasma cGMP levels become highly correlated with a putative ANP concentration in the "effect compartment."

Although nonlinear, this dose-dependent relationship between cGMP and ANP was also observed in three additional studies. First, following "stepped infusions" of ANP (three successive 30-min constant infusions at 4, 8, and 16 pmol/kg/min), plasma ANP levels increased up to 19 ± 2, 58 ± 9, and 74 ± 11 pmol/liter, whereas peak plasma cGMP levels were, respectively, 6 ± 1, 16 ± 2, and 15 ± 3 nmol/liter (Cusson *et al.*, 1987). However, data were only available in three of the seven healthy volunteers due to ANP-induced hypotension. In a second study, ANP was infused at 4 pmol/kg/min for 3 h ($n = 5$) and plasma ANP and cGMP levels increased to 25 ± 7 pmol/liter and 13 ± 2 nmol/liter, respectively (Cusson *et al.*, 1989). In the third investigation, ANP was infused at 0.5 pmol/kg/min for 12 h, and plasma ANP and cGMP levels rose to 12 ± 3 pmol/liter and 4.0 ± 0.8 nmol/liter, respectively.

Certainly, the different durations of infusion could have introduced a confounding factor in the evaluation of that relationship. When the 30-min and 3-h ANP infusions at 4 pmol/kg/min were compared, it became apparent that infusion duration (hence, the cumulative dose), and not only peak level, was an important determinant of the plasma cGMP values achieved. After 30 min, plasma ANP and cGMP concentrations were 19 ± 2 pmol/liter and 6 ± 1 nmol/liter, respectively (Cusson *et al.*, 1987), whereas in the 3-h study, plasma ANP increased to about a similar level (25 ± 7 pmol/liter), but plasma cGMP rose much more (13 ± 2 nmol/liter) (Cusson *et al.*, 1989).

With regard to urinary cGMP excretion, a dose-dependent relationship with ANP infusion rate was observed (Cusson *et al.*, 1987). In healthy volunteers given stepped infusions at 4, 8, and 16 pmol/kg/min, the corresponding urinary cGMP excretion rates were 0.9 ± 0.3, 3.7 ± 0.5, and 6.3 ± 1.7 μmol/min, respectively. As described above for plasma cGMP, the duration of ANP infusion was an important determinant of the urinary cGMP excretion rate. With the 3-h ANP infusion of 4 pmol/kg/min, urinary cGMP excretion was greater than after 30-min infusion of the same dose (4 vs 1 μmol/min) (Cusson *et al.*, 1989). Finally, when ANP was infused at the rate of 0.5 pmol/kg/min for 12 h, urinary cGMP excretion ranged from 0.6 to 0.8 μmol/min (about twofold more than with the vehicle infusion). Thus, the urinary cGMP excretion rate is related to the cumulative ANP dose (infusion rate and duration of infusion). Again, this is perhaps due to the irreversible activation of particulate guanylate cyclase by ANP (Tremblay *et al.*, 1986).

Interestingly, 48-h ANP infusions at the constant rate of about 0.5 pmol/kg/min in African green vervet monkeys did not increase plasma cGMP levels despite evidence of ANP actions such as decreased blood pressure and aldosterone levels (Hamet *et al.*, 1989b). This could have been due to the absence of a rise in extracellular cGMP, from either a lack of its egression (Hamet *et al.*, 1989a) or its accelerated metabolic clearance (Hamet *et al.*, 1975).

The nonlinear relationship between ANP and extracellular cGMP could also be related to the fact that ANP is measured as a circulating hormone, whereas cGMP is a marker of intracellular production, part of which is bound to impact on enzymes such as cGMP protein kinase.

Although nonlinear, cGMP increases in plasma reflect selective stimulation of particulate guanylate cyclase. We have demonstrated previously that infusion of a nitric oxide agonist, sodium nitroprusside (SNP), a stimulator of soluble guanylate cyclase, *does not* increase plasma cGMP and only modestly elevates its urinary excretion. ANP infused in patients produces expected elevations of plasma and urine cGMP levels even though the hemodynamic effects of SNP and ANP are similar (Roy *et al.*, 1989). This may be due to the fact that soluble guanylate cyclase is mostly contained in vascular smooth muscle cells and that the ANP-induced egression of cGMP in fact reflects stimulation of the particulate form of the enzyme in endothelial cells. This topology has to be considered when interpreting data of circulating cGMP in physiology and pathophysiology.

III. Cyclic GMP Measurements in Essential Hypertension

A. Baseline Values

Baseline (supine, at rest) plasma cGMP concentrations in untreated essential hypertensives are comparable to values found in healthy, aged-matched volunteers. Plasma cGMP levels in 97 patients ages 21 to 71 years, averaged 5.0 ± 0.3 nmol/liter with blood pressure ranges of 122–228/90–134 mm Hg, whereas 77 healthy volunteers ages 22 to 86 years had values of 4.7 ± 0.2 nmol/liter and 88–150/54–90 mm Hg, respectively (Genest *et al.*, 1989). It can be pointed out that although plasma ANP levels rise with age in normotensive subjects, its cellular marker, cGMP, increases with age in hypertensive patients (Table I).

B. Effect of Blood Volume Expansion

Blood volume expansion elevates plasma cGMP levels in patients with essential hypertension similarly to changes described previously in healthy volunteers. When blood volume expansion was produced exogenously in seven male essential hypertensives ages 26 to 63 years by intravenous infusion of hypotonic, isotonic, or colloid solutions, as was done in healthy volunteers, hematocrit was reduced in a stepwise manner by 3, 8, and 14%, respectively. Plasma cGMP levels rose from 4.2 ± 0.6 to 5.6 ± 0.8 nmol/liter (p = NS), from 4.2 ± 0.7 to 5.8 ± 1.2 nmol/liter (p = NS) and from 4.6 ± 0.7 to 7.8 ± 1.4 nmol/liter ($p < 0.01$), respectively. Relative to baseline values, these rises in plasma cGMP concentration averaged about 35, 40, and 75% respectively, and were associated with increases in plasma ANP levels of about 25, 65, and 250% respectively (J. R. Cosson *et al.*, unpublished data). Overall, these changes were quite similar to those seen in healthy volunteers (see 2.2). In a more recent study, both plasma cGMP and ANP levels increased by about 100%, following endogenous blood volume expansion with HOI (Larochelle *et al.*, 1992).

C. Effect of Pressor Doses of Phenylephrine

Phenylephrine, an α-agonist, was administered to five patients with mild essential hypertension at constant rates of 0.2 and 1.0 μg/kg/min over 4 h, to increase diastolic blood pressure by less than 5 and up to 15 mm Hg, respectively (Closas *et al.*, 1988). As in the case of healthy volunteers (see III,D), these infusions, aimed at assessing the effect of an acute increase in blood pressure on ANP secretion, were compared to vehicle infusion. In these untreated hypertensive patients with baseline resting supine blood pressure in the range 135/85 mm Hg, plasma ANP and cGMP levels remained unchanged after the lower dose infusion, but rose by about 300 and 115%, respectively, with the higher dose.

D. Effect of Intravenous ANP Administration

In keeping with our data in healthy volunteers, and in addition to linking plasma cGMP and ANP levels, findings in our three studies, where ANP was infused intravenously, suggest that more cGMP is produced in response to ANP in patients with essential hypertension. This is shown in Table VII. Similar findings have been reported in hypertensive animals (Pang *et al.*, 1985). However, a similar kind of nonlinear E-max relationship was also found in hypertension (du Souich *et al.*, 1989). Moreover,

Table VII
Increased cGMP Responses to Intravenous ANP Infusions in Essential hypertension

Reference	ANP dose	Subjects	Peak plasma [ANP] (pmol/liter)	Peak plasma [cGMP] (nmol/liter)
Cusson *et al.* (1987)	4 pmol/kg/min	HV (*n* = 3)	19 ± 2	6 ± 1
	× 30 min	EH (*n* = 5)	22 ± 3	11 ± 1
	8 pmol/kg/min	HV (*n* = 3)	58 ± 9	16 ± 2
	× 30 min	EH (n = 5)	59 ± 12	21 ± 3
	16 pmol/kg/min	HV (*n* = 3)	74 ± 11	15 ± 3
	× 30 min	EH (*n* = 5)	74 ± 16	30 ± 4
Cusson *et al.* (1989)	4 pmol/kg/min	HV (*n* = 5)	25 ± 7	13 ± 2
	× 3 h	EH (*n* = 4)	24 ± 6	16 ± 4
Cusson *et al.* (1990)	0.5 pmol/kg/min	HV (*n* = 6)	12 ± 3	4.0 ± 0.8
	× 12 h	EH (*n* = 6)	12 ± 2	5.0 ± 0.5

Note. HV, healthy volunteers; EH, essential hypertensives.

in hypertensive patients, there is a trend toward greater urinary cGMP excretion. For example, our own data obtained through the different protocols of ANP administration described previously, showed that urinary cGMP excretion rates were increased by 20 to 70% in hypertensive patients relative to healthy volunteers (Cusson *et al.*, 1987, 1989, 1990). Our most recent data demonstrated an increase of particulate guanylate cyclase in glomeruli and lung in spontaneously hypertensive rats as opposed to several normotensive strains. This increase in enzyme activity was revealed by quantitative polymerase chain reaction (using a mutated exogenous template as competitor) due to excessive mRNA accumulation of Type A ANP receptors (J. Tremblay, personal communication).

In summary, data obtained in patients with essential hypertension, whether following endogenous ANP release or after exogenous ANP administration, suggest (*a*) that the production of cGMP is related to plasma ANP concentration in a nonlinear fashion, and (*b*) that the target cells are hyperresponsive to ANP in essential hypertension.

IV. Use of Cyclic GMP Measurements as a Tool in Other Diseases

In contrast to the general clinical research arena, cGMP measurements have not been studied in great detail as a diagnostic tool or to monitor disease status other than cardiovascular diseases.

Besides hypertension, heart failure has recently received much more attention in relation to cGMP measurements. Not only has it been shown that plasma cGMP concentration is correlated with intracardiac pressures (as is plasma ANP level), but plasma cGMP and ANP are closely correlated in patients with cardiac diseases and heart failure, with Pearson's *r* values being greater than 0.70 (Dussaule *et al.*, 1988; Nakaoka *et al.*, 1988; Hauptlorenz and Puschendori, 1989; Stangl *et al.*, 1990; Vorderwinkler *et al.*, 1991). This is similar to the data obtained in cardiomyopathic hamsters (Cantin and Genest, 1985). Plasma cGMP levels are also relatively sensitive to changes in clinical status with treatment of heart failure (Dussaule *et al.*, 1988; Nakaoka *et al.*, 1988; Stangl *et al.*, 1990). Similarly, urinary cGMP excretion is increased in heart failure, and decreases with clinical improvement (Baudouy *et al.*, 1991).

Plasma cGMP is elevated in end-stage renal failure patients on chronic hemodialysis (Hamet *et al.*, 1975; Lauster *et al.*, 1990), just as is plasma ANP. Plasma cGMP (and plasma ANP) levels are also reported to be increased in cirrhotic patients (Jespersen *et al.*, 1990).

Plasma and urinary cGMP have also been measured in a variety of other disorders. For example, in obstructive sleep-apnea patients, the urinary cGMP excretion rate is increased and continuous positive nasal air pressure improves both health and cGMP levels (Krieger *et al.*, 1989). Plasma and urinary cGMP seem to be elevated in certain malignancies such as ovarian cancer (Turner *et al.*, 1990; Peracchi *et al.*, 1990). They are also increased in patients with schizophrenia (Okada *et al.*, 1991). On the other hand, plasma cGMP levels are not elevated in patients with various other pulmonary diseases, and in patients with some gastroenterological and rheumatological diseases (Vorderwinkler *et al.*, 1991). However, the significance of all these findings clearly remains to be elucidated.

V. Conclusions

In summary, the reviewed data show that plasma and urinary levels of cGMP are correlated with plasma ANP levels within a wide range, from physiological to pharmacological concentrations. cGMP measurement in plasma and urine can therefore be seen as a potentially useful tool to monitor the effects of exogenous ANP and the results of physiological maneuvers where blood volume and atrial stretch are expected to change.

Acknowledgments

Data from our laboratories reviewed in this manuscript originated from studies completed since 1986 and supported by the Medical Research Council of Canada and the Canadian Heart and Stroke Foundation. We gratefully acknowledge the expert technical assistance of Ms Suzanne Cossette (cGMP measurements), and Ms Dominique Falstrault (ANP measurements), as well as the excellent research nursing of Ms Martine Bouchard, Ms France Boulianne, Ms Marie-Ange Boutin, Ms Lucette Gauthier, Ms Mireille Kirouac, Ms Suzanne Paris, and Ms Colette Vanier. We also thank Ms Louise Murray for typing the manuscript. The editorial contribution of Mr. Ovid Da Silva is also appreciated.

References

Baudouy, P. Y., Abassade, P., Valleteau de Moulliac, M., and Michel, J. B. (1991). Evaluation de l'excrétion urinaire du GMP cyclique en clinique cardiologique. *Arch. Mal. Coeur Uaiss* **84,** 777–784.

Bell, G. M., Atlas, S. A., Pecker, M., Sealey, J. E., James, G., and Laragh, J. H. (1990). Diurnal and postural variations in plasma atrial natriuretic factor, plasma guanosine 3′ : 5′-cyclic monophosphate and sodium excretion. *Clin. Sci.* **79,** 371–376.

Cantin, M., and Genest, J. (1985). The heart and the atrial natriuretic factor. *Endocr. Rev.* **6,** 107–127.

Closas, J., Genest, J., Larochelle, P., Cusson, J., Gutkowska, J., Hamet, P., De Léan, A., Thibault, G., and Cantin, M. (1988). Effets de la phényléphrine sur le facteur natriurétique de l'oreillette et l'axe rénine-aldostérone chez les sujects normaux et les sujets hypertendus essentiels. *Arch. Mal. Coeur Uaiss* **81,** 75–78.

Cusson, J. R., Hamet, P., Gutkowska, J., Kuchel, O., Genest, J., Cantin, M., and Larochelle, P. (1987). Effects of atrial natriuretic factor on natriuresis and cGMP in patients with essential hypertension. *J. Hypertens.* **5,** 435–443.

Cusson, J. R., du Souich, P., Hamet, P., Schiffrin, E. L., Kuchel, O., Tremblay, J., Cantin, M., Genest, J., and Larochelle, P. (1988). Effects and pharmacokinetics of bolus injections of atrial natriuretic factor in normal volunteers. *J. Cardiovasc. Pharmacol.* **11,** 635–642.

Cusson, J. R., Thibault, G., Kuchel, O., Hamet, P., Cantin, M., and Larochelle, P. (1989). Cardiovascular, renal and endocrine responses to low doses of atrial natriuretic factor in mild essential hypertension. *J. Hum. Hypertens.* **3,** 89–96.

Cusson, J. R., Thibault, G., Cantin, M., and Larochelle, P. (1990). Prolonged low dose infusion of atrial natriuretic factor in essential hypertension. *Clin. Exp. Hypertens.* **A12**(1); 111–135.

du Souich, P., Larochelle, P., Marleau, S., and Cusson, J. (1989). The dose-response curve in indirectly acting drugs. *In* "Dose-Response Relationship in Clinical Pharmacology" (L. Lasagna, S. Erill and C. A. Naranjo, eds.), pp. 95–114. Elsevier, Amsterdam.

Dussaule, J. C., Vahanian, A., Michel, P. L., Soullier, I., Czekalski, S., Acar, J., and Ardaillou, R. (1988). Plasma atrial natriuretic factor and cyclic GMP in mitral stenosis treated by balloon valvulotomy. *Circulation* **78,** 276–285.

Epstein, M., Loutzenhiser, R., Friendland, E., Aceto, R. M., Camargo, M. J. F., and Atlas, S. A. (1987). Relationship of increased plasma atrial natriuretic factor and renal sodium handling during immersion-induced central hypervolemia in normal humans. *J. Clin. Invest.* **79,** 738–745.

Furtwängler, W., Balogh, D., Pomaroli, A., Koller, J., Wieser, C. H., and Mair, P. (1990). Plasmaspiegel des Atrialen-Natriuretischen-Peptids (ANP), des syklischen-Guanosin-Mono-Phosphats (cGMP) sowie des Renins nach Gabe von 7,5% NaCl + 6% Hydroxyäthylstärke (HH) oder Ringer-Laktat(RL). *Anaesthesist* **39,** 499–504.

Gauer, O. H., and Henry, J. P. (1976). Neurohormonal control of plasma volume. *Int. Rev. Physiol.* **9,** 145–190.

Genest, J., Larochelle, P., Cusson, J. R., and Cantin, M. (1989). The mechanisms of hypertension. Sodium and the atrial natriuretic factor. *Clin. Exp. Hypertens.* **B8**(1), 67–93.

Gerbes, A. L., Arendt, R. M., Gerzer, R., Schnizer, W., Jüngst, D., Paumgartner, G., and Wernzes, H. (1988). Role of atrial natriuretic factor, cyclic GMP and the renin-aldosterone system in acute volume regulation of healthy subjects. *Eur. Clin. Invest.* **18,** 425–429.

Gerzer, R., Witzgall, H., Tremblay, J., Gutkowska, J., and Hamet, P. (1985). Rapid increase in plasma and urinary cyclic GMP after bolus injection of atrial natriuretic factor in man. *J. Clin. Endocrinol. Metab.* **61**(6), 1217–1219.

Gutkowska, J., Bonan, R., Roy, D., Bourassa, M., Garcia, R., Thibault, G., Genest, J., and Cantin, M. (1986). Atrial natriuretic factor in human plasma. *Biochem. Biophys. Res. Commun.* **139,** 287–295.

Hamet, P., Stouder, D. A., Ginn, H. E., Hardman, J. G., and Liddle, G. W. (1975). Studies of the elevated extracellular concentration of cyclic AMP in uremic man. *J. Clin. Invest.* **56,** 339–345.

Hamet, P., Pang, S. C., and Tremblay, J. (1989a). Atrial natriuretic factor induced egression of cyclic guanosine 3′ : 5′ monophosphate in cultured vascular smooth muscle and endothelial cells. *J. Biol. Chem.* **264,** 12364–12369.

Hamet, P., Testaert, E., Palmour, R., Larochelle, P., Cantin, M., Gutkowska, J., Langlois, Y., Ervin, F., and Tremblay, J. (1989b). Effect of prolonged infusion of ANF in normotensive and hypertensive monkeys. *Am. J. Hypertens.* **2,** 690–695.

Hauptlorenz, S., and Puschendori, B. (1989). Influence of β-blockade on ejection fraction, ANP, and cGMP in patients after myocardial infarction. *JAMA, J. Am. Med. Assoc.* **262**(21), 2996.

Jespersen, B., Jensen, L., Sorensen, S. S., and Pedersen, E. B. (1990). Atrial natriuretic factor, cyclic 3′,5′-guanosine monophosphate and prostaglandin E_2 in liver cirrhosis: Relation to blood volume and changes in blood volume after furosemide. *Eur. J. Clin. Invest.* **20,** 632–641.

Krieger, J., Schmidt, M., Sforza, F., Lehr, L., Imbs, J. L., Coumaros, G., and Kurtz, D. (1989). Urinary excretion of guanosine 3′ : 5′-cyclic monophosphate during sleep in obstructive sleep apnoea patients with and without nasal continuous positive airway pressure treatment. *Clin. Sci.* **76,** 31–37.

Larochelle, P., Cusson, J. R., Hamet, P., Schiffrin, E. L., and du Souich, P. (1994). Renal effects of immersion in essential hypertension (EH). *Am. J. Hypertens.* **7** (in press).

Larose, P., Meloche, S., du Souich, P., De Léan, A., and Ong, H. (1985). Radioimmunoassay of atrial natriuretic factor: human plasma levels. *Biochem. Biophys. Res. Commun.* **130,** 553–558.

Lauster, F., Gerzer, R., Weil, J., Fülle, H. J., and Schiffl, H. (1990). Assessment of dry body-weight in haemodialysis patients by the biochemical marker cGMP. *Nephrol., Dial., Transplant.* **5,** 356–361.

Lewis, H. M., Wilkins, M. R., Selwyn, B. M., Yelland, U. J., Griffith, M. E., and Bhoola, K. D. (1988). Urinary guanosine 3′ : 5′-cyclic monophosphate but not tissue kallikrein follows the plasma atrial natriuretic factor response to acute volume expansion with saline. *Clin. Sci.* **75,** 489–494.

Nakaoka, H., Imataka, K., Kitahara, Y., Fujii, J., Ishibashi, M., and Yamaji, T. (1988). Relationship between plasma levels of atrial natriuretic peptide and cyclic guanosine monophosphate in patients with heart diseases. *Jpn. Circ. J.* **52,** 30–33.

Okada, F., Tokumitsu, Y., Honma, M., and Yi, M. (1991). Plasma cyclic nucleotide responses to psychological stress in patients with schizophrenia. *Biol. Psychiatry* **29,** 613–617.

Pang, S. C., Hoang, M. C., Tremblay, J., Cantin, M., Garcia, R., Genest, J., and Hamet, P. (1985). Effect of natural and synthetic atrial natriuretic factor on arterial blood pressure, natriuresis and cyclic GMP excretion in spontaneously hypertensive rats. *Clin. Sci.* **69,** 721–726.

Peracchi, M., Bamonti-Catena, F., Bareggi, B., Calori, R., and Maiolo, A. T. (1990). Plasma cyclic nucleotide levels in patients with refractory anaemia with excess of blasts. *Blut* **60,** 177–180.

Richards, A. M., Tonolo, G., Fraser, R., Morton, J. J., Leckie, B. J., Ball, S. G., and Robertson, J. I. S. (1987). Diurnal change in plasma atrial natriuretic peptide concentrations. *Clin. Sci.* **73,** 489–495.

Richman, R. A., Kopf, S. G., Hamet, P., and Johnson, R. A. (1980). Preparation of cyclic nucleotide antisera with thyroglobulin cyclic nucleotide conjugates. *J. Cyclic Nucleotide Protein Phosphorylation Res.* **6,** 461–468.

Roy, L. F., Ogilvie, R. I., Larochelle, P., Hamet, P., and Leenen, F. H. H. (1989). Cardiac and vascular effects of atrial natriuretic factor and sodium nitroprusside in healthy men. *Circulation* **79,** 383–392.

Sagnella, G. A., Markandu, N. D., Shore, A. C., and MacGregor, G. A. (1985). Effects of changes in dietary sodium intake and saline infusions on immunoreactive atrial natriuretic peptide in human plasma. *Lancet* **2,** 1208–1211.

Sagnella, G. A., Singer, D. R. J., Markandu, N. D., MacGregor, G. A., Shirley, D. G., Tremblay, J., and Hamet, P. (1990). Atrial natriuretic peptide - cyclic GMP coupling and urinary sodium excretion during acute volume expansion in man. *Can. J. Physiol. Pharmacol.* **68,** 535–538.

Shenker, Y., Sider, R. S., Ostafin, E. A., and Grekin, R. J. (1985). Plasma levels of immunoreactive atrial natriuretic factor in healthy subjects and in patients with edema. *J. Clin. Invest.* **76,** 1684–1687.

Stangl, K., Baumann, G., Weil, J., Gerzer, R., Kerscher, M., and Blömer, H. (1990). Akute Senkung erhöhter Spiegel des atrialen natriuretischen Peptids (ANP) und des zyklischen Guanosinmonophosphats (cGMP) bei Patienten mit chronischer Herzinsuffizienz durch beta-adrenerge Stimulation mit Dopexamin-Hydrochlorid. *Z. Kardiol.* **79,** 417–423.

Tremblay, J., Gerzer, R., Pang, S. C., Cantin, M., Genest, J., and Hamet, P. (1986). ANF stimulation of detergent-dispersed particulate guanylate cyclase from bovine adrenal cortex. *FEBS Lett.* **181,** 17–22.

Turner, G. A., Greggi, S., Duthrie, D., Panici, B. P., Ellis, R. D., Scambia, G., and Mancuso, S. (1990). Monitoring ovarian cancer using urine cyclic GMP. *Eur. J. Gynaec. Oncol.* **11,** 421–427.

Vorderwinkler, K. P., Artner-Dworzak, E., Jakob, G., Mair, J., Diensti, F., Pichler, M., and Pucshendorf, B. (1991). Release of cyclic guanosine monophosphate evaluated as a diagnostic tool in cardiac diseases. *Clin. Chem.* (*Winston-Salem, N.C.*) **37**(2), 186–190.

Wehling, M., Müller, T., Heim, J. M., Lorenz, R., Witzgall, H., Weil, J., and Gerzer, R. (1989). Effects of clonidine and dihydralazine on atrial natriuretic factor and cGMP in humans. *J. Appl. Physiol.* **67,** 938–944.

Future Directions

Ferid Murad

Molecular Geriatrics Corporation
Lake Bluff, Illinois 60044

Although much has been learned about cyclic GMP synthesis, metabolism, and function since its discovery in the early 1960s, much remains to be done. For example, how many isoforms of guanylyl cyclase, cyclic nucleotide phosphodiesterase, and nitric oxide synthase truly exist? How many of the isoforms are discrete gene products, how are they transcriptionally regulated, and are there alternate splicing mechanisms that are also regulatable? For example, some years ago we found that proliferating tissues generally have greater amounts of the particulate guanylyl cyclase activity than the control or nonproliferating tissues (Kimura and Murad, 1975a,b,c). This altered subcellular distribution of guanylyl cyclase was usually associated with increased levels of cyclic GMP. Does this suggest that cyclic GMP participates in cellular proliferation or its regulation in some capacity? Relatively little has been done in comparable studies with the cyclic nucleotide phosphodiesterases or nitric oxide synthases. Are there other mechanisms to alter the subcellular distribution of these enzyme isoforms and perhaps their physiological effects and functions? For example, what are the post-translational modifications that could occur that may regulate activity or cellular compartmentation of these enzymes and their isoforms?

Although we and others have reported that other free radicals in addition to nitric oxide can activate guanylyl cyclase, how general is this activation mechanism and is the regulation by other free radicals and reactive species of oxygen physiologically or pathologically relevant? Also, is the formation of cyclic AMP by activated guanylyl cyclase physiologically relevant?

Does this occur in any or many cells and tissues? What data currently exist in the literature where increases in cyclic AMP levels were due to activation of guanylyl cyclase rather than altered adenylyl cyclase or cyclic nucleotide phosphodiesterase?

In most systems the second messengers described to date are quite interactive with each other. For example, cyclic GMP and cyclic AMP can compete for hydrolysis by various phosphodiesterases. It has been suggested that the effects of some hormones that increase cyclic GMP are due to secondary decreases in cyclic AMP since cyclic GMP may activate some isoforms of cyclic nucleotide phosphodiesterase. These cyclic nucleotides can also regulate cytosolic calcium concentrations through several possible mechanisms including inositol tris-phosphate formation and metabolism. The altered cytosolic calcium may regulate synthesis or hydrolysis of either cyclic nucleotide. Both cyclic nucleotides can also activate both cyclic GMP- and cyclic AMP-dependent protein kinases, usually with K_a values that are about 30- to 100-fold different. Perhaps conditions or situations may exist where one cyclic nucleotide can activate the others kinase; i.e., some cyclic GMP effects could be mediated, through cyclic AMP-dependent protein kinase regulation or via altered cyclic AMP accumulation due to its effects on the different isoforms of cyclic nucleotide phosphodiesterase. There may also be situations where the opposite scenario is the case. Although it should be obvious from the various isoforms of the cyclases and phosphodiesterases that there will be discrete intracellular compartments of these cyclic nucleotides with perhaps different physiological and biochemical effects, this hypothesis have never been definitively proven. Although there has been an extensive search by a number of laboratories for unique or specific protein substrates for the cyclic GMP-dependent and cyclic AMP-dependent protein kinases, only a few such specific substrates have, in fact, been reported. Perhaps the specificity of the phosphorylation is more with the protein's subcellular compartmentation and proximity to a kinase rather than a markedly different affinity or K_m for the kinase. What other macromolecules and potential "receptors" do cyclic GMP interact with in addition to cyclic nucleotide-dependent protein kinases and some isoforms of cyclic nucleotide phosphodiesterase? Will these other macromolecular targets be important mechanisms for signal transduction and information transfer?

Why do some cells and tissues extrude or release cyclic nucleotides into the medium or extracellular space greater than other cells? Are these transport processes active, passive, or regulatable? Could extracellular cyclic nucleotides transmit information to other cells and tissues locally or at a distant site and function as a paracrine substance or hormone?

It has also been suspected that extracellular cyclic nucleotides do not transmit information, particularly since their concentrations in plasma are relatively low (picomolar to nanomolar) and much higher concentrations are required for effects on cells or tissues when added exogenously due to their net charge and poor permeability. There are undoubtedly situations in synaptic clefts, tight junctions, and interstitial fluid where cyclic nucleotide concentrations are much higher than plasma concentrations and under these conditions information transfer could be expected. How can extracellular concentrations of cyclic nucleotides in various body fluids such as plasma, urine, spinal fluid, saliva, and milk be used to evaluate tissue function and hormonal responses in different metabolic or endocrinologic diseases? To date, this has been useful in some calcium disorders including parathyroid diseases. Can provocative diagnostic tests be designed to measure cyclic nucleotides release from a tissue or organ in order to test its function after an appropriate stimulus?

Although the list of functions for cyclic GMP has grown considerably in the past 10 to 15 years with effects on smooth muscle relaxation, intestinal secretion, phototransduction, platelet aggregation, platelet adhesion, endocrine secretion, and neuronal regulation, the list is undoubtedly in its infancy. To date, there is a relatively short list of small intracellular second messengers that include cyclic AMP, cyclic GMP, nitric oxide, calcium, inositol tris-phosphate, diacylgycerol, eicosanoids, and a few others. Perhaps we should assume that 10 to 20% of all the physiological and biochemical effects regulated by such second messenger systems will be regulated by cyclic GMP and/or nitric oxide. Will there be effects of cyclic GMP on transcriptional control, protein synthesis, cell proliferation, etc.? Some years ago, we proposed that since free radicals could regulate cyclic GMP synthesis, perhaps cyclic GMP could act as a "feedback regulator" to control critical redox pathways.

Presumably, as we expand our knowledge base even further with the regulation of cyclic GMP synthesis, metabolism, and function, we should become more clever and predictive in rationally designing useful therapeutic agents whose mechanisms of action are mediated through these processes. Undoubtedly the effects of some methylxanthines such as theophylline on airway and vascular smooth muscle relaxation are mediated through cyclic nucleotide phosphodiesterase inhibition and cyclic GMP accumulation rather than cyclic AMP accumulation. Many such examples of useful therapeutic agents whose effects are due to altered cyclic GMP and/or nitric oxide formation and metabolism are expected in the future. For some years it has been suspected that the formation of superoxide anion and other reactive oxygen species, such as hydroxyl free radical, has important effects in inflammation, vascular and tissue injury, atherogenesis, and lipid peroxidation. Superoxide can interact with nitric oxide to form another reactive species, peroxynitrite. Perhaps the deleterious effects of some reactive oxygen species are due to the removal of nitric oxide and its beneficial effects. Alternatively, some of their effects may be mediated through the formation of peroxynitrite.

There are undoubtedly many more important questions than can and should also be asked. It has been possible to design experiments related to cyclic GMP synthesis, metabolism, and function in recent years that usually work and, in most cases, provide expected and predictable data. This has been a major achievement from where the field was just 10 to 20 years ago, when most of the work was a descriptive "shotgun" approach. The known pathways and reagents currently available that modify the cyclic GMP cascade and system indicate to many of us in the field that there have been many significant advances in recent years and a much greater expansion of knowledge is expected in the near future.

References

Kimura, H., and Murad, F. (1975a). Two forms of guanylate cyclase in mammalian tissues and possible mechanisms for their regulation. *Metab., Clin. Exp.* **24,** 439–445.

Kimura, H., and Murad, F. (1975b). Increased particulate and decreased soluble guanylate cyclase activity in regenerating liver, fetal liver, and hepatoma. *Proc. Natl. Acad. Sci. U.S.A.* **72,** 1965–1969.

Kimura, H., and Murad, F. (1975c). Subcellular localization of guanylate cyclase. *Life Sci.* **17,** 837–844.

Index

Action potentials, cyclic nucleotide effects, 231–232
Aldosterone, inhibition of production, atrial natriuretic peptide-mediated, 103–104
Angiotensin, Ca^{2+} channel regulation, 239–241
ATP, regulation of Ca^{2+} channels, 234–239
Atrial natriuretic peptide, 172–173
  inhibition of aldosterone production, 103–104
  intravenous, effect on cGMP, 311–313
    in essential hypertension, 314–315
  receptors, 173–174
Atrial natriuretic peptide receptors, 8–9, 69–74
  cloning, 71–72
  coupling to activation of particulate guanylate cyclase, 72–74
  heterogeneity of subunit structure, 70
  pharmacological heterogeneity, 69–70
  on transmembrane protein with guanylate cyclase residue, 70–71
Autoinhibitory domain, cGMP kinase, 130–135
Autophosphorylation sites, cGMP kinase, 132
Azide, conversion to nitric oxide, 22–23

Bacteria, cGMP in, 286–287
Bicarbonate, secretion, cGMP effect, 261

Calcium
  cGMP effects on
    endoplasmic reticulum Ca^{2+} pump, 200–201
    Na^{+}–Ca^{2+} exchange mechanisms, 201–203
    plasmalemmal Ca^{2+} pump, 196–200
    receptor-operated Ca^{2+} entry, 205–207
    voltage-dependent Ca^{2+} channels, 203–205
  channels
    differences between slow and fast, 218, 220
    voltage-dependent, 203–205
  intracellular levels, regulation by cGMP kinase, 152–155
  intracellular signaling and, 182–183
  receptor-operated entry, 205–207
  slow channels, 217–246
    action potential induction, 221
    activity, changes over time, 237–238
    current types, 236–237
    schematic model, 218–219
Calmodulin–protein kinase, 229
Cardiac muscle, 220–230
  calmodulin–protein kinase and protein kinase C, 229
  cAMP stimulation, 220–222
  cGMP inhibition of current, 224–229
  phosphorylation hypothesis, 222–224
Cellular functions, agonists, 195
Chlorine
  absorption
    cGMP effect, 257–259
    inhibition mechanisms, 265–266
  channel, cystic fibrosis transmembrane conductance regulator, 270–273
  secretion
    cGMP effect, 259–261
    stimulation mechanisms, 267
Cloning
  atrial natriuretic peptide receptors, 71–72
  Escherichia coli, heat-stable enterotoxin receptors, 77
  guanylyl cyclase isoforms, 7–15
Copper, bound to guanylate copper, role, 45–46
Cyclase–kinase–phosphatase family, members, 14
Cyclic AMP, Ca^{2+} slow channel
  inhibition, 230–234
  stimulation, 220–222

Cyclic GMP
  Ca^{2+}/CaM-dependent cyclic nucleotide phosphodiesterase, regulation, 95–96
  control of steady state levels, 88
  cyclic nucleotide hydrolysis regulation, 87–107
  functions, 323
  history, 104
  hydrolysis control by cyclic nucleotide phosphodiesterase, 88–89
  inhibition of slow Ca^{2+} current, 224–229
  intracellular signaling and, 182–183
  isoforms, 321–322
  measurements, *see* Measurements
  mechanisms, 184
  nitric oxide–heme complex role, 55–56
  nitrosyl-heme role, 56–57
  phenobarbitol effects, 57
  physiological responses to, 87
  signal transduction system, 196
  synthesis, 2–3
    endothelial-derived relaxing factor effects, 26–27
    in intestine, 254–257
    nitric oxide effects, 3–4, 21–26
Cyclic nucleotide, extracellular, 322–323
Cyclic nucleotide phosphodiesterase, 87–107
  basis for family designation, 89–91
  Ca^{2+}/CaM-dependent family, 91–96
    cGMP regulation, 95–96
    isoforms, 91–92
    kinetic properties, 92–93
    structure and domain organization, 93
    tissue and cellular distribution, 94–95
  conserved motifs, 89–91
  GMP-inhibited family, 104–106
  GMP-specific family, 96–99
    cellular distribution and functions, 97–98
    properties, 97
    regulation of activity, 98
    structural features, 99
  GMP-stimulated family, 99–104
    domain organization, 101
    isoforms, 99
    kinetic properties, 99–100
    regulation, 102–104
    tissue distribution, 101–102
  multiple, cGMP hydrolysis control, 88–89
Cystic fibrosis transmembrane conductance regulator, chlorine channel, 270–273

Dictyostelium discoideum, cGMP in, 287–291
Dimerization domain, cGMP kinase, 129–130
DNA
  complementary
    guanylate cyclase, particulate, 8, 71
    soluble, α-subunit, 11
Drosophila, cGMP kinase, 123–124

Electrolytes, secretion in intestines, cGMP-induced, 273–275
Endothelium-derived relaxing factor, 175–176
  effects on cGMP formation, 26–27
Enzymes, nitric oxide-generating, in smooth muscle, 178–179
Epithelial cells, cGMP effects, 257–262
Escherichia coli
  heat-stable enterotoxin, 3, 253
    particulate guanylate cyclase regulation, 67–82
  heat-stable enterotoxin receptor, 74–80
    cloning, 77
    coupling, activation of particulate guanylate cyclase and, 80
    high-affinity, 75–76
    low-affinity, 75
    pharmacological heterogeneity, 74–76
    purification, intestinal mucosa receptors, 78–79
    relationship with particulate guanylate cyclase in intestinal cells, 79–80
    subcellular distribution heterogeneity, 76–77
    subunit structure heterogeneity, 76
Essential hypertension, cGMP measurements, 313–315
Exon–intron organization, cGMP kinase, 123–124

Fluids, secretion in intestines, cGMP-induced, 273–275
Fungi, cGMP in, 291–292

G-protein, calcium, slow channels regulation, 243–245
Guanylate cyclase
activation
nitric oxide, 37–38, 55–56
phenylhydrazine, 36–40, 46–50
phenyl radical, 55–56
protoporphyrin IX, 37–43
kinetics, 43–45
activity, 295–297
catalytic, 55
free radical activation mechanism and, 23
regulation by porphyrins and metalloporphyrins, 54–58
thiol effects, 38–40
atrial natriuretic peptide, 172–173
cytosolic, 41–42
endothelium-derived relaxing factor, 175–176
enzyme structure, 179–180
nitric oxide-generating enzymes in smooth muscle, 178–179
nitrovasodilator metabolism, 174–175
nonadrenergic noncholinergic nerves, 175–178
in *Dictyostelium discoideum*, 289–291
effects of structurally modified protoporphyrin IX, 50–53
heme-containing, 43–45
heme-deficient cytosolic, spectral properties, 60
expression, 35
interaction with nitric oxide, 36
intestinal form, 254–255
isoforms
cloning, 7–15
particulate, 7–11
soluble, 11–12
structures, 13–14
localization in intestine, 255–257
nitrovasodilator metabolism, 174–175
particulate, 172–174
soluble, 11–12
activation, 24
regulation, 35–62, *see also* Metalloporphyrins; Porphyrins
nitric oxide–cGMP signal transduction system, 19–30
nitric oxide–heme exchange with hemoproteins, 58–61
particulate, 67–82, 172–174
activation
coupling of atrial natriuretic peptide receptors to, 72–74
Escherichia coli, heat-stable enterotoxin receptors coupling and, 80
atrial natriuretic peptide receptors, 69–74
cDNA, 71
Escherichia coli, heat-stable enterotoxin receptors and, 74–80
activation, 74–75
on transmembrane protein with atrial natriuretic peptide receptors, 70–71
relationship with *Escherichia coli,* heat-stable enterotoxin receptors in intestinal cells, 79–80
porphyrin binding site, 54–55
role, copper bound to, 45–46
Guanylyl cyclase, *see* Guanylate cyclase
Guanylyl cyclase-activating factor synthase, *see* Nitric oxide synthases

Heme
binding, 36
required by nitric oxide, 40–43
synthesis, induction by phenobarbital, 57
Hemoproteins
nitric oxide–heme exchange with guanylate cyclase, 58–61
reaction with phenylhydrazine, 49–50
Hydralazine, guanylate cyclase activation, 46–50
Hypertension, essential, cGMP measurements, 313–315

Inositol phosphates, cGMP effects
intestinal transport, 253–275
production, 207–209
inhibition mode, 208–209
mechanisms, 207–208
Intestinal transport, cGMP effects
in mammals, 257–262
Cl and HCO_3 secretion, 259–261
localization, 261–262
Na and Cl absorption, 257–259
nonepithelial cells, 262
in winter flounder, 263
Intestine
cGMP
action, 263–267
Cl secretion stimulation, 267
inhibition of Na and Cl absorption, 265–266
mechanisms, 263–265
pathways, 264
synthesis in, 254–257
cGMP-dependent protein kinase, 267–270
electrolytes and fluids, secretion, cGMP-induced, 273–275
Intracellular signaling, cGMP and calcium and, 182–183

Measurements
in essential hypertension, 313–315
in healthy humans, 305–313
blood volume expansion, 309–310
intravenous atrial natriuretic peptides effect, 311–313
measurement conditions, 306–309
phenylephrine effect, 311
as tool in diseases, 315–316
Metalloporphyrins, guanylate cyclase and regulation of activity, 54–58
structure–activity relationships, 50–54
Mg^{2+}/ATP, cGMP kinase subdomain for, 144–145
Microheterogeneity, cGMP kinase, 126–128
Muscarinic agonists, inhibition of Ca^{2+} slow channel, 227, 229

Natriuretic peptides, *see also* Atrial natriuretic peptides
particulate guanylate cyclase regulation, 67–82
Nerves, nonadrenergic noncholinergic, 176–178
Nitric oxide
effects on cGMP synthesis, 3–4, 21–26
formation, 27–30
functions, 20–21
guanylate cyclase activation, 37–38
heme requirement, 40–43
interaction with guanylyl cyclase, 36
Nitric oxide–cGMP signal transduction system, 19–30
scheme, 28
Nitric oxide–heme complex, 55–56
exchange between hemoproteins and guanylate cyclase, 58–61
Nitric oxide synthases, 27–30
isoforms, 27, 29
Nitroprusside, 38–39
Nitrosoguanidines, 38–39
S-Nitrosothiols, 38–39
Nitrosyl-hemoproteins, high-molecular-weight, 58
Nitrovasodilators
conversion to nitric oxide, 22–23
metabolism, 174–175
Nonepithelial cells, cGMP effects, 262
Nucleotides, cyclic, effects on action potentials, 231–232

Paramecium, see Protozoans
Particulate guanylate cyclase–cGMP second messenger system, 81–82
Phenobarbital, heme synthesis induction, 57
Phenylephrine, effect on cGMP, 311
in essential hypertension, 314
Phenylhydrazine
guanylate cyclase activation, 36–37
mechanism, 46–50
reaction with hemoproteins, 49–50
Phosphodiesterase, 322
cGMP-binding, 148

in *Dictyostelium discoideum*, 289–291
in protozoas, 297
Phosphorylation
hypothesis, Ca^{2+} slow channel, 222–224
sites, cGMP kinase, 127–128
Platelet aggregation, inhibition, by cGMP kinase, 152
Porphyrins
binding to guanylyl cyclase, 36
guanylate cyclase
binding site, 54–55
regulation, 54–58
structure–activity relationships, 50–54
Protein kinase, cGMP-dependent, 115–159
amino acid sequences, 118–119
autoinhibitory domain, 130–135
autophosphorylation sites, 132
interaction with cGMP-binding sites, 142–143
interaction with other domains, 135
monomeric kinases, 132–133
carboxyl-terminal domain, 150
catalytic domain, 128, 144–150
interaction with cGMP-binding sites, 142–143
subdomains for Mg^{2+}/ATP and protein substrate binding, 144–150
substrate specificity, 145–150
catalytic efficiency, 146
cGMP binding domains, 136–144
analog specificities, 143–144
characteristics, 136–137
interaction with autoinhibitory and catalytic domain, 142–143
kinetically distinct sites, 141–142
specificity determinants, 140–141
structural features, 136, 138–140
cross-activation, 156–159
dimerization domain, 129–130
functional domains, 125
inhibitor potency, 148–150
intestinal form, 267–270
structural models, 268
isozymes, 118–124
chromosomal location, 123–124
mRNA size and distribution, 121–123
type I, 120
type II, 120–121
microheterogeneity, 126–128
nonmammalian, 124
phosphorylation sites, 127–128
physiological function, 150–156
intracellular calcium regulation, 152–155
platelet aggregation inhibition, 152
smooth muscle regulation, 151–152
in protozoa, 297
smooth muscle relaxation, 182
structure, 124–126
tissue distribution, 117–118
Protein kinase A, 272–273
Protein kinase C, 229
calcium, slow channels regulation, 242–243
Protein kinase-G II, 265, 269–270
Protein kinase inhibitor, potency, 148–150
Protoporphyrin IX
guanylate cyclase activation, 37–43
heme requirement, 40–43
kinetics, 43–45
thiol effects, 38–40
propionic acid residues, 53
structural modifications, effects on guanylate cyclase, 50–53
Protozoans, cGMP in, 292–297
guanylyl cyclases, 295–297
intracellular formation regulation, 292–295
phosphodiesterase and cGMP-dependent protein kinase, 297
Pyridine hemochrome, 60

RNA, messenger
α and β subunits, guanylyl cyclase, 12
size and distribution, cGMP kinase, 121–123

Sea urchin, guanylyl cyclase cDNA, 8
Skeletal muscle fibers, calcium, slow channels, 245–246
Slime molds, cGMP in, 287–292

Smooth muscle
  regulation, by cGMP kinase, 151–152
  relaxation, 171–184, *see also* Guanylyl cyclase
    cGMP, calcium, and intracellular signaling, 182–183
    cGMP kinase, 182
    correlation with cGMP levels, 180–181
    nitric oxide effects, 21–26
    nitric oxide-generating enzymes, 178–179
    nonadrenergic noncholinergic nerves, 176–178
  vascular cells, 230–241
    angiotensin regulation, 239–241
    ATP regulation, 234–239
    G-protein regulation, 243–245
    inhibition by cAMP and cGMP, 230–234
    properties in, 222
    protein kinase C regulation, 242–243
Sodium–Calcium exchange mechanisms, 201–203
Sodium, absorption
  cGMP effect, 257–259
  inhibition mechanisms, 265–266

Tetrahymena, see Protozoans
Thiols, effects on guanylate cyclase activity, 38–40
Transmembrane protein, particulate residue on, 70–71

Contents of Previous Volumes

Volume 21

Glycosphingolipids That Can Regulate Nerve Growth and Repair
A. Claudio Cuello

New Approaches to Vaccination
Charles Flexner

Allosteric Modulation of *N*-Methyl-D-Aspartate Receptors
Ian J. Reynolds and Richard J. Miller

Erythropoietin: Regulation of Erythropoiesis and Clinical Use
Emmanuel N. Dessypris and Sanford B. Krantz

DNA Topoisomerases as Anticancer Drug Targets
Erasmus Schneider, Yaw-Huei Hsiang, and Leroy F. Liu

Multidrug Resistance and Chemosensitization: Therapeutic Implications for Cancer Chemotherapy
Elias Georges, Frances J. Sharom, and Victor Ling

Peptides: Chemistry, Biology, and Pharmacology
Amrit K. Judd and Gary K. Schoolnik

Volume 22

Acyclovir: Mechanism of Antiviral Action and Potentiation by Ribonucleotide Reductase Inhibitors
John E. Reardon and Thomas Spector

Rational Approaches to Osteoporosis Therapy
Robert Marcus

Molecular Asymmetry and Its Pharmacological Consequences
Kenneth M. Williams

Blood–Brain Barrier: Transport Studies in Isolated Brain Capillaries and in Cultured Brain Endothelial Cells
Yoshinobu Takakura, Kenneth L. Audus, and Ronald T. Borchardt

Protein Kinase Inhibitors: Probes for the Functions of Protein Phosphorylation
John E. Casnellie

Renin Inhibitors
Hollis D. Kleinert, William R. Baker, and Herman H. Stein

The Capactitative Model for Receptor-Activated Calcium Entry
James W. Putney, Jr.

Calcium Channel Antagonists in the Prevention of Neurotoxicity
Stuart A. Lipton

New Directions in the Delivery of Drugs and Other Substances to the Central Nervous System
Yvette Madrid, Laura Feigenbaum Langer, Henry Brem, and Robert Langer

Hormonal Regulation of Cytochrome *P*-450 Gene Expression
Johan Lund, Peter G. Zaphiropoulos, Agneta Mode, Margaret Warner, and Jan-Ake Gustafsson

Volume 23

Advanced Glycosylation: Chemistry, Biology, and Implications for Diabetes and Aging
Richard Bucala and Anthony Cerami

Complex Carbohydrates in Drug Development
Ronald L. Schnaar

New Developments in Enteric Bacterial Toxins
David A. Bobak and Richard L. Guerrant

Superoxide Dismutase: Pharmacological Developments and Applications
Bassam A. Omar, Sonia C. Flores, and Joe M. McCord

Pharmacological Implications of Interleukin-5 in the Control of Eosinophilia
Colin J. Sanderson

Advances in Antiarrhythmic Drug Therapy
Donald C. Harrison and Michael B. Bottorff

New Developments in Thrombolytic Therapy
Stephen F. Badylak, Jack Henkin, Sandra F. Burke, and Arthur A. Sasahara

Therapy of Hematopoietic Disorders with Recombinant Colony-Stimulating Factors
Robert S. Negrin and Peter L. Greenberg

Mechanisms of Xenobiotic-Induced Renal Carcinogenicity
Wolfgang Dekant and Spyridon Vamvakas

Volume 24

Antibody Engineering Using *Escherichia coli* as Host
E. Sally Ward

Insulin Mediators and the Mechanism of Insulin Action
Guillermo Romero and Joseph Larner

Activation of Latent Transforming Growth Factor β
Robert Flaumenhaft, Soichi Kojima, Mayumi Abe, and Daniel B. Rifkin

Structure and Function of P-Glycoprotein in Normal Liver and Small Intestine
Zenaida C. Gatmaitan and Irwin M. Arias

Antibody-Directed Enzyme Prodrug Therapy (ADEPT)
Kenneth D. Bagshawe

Mechanisms and Therapeutic Potential of Vanilloids (Capsaicin-like Molecules)
Arpad Szallasi and Peter M. Blumberg

Multidrug Resistance in Cancers of Childhood: Clinical Relevance and Circumvention
Helen S. L. Chan, Paul S. Thorner, George Haddad, Gerrit DeBoer, Brenda L. Gallie, and Victor Ling

Phospholipase D: Regulation and Functional Significance
Neil T. Thompson, Lawrence G. Garland, and Robert W. Bonser

Pharmacology of Nerve Growth Factor in the Brain
Franz Hefti and Paul A. Lapchak

Molecular Mechanisms in Acute Lung Injury
Peter A. Ward and Michael S. Mulligan

Volume 25

Pharmacology of Interleukin-1 Actions in the Brain
Nancy J. Rothwell and Giamal Luheshi

Interleukin-1
Charles A. Dinarello

Modulation of Cytokine Function: Therapeutic Applications
Brian Henderson and Stephen Poole

Integrins, ICAMS, and Selectins: Role and Regulation of Adhesion Molecules in Neutrophil Recruitment to Inflammatory Sites
Takashi Kei Kishimoto and Robert Rothlein

Immune Modulating Therapies for Idiopathic Inflammatory Bowel Diseases
Douglas S. Levine

Selectins in Leukocyte Extravasation: Function of a Common Epitope on L- and E-Selectin
Mark A. Jutila

Endogenous Cardiac Glycosides
Ralph A. Kelly and Thomas W. Smith

Antisense Catalytic RNAs as Therapeutic Agents
Daniela Castanotto, John J. Rossi, and Nava Sarver

Gene-Mimetic Substances: Drugs Designed to Intervene in Gene Expression
Jack S. Cohen

Progress toward Understanding the Cannabinoid Receptor and Its Second Messenger Systems
Billy R. Martin, Sandra P. Welch, and Mary Abood

HIV Protease as an Inhibitor Target for the Treatment of AIDS
Paul L. Darke and Joel R. Huff

ISBN 0-12-032926-3